Advances in Intelligent and Soft Computing 88

Editor-in-Chief: J. Kacprzyk

T0074375

Advances in Intelligent and Soft Computing

Editor-in-Chief

Prof. Janusz Kacprzyk
Systems Research Institute
Polish Academy of Sciences
ul. Newelska 6
01-447 Warsaw
Poland
E-mail: kacprzyk@ibspan.waw.pl

Further volumes of this series can be found on our homepage: springer.com

Yves Demazeau, Michal Pěchouček,
Juan M. Corchado, and Javier Bajo (Eds.)

Advances on Practical Applications of Agents and Multiagent Systems

9th International Conference on Practical
Applications of Agents and Multiagent Systems

 Springer

Editors

Prof. Yves Demazeau
Laboratoire d'Informatique de Grenoble
Centre National de la Recherche
Scientifique
Maison Jean Kuntzmann
110 av. de la Chimie
F-38041 Grenoble
France

Prof. Michal Pĕchoucĕk
Czech Technical University
Department of Cybernetics
Technická 2
16627 Prague 6
Czech Republic

Prof. Juan M. Corchado
Universidad de Salamanca
Departamento de Informática
y Automática
Facultad de Ciencias
Plaza de la Merced S/N
37008 Salamanca
Spain
E-mail: corchado@usal.es

Prof. Javier Bajo Pérez
Universidad Pontificia de Salamanca
Escuela Universitaria de Informática
Compañía 5
37002 Salamanca
Spain

ISBN 978-3-642-19874-8   e-ISBN 978-3-642-19875-5

DOI 10.1007/978-3-642-19875-5

Advances in Intelligent and Soft Computing   ISSN 1867-5662

Library of Congress Control Number: 2011923222

Typeset & Cover Design: Scientific Publishing Services Pvt. Ltd., Chennai, India.

Printed on acid-free paper
5 4 3 2 1 0
springer.com

Preface

Research on Agents and Multi-Agent Systems has matured during the last decade and many effective applications of this technology are now deployed. An international forum to present and discuss the latest scientific developments and their effective applications, to assess the impact of the approach, and to facilitate technology transfer, has become a necessity.

PAAMS, the International Conference on Practical Applications of Agents and Multi-Agent Systems is the international yearly tribune to present, to discuss, and to disseminate the latest developments and the most important outcomes related to real-world applications. It provides a unique opportunity to bring multi-disciplinary experts, academics and practitioners together to exchange their experience in the development of Agents and Multi-Agent Systems.

This volume presents the papers that have been accepted for the 2011 edition. These articles capture the most innovative results and this year's trends: Finance and Trading, Information Systems and Organisations, Leisure Culture and Interactions, Medicine and Cloud Computing, Platforms and Adaptation, Robotics and Manufacturing, Security and Privacy, Transports and Optimisation paper has been reviewed by three different reviewers, from an international committee composed of 75 members from 24 different countries. From the 81 submissions received, 15 were selected for full presentation at the conference, and 24 were accepted for short presentation at the conference.

We would like to thank all the contributing authors, as well as the members of the Program Committee and the Organizing Committee for their hard and highly valuable work. Their work has helped to contribute to the success of the PAAMS'11 event. Thanks for your help, PAAMS'11 wouldn't exist without your contribution.

Yves Demazeau   Juan Manuel Corchado
Michal Pěchouček   Javier Bajo
PAAMS'11 Program Co-chairs   PAAMS'11 Organizing Co-chairs

Organization

General Co-chairs

Yves Demazeau   Centre National de la Recherche Scientifique (France)

Michal Pěchouček (Co-chair)   Czech Technical University in Prague (Czech Republic)

Juan M. Corchado   University of Salamanca (Spain)

Javier Bajo   Pontifical University of Salamanca (Spain)

Advisory Board

Juan Pavón   Universidad Complutense de Madrid (Spain)

Frank Dignum   Utrecht University (The Netherlands)

Program Committee

Yves Demazeau (Co-chair)   Centre National de la Recherche Scientifique (France)

Michal Pěchouček (Co-chair)   Czech Technical University in Prague (Czech Republic)

Carole Adam   University of Grenoble (France)

Frederic Amblard   University of Toulouse (France)

Francesco Amigoni   Politecnico di Milano (Italy)

Luis Antunes   University of Lisbon (Portugal)

Javier Bajo   Pontifical University of Salamanca (Spain)

Zbigniew Banaszak   Warsaw University of Technology (Poland)

Fabio Luigi Bellifemine   Telecom Italia (Italy)

Olivier Boissier   Ecole Nationale Superieure des Mines de Saint Etienne (France)

Juan A. Botía   University of Murcia (Spain)

Vicente Botti   Polytechnic University of Valencia (Spain)

Stefano Bromuri   University of Applied Sciences Western Switzerland (Switzerland)

Sven Brueckner   Jacobs Technology Inc. (USA)

Valerie Camps   University of Toulouse (France)

Longbing Cao   University of Technology Sydney (Australia)

Lawrence Cavedon   RMIT Melbourne (Australia)
Pierre Chevaillier   University of Brest (France)
Juan Manuel Corchado   University of Salamanca (Spain)
Keith Decker   University of Delaware (USA)
Frank Dignum   Utrecht University (The Netherlands)
Virginia Dignum   Utrecht University (The Netherlands)
Alexis Drogoul   Institut de Recherche pour l Developpement
                            (Vietnam)
Julie Dugdale   University of Grenoble (France)
Edmund Durfee   University of Michigan (USA)
Partha S Dutta   Rolls-Royce (United Kingdom)
Amal Elfallah   University of Paris 6 (France)
Torsten Eymann   University of Bayreuth (Germany)
Rubén Fuentes   University Computense de Madrid (Spain)
Francisco Garijo   Telefonica I+D (Spain)
Khaled Ghedira   National School of Computer Sciences
                            (Tunisia)
Sylvain Giroux   Unversity of Sherbrooke (Canada)
Pierre Glize   University of Toulouse (France)
Vladimir Gorodetski   University of Saint Petersburg (Russia)
Dominic Greenwood   Whitestein Technologies (Switzerland)
Kasper Hallenborg   University of Southern Denmark (Denmark)
Benjamin Hirsch   Technical University Berlin (Germany)
Tom Holvoet   Catholic University of Leuven (Belgium)
Shinichi Honiden   National Institute of Informatics Tokyo
                            (Japan)
Jomi Hübner   Federal University of Santa Catarina (Brazil)
Toru Ishida   University of Kyoto (Japan)
Takayuki Ito   Massachussets Institute of Technology (USA)
Vicente Julian   Polytechnic University of Valencia (Spain)
Achilles Kameas   University of Patras (Greece)
Jeffrey O. Kephart   IBM Research (USA)
Franziska Kluegl   University of Örebro (Sweden)
Matthias Klusch   DFKI (Germany)
Martin Kollingbaum   University of Aberdeen (United Kingdom)
Jaroslaw Kozlak   University of Science and Technology in
                            Krakow (Poland)
Zakaria Maamar   Zayed University (United Arab Emirates)
Rene Mandiau   University of Valenciennes (France)
Philippe Mathieu   University of Lille (France)
Eric Matson   Purdue University (USA)
Fabien Michel   University of Reims (France)
José M. Molina   Universidad Carlos III de Madrid (Spain)
Mirko Morandini   University of Trento (Italy)
Joerg Mueller   Clausthal University of Technology
                            (Germany)

Peter Novak	Czech Technical University in Prague (Czech Republic)
Eugenio Oliveira	University of Porto (Portugal)
Sascha Ossowski	University of Rey Juan Carlos (Spain)
Julian Padget	University of Bath (United Kingdom)
Adolfo López Paredes	University of Valladolid (Spain)
Juan Pavón	Universidad Complutense de Madrid (Spain)
Jeremy Pitt	Imperial College of London (United Kingdom)
Juan Rodriguez Aguilar	Artificial Intelligence Research Institute (Spain)
Simon Thompson	British Telecom IIS Research Centre (United Kingdom)
Paolo Torroni	University of Bologna (Italy)
Rainer Unland	University of Duisburg (Germany)
Domenico Ursino	University of Reggio Calabria (Italy)
Birna van Riemsdijk	University of Delft (The Netherlands)
Jacques Verriet	Embedded Systems Institute (The Netherlands)
Jiri Vokrinek	Czech Technical University in Prague (Czech Republic)
Danny Weyns	Catholic University of Leuven (Belgium)
Niek Wijngaards	Thales, D-CIS lab (The Netherlands)
Gaku Yamamoto	IBM and Tokyo Institute of Tecnology (Japan)

Organizing Committee

Juan M. Corchado (Chairman)	University of Salamanca (Spain)
Javier Bajo (Co-Chairman)	Pontifical University of Salamanca (Spain)
Juan F. De Paz	University of Salamanca (Spain)
Sara Rodríguez	University of Salamanca (Spain)
Dante I. Tapia	University of Salamanca (Spain)
Emilio Corchado	University of Salamanca (Spain)
Fernando de la Prieta Pintado	University of Salamanca (Spain)
Davinia Carolina Zato Domínguez	University of Salamanca (Spain)

Contents

Information and Organisations

Robotics and Manufacturing

Security and Privacy

Platforms and Adaptation

Transports and Optimisation

Leisure, Culture, Interactions

Medicine and Cloud Computing

Finance and Trading

A J-MADeM Agent-Based Social Simulation to Model Urban Mobility

Francisco Grimaldo, Miguel Lozano, Fernando Barber,
and Alejandro Guerra-Hernández

Abstract. The mobility models followed within metropolitan areas, mainly based on the massive use of the car instead of the public transportation, will soon become unsustainable unless there is a change of citizens' minds and transport policies. The main challenge related to urban mobility is that of getting free-flowing greener cities, which are provided with a smarter and accessible urban transport system. In this paper, we present an agent-based social simulation approach to tackle this kind of social-ecological systems. The Jason Multi-modal Agent Decision Making (J-MADeM) library enable us to model and implement the social decisions made by each habitant about how to get to work every day, e.g., by train, by car, sharing a car, etc. In this way, we focus on the decision making aspects of this problem at a micro level, instead of focussing on spatial or other macro issues. The first results show the different outcomes produced by societies of individualist and egalitarian agents, in terms of the average travel time, the use of the urban transportation and the amount of CO_2 emitted to the environment.

1 Introduction and Related Work

The mobility models followed within metropolitan areas, mainly based on the massive use of the car instead of the public transportation, will soon become unsustainable unless there is a change of citizens' minds and transport policies. This fact

Francisco Grimaldo · Fernando Barber · Miguel Lozano
Departament d'Informàtica, Universitat de València, Av. Vicent Andrés Estellés s/n,
(Burjassot) València, Spain 46100
e-mail: `francisco.grimaldo@uv.es`, `miguel.lozano@uv.es`,
        `fernando.barber@uv.es`

Alejandro Guerra-Hernández
Departamento de Inteligencia Artificial, Universidad Veracruzana,
Facultad de Física e Inteligencia Artificial, Sebastián Camacho No. 5,
Xalapa, Ver., México, 91000
e-mail: `aguerra@uv.mx`

Y. Demazeau et al. (Eds.): Adv. on Prac. Appl. of Agents and Mult. Sys., AISC 88, pp. 1–11.
springerlink.com

has been highlighted, for instance, by the European Commission through the Green Paper on Urban Mobility [3]. Cities all over the world face similar problems, e.g., congestion, road safety, security, pollution, climate change due to CO_2 emissions, etc. Since these problems are increasing constantly, humankind pays a high price both in economic and environmental terms, as well as for the health and quality of life of citizens. This money would be better spent for developing more efficient transport systems.

The main challenge related to urban mobility is that of getting free-flowing greener cities, which are provided with a smarter and accessible urban transport system. Agent-based social simulation (ABSS) has been proposed as a suitable manner to tackle this kind of Social-Ecological systems and Environmental Management [14], as well as in Economics [19], Anthropology [15], and Ecology [11]. ABSS provides a framework for implementing techniques that fulfill the requirements of environmental modelling. First, ABSS allows to couple the model of the environment to the social entities that it includes. For example, it makes possible to model aspects such as the roles of social interaction and the disaggregated adaptive human decision-making; and second, it enables the study the relationships between the micro-macro levels of decision making, and the emergence of collective behavior as the response to changes in the environment or in the environmental management policies.

Social and organizational models are being studied under the scope of multi-agent systems (MAS) in order to regulate the autonomy of self-interested agents. Nowadays, the performance of a MAS is determined not only by the degree of de-liberativeness but also by the degree of sociability. In this sense, sociability points to the ability to communicate, cooperate, collaborate, form alliances, coalitions and teams. The assignment of individuals to an organization generally occurs in Human Societies [16], where the organization can be considered as a set of behavioural constraints that agents adopt, e.g., by the role they play [6].

The definition of a proper MAS organization is not an easy task, since it involves dealing with three dimensions: functioning, structure, and norms [13]. From the functioning perspective, systems focus on achieving the best plans and cover aspects such as: the specification of global plans, the policies to allocate tasks to agents, the coordination of plans, etc. [5]. From the structural perspective, systems focus on defining the organizational structures (roles, relations among roles, groups of roles, etc.) that establishes the obligations/permissions of their agents [8]. Very few models deal with both previous dimensions to support agent decision making about organizations, e.g., MOISE+ [13]. For the sake of simplicity, the third dimension is not discussed here.

Social reasoning has been extensively studied in MAS in order to incorporate social actions to cognitive agents [4]. As a result of these works, agent interaction models have evolved to social networks that try to imitate the social structures found in real life [12]. Social dependence networks allow agents to cooperate or to per-form social exchanges attending to their dependence relations [18]. Trust networks can define different delegation strategies by means of representating the attitude towards the others through the use of some kind of trust model, e.g., reputation

[7]. Agents in preference networks express their preferences normally using utility functions so that personal attitudes can be represented by the differential utilitarian importance they place on the others' utilities. Following this preferential approach, the MADeM (Multi-modal Agent Decision Making) model [9] is a market-based mechanism for social decision making, capable of simulating different kinds of social welfares (e.g. elitist, utilitarian), as well as social attitudes of their members (e.g. egoism, altruism).

In this paper we present an ABSS approach to model the mobility within a metropolitan area. Other platforms have face this problem following a similar approach (e.g. UrbanSim [20]) but they usually focus on spatial issues to simulate large-scale urban areas, using gridcell, zonal, or parcel geographies. Instead, we propose focusing on the agent's decision making. At this micro level, the J-MADeM library has been used to model and implement the social decisions made by each habitant about how to get to work every day, e.g., by train, by car, sharing a car, etc. Although there is still work in progress, the main goal of the proposed system is to be used for research and as a decision support platform for metropolitan planning.

The rest of the paper is organized as follows: The next section reviews the J-MADeM library, which allows programming MADeM decisions at the agent level. In section 3 we introduce the urban mobility simulation framework as well as the definition of a simple "travel to work" scenario. Section 4 shows the different outcomes produced by a society of individualist and a society of egalitarian agents in terms of the average travel time, the use of the urban transportation and the amount of CO_2 emitted to the environment. Finally, in section 5 we state the conclusions and discuss about future work.

2   The J-MADeM Library

J-MADeM [10] is a full-fledge AgentSpeak(L) [17] library that implements the Multi-modal Agent Decision Making (MADeM) [9] model in *Jason* [1], the well known extended java based interpreter for this agent oriented programming language. The MADeM model provides agents with a general mechanism to make socially acceptable decisions. In this kind of decisions, the members of an organization are required to express their preferences with regard to the different solutions for a specific decision problem. The whole model is based on the MARA (Multi-Agent Resource Allocation) theory [2], therefore, it represents each one of these solutions as a set of resource allocations. MADeM can consider both tasks and objects as plausible resources to be allocated, which it generalizes under the term *task-slots*. MADeM uses first-sealed one-round auctions as the allocation procedure and a multi-criteria winner determination problem to merge the different preferences being collected according to the kind of agent or society simulated.

The J-MADeM library provides an agent architecture that *Jason* agents can use to carry out their own MADeM decisions; an ontology to express MADeM data as beliefs and rules; and a plan library to execute MADeM processes. The agent architecture `jmadem.MADeMAgArch`, implements in Java a set of actions (Table 1) performing the basic operations of the model. As usual in *Jason*, actions are prefixed by the name of the library, e.g., to set the welfare of the society as a nash equilibrium, the action jmadem.set_welfare(nash) is executed in a plan. Other MADeM parameters are defined in the same way. Although this Java based actions are often more efficient than the AgentSpeak(L) plans and rules, they hide information to the agents. For instance, the action *construct_allocations*/4 basically computes the cartesian product of the slots domains, so that some kind of filtering at the Java level is required to obtain "legal" allocations; but the agents do not know what a "legal" allocation is, to the detriment of the agent metaphor, e.g., they can not reason about legal allocations, nor communicate about them.

Table 1 Actions defined in the J-MADeM library.

Action	Description
add_utility_function("P.U")	P is a Java package name and U the utility function name.
add_utility_function(U,N)	U is a utility name and N is fully qualified name of the function Java class.
construct_allocations(T,S,E,Al)	$T = t(S_1,\ldots,S_n)$ is a function denoting a task t of n slots, $S \subseteq \{S_1,\ldots,S_n\}$ is a set of task slots to be allocated, $E = [[e_1,\ldots,e_j],\ldots]$ elements in the domain of each slot, Al is the computed list of allocations
launch_decision(A,AL,U,DId)	A is a set of agents, AL is a set of allocations, U is a list of utilify functions, and DId is the output parameter.
launch_decision1(A,AL,U,DId)	As above, but it returns only 1 solution.
remove_utility_function(U,N)	U and N are as above.
reset_personal_weights(PW)	$PW = [jmadem\_personal\_weight(A,\_),\ldots]$.
reset_utility_weights(UW)	$UW = [jmadem\_utility\_weight(U,\_),\ldots]$.
set_list_of_personal_weights(PW)	$PW = [jmadem\_personal\_weight(A,W),\ldots]$, where A is an agent and $W \in \Re$ his personal weight.
set_list_of_utility_weights(UW)	$UW = [jmadem\_utility\_weight(U,W),\ldots]]$, where U is an utility name and $W \in \Re$ its weight.
set_personal_weight(A,W)	A is an agent and $W \in \Re$ is his weight.
set_remove_MADeM_data(V)	If V is *true* MADeM data is deleted at the Java level, once the decision is done.
set_timeout(T)	T is a numerical value in milliseconds (1000 by default).
set_utility_weight(U,W)	U is a utility name and $W \in \Re$ is its weight.
set_welfare(W)	$W \in \{utilitarian, egalitarian, elitist, nash\}$ is the welfare.

In order to provide a full-fledge AgentSpeak(L) layer in the library, J-MADeM agents use an ontology (Table 2) to define the data of a decision process declaratively, as beliefs and rules. In this way, data is accessible to Test Goals and Speech Acts with *Ask*-like performatives. Utilities and filters can also be defined as beliefs or rules. For instance, considering allocations of the form showed below in equation 1, the rule:

```
jmadem_utility(dummyUF,_,Alloc,0) :-
    .my_name(Myself) &
    owns(Myself,Vehicle) &
    .member(travel_by(_,Vehicle), Alloc) &
    not .member(travel_by(Myself,Vehicle),Alloc).
```

Table 2 The ontology used by J-MADeM agents.

Belief formula	Description
jmadem_list_of_personal_weights(PW)	*PW* is a list of personal weight, as defined below.
jmadem_list_of_utility_weights(UW)	*UW* is a list of utility weights, as defined below.
jmadem_filter(F,Al)	*F* is the name of the filter
	Al is an allocation to be filtered.
jmadem_personal_weight(A,W)	*A* is an agent and $W \in \Re$ his weight.
jmadem_timeout(T)	*T* is the timeout in millisecond (1000 by default).
jmadem_utility(U,N)	*U* is the utility function name and
	N is the name of the java class.
jmadem_utility(U,A,Al,V)	*U* is the utility function name,
	A is the auctioneer agent,
	Al is an allocation, and
	V is the utility value assigned to *Al* according to *U*.
jmadem_utility_weight(U,W)	*U* is an utility name and $W \in \Re$ is its weight.
jmadem_welfare(W)	$W \in \{utilitarian, egalitarian, elitist, nash\}$.

expresses that an agent is not interested in sharing his vehicle if he is not travelling by too, following the utility function dummyUF. And:

```
jmadem_filter(dummyFilter,Alloc) :-
    .my_name(Myself) &
    owns(Myself,Vehicle) &
    .member(travel_by(_,Vehicle), Alloc) &
    not .member(travel_by(Myself,Vehicle),Alloc).
```

defines a filter to delete such instances from the set of allocations computed by the agent. In addition, J-MADeM provides a library of plans jmadem.asl to call MADeM processes as Achieve Goals. The trigger events recognized by these plans are listed in Table 3. Utilities and filters can also be defined as plans. For instance, the utility function in the previous example would be defined as a plan as follow:

```
+!jmadem_utility(dummyUF,_,Alloc,0) :
    .my_name(Myself)&
    owns(Myself,Vehicle) &
    .member(travel_by(_,Vehicle), Alloc) &
    not .member(travel_by(Myself,Vehicle),Alloc).
```

Table 3 Trigger Events used by J-MADeM agents.

Trigger Event	Description
+!jmadem_get_utility_function_names(U)	*U* is a list of utility names.
+!jmadem_construct_allocations(T,E,Al)	*T* is a set of task slots,
	E is a logic formula to compute the elements of the allocation, and
	Al is the resulting set of allocations.
+!jmadem_filter_allocations(F,Al,FAls)	*F* is a filter,
	Al is a set of allocations,
	FAls is a set of filtered allocations.
+!jmadem_launch_decision(A,Al,U,DId)	*A* is a set of agents,
	Al is a set of allocations,
	U is a list of utility function names,
	DId is a decision identifier.
+!jmadem_launch_decision1(A,Al,U,DId)	As above, but for 1 solution.

Then, Speech Acts with *AskHow*-like performatives can be used to exchange utilities and filters defined as plans. Interestingly, there is a plan for constructing allocations after the beliefs of an agent, finding all the allocations that satisfies a logical query E defined by the programmer. Thus "legal" allocations are computed directly. Alternatively, allocations can be further filtered by means of the achieve goal !jmadem_filter_allocations.

3   Urban Mobility Simulation Framework

In this section we introduce an urban mobility simulation framework developed over Jason that allows to model the mobility within a metropolitan area. This multi-agent system is highly configurable through XML configuration files, thus, it can be applied to different scenarios. For instance, the user can specify how many towns surround the city as well as the roads that interconnect them. For each of these entities, concrete parameters can be set such as: the number of habitants, the income per capita distribution, the transports available (e.g. car, train, bus), the length and flow of the roads, etc. The environment is based on a very simple traffic simulator that returns the real times and consumptions of each vehicle. On the other hand, each citizen is represented by an agent that uses the J-MADeM library to make decisions that balance individual and social preferences.

As a proof of concept, in this paper we present the "travelling to work" scenario. This scenario represents a 20 Km long road connecting a residential town to a city. Every morning, the habitants of this town must travel to the city to reach their workplaces. Each habitant owning a car can drive alone to work but he/she can also share the car with other habitants, thus lowering the expenses and the CO_2 emissions. Besides, there is the possibility to travel by train, which in the experiments is considered to emit no CO_2 and to cost 1 €/trip. Cars travel at an average speed of 100 Km/h and the train does at 60 Km/h, including all possible stops. However, as the road has a limited flow, when too many cars try to enter the city at the same time they will create a traffic jam, which may produce long delays. We have also set to 5 minutes the delay associated with both catching the train and picking-up each extra passenger in a shared car.

J-MADeM has been used in this scenario to model the main decision that habitants make every morning. That is, which transport to use for travelling to work: alone in their own car, sharing a car or by train. Citizens are randomly organized in decision groups meaning their family, friends, neighbors, etc. As we have fixed the maximum capacity of cars to 4 people, this is also the size of the decision groups. Therefore, the allocations used to represent each travel alternative for each group in this scenario are as follows:

$$alloc_i = [travel\_by(agent_1, vehicle_1), ..., travel\_by(agent_4, vehicle_4)] \qquad (1)$$

where $agent_i$ are the group members and $vehicle_i \in \{car_1, ..., car_4, train\}$ is the transport chosen by each member (logically, car_i belongs to $agent_i$). It should be noticed that, even though every habitant can travel in any car, it is a must that the owner of a car also travels in the car to be a valid allocation.

J-MADeM then collects the preferences of the group about every possible alternative. To express their personal preferences according to different points of view, habitants compute the utility functions defined in equation 2. Function UF_{eco} represents economy and it calculates the monetary cost of each allocation. Function UF_{tmp} informs about the travel time associated to the allocation and, finally, function UF_{CO_2} models its ecological impact in terms of the kilograms of CO_2 emitted to the environment. Consumption and travel times are estimated by remembering previous travel experiences with a similar vehicle-partners configuration. Utility functions represent costs in euros to be able to properly combine them in the J-MADeM process. Finally, J-MADeM selects the winner allocation, which is passed to the traffic simulator in the environment. For the winner determination, we use the Utilitarian collective utility function as an appropiated social welfare to reflect the aggregate impact of the kind of allocations considered.

$$UF_{eco}(alloc_i) = (Consumption(alloc_i) * PricePerLitre)/Partners(alloc_i)$$
$$UF_{tmp}(alloc_i) = Time(alloc_i) + Partners(alloc_i) * PickUpTime \qquad (2)$$
$$UF_{CO_2}(alloc_i) = (Consumption(alloc_i) * CO2PerLitre)/Partners(alloc_i)$$

Other works [14] have assumed that agents use different world views to interpret the climate change and, consequently, they have distinguished different types of policies based on cultural perspectives:

- *Hierarchical*: It assumes that nature is stable in most cases but it can collapse if we go beyond the limits of its capacity.
- *Egalitarians*: It assumes that the nature is highly unstable and the least human intervention may lead to a collapse.
- *Individualist*: It assumes that the nature provides plenty of resources and it will remain stable under human interventions. Essentially, it encourages strategies that maximize the economic growth.

In order to model the individualist and the egalitarian perspectives in the "travelling to work" scenario, we use the weights that J-MADeM allows to associate to each utility function. Thus, we can simulate the behavior of an individualist and an egalitarian society by using the utility weights in table 4

Table 4 Utility weights used for defining the Individualist and the Egalitarian societies

Type	UF_{eco}	UF_{tmp}	UF_{CO_2}
Individualist	1	0.1	0.1
Egalitarian	1	0.5	0.4

4   Results

This section shows the first results obtained when running the "travelling to work" scenario with 32 habitants for a period of 100 cycles or days. Following table 4, we have simulated the behavior of two types of habitants: a society of individualist and a society of egalitarian agents.

Figure 1 shows the outcomes produced by these societies in terms of the average travel time, the use of the urban transportation and the amount of CO_2 emitted to the environment. In the top left-hand corner, figure 1.a shows the average car/train travel time (in seconds) for both egalitarian and individualist populations. In the top right-hand corner, figure 1.b shows the total amount of kilograms of CO_2 emitted to the environment each day (cycle). In the bottom left-hand corner, figure 1.c shows the average number of passengers per car and, finally, figure 1.d shows the number of habitants travelling by each type of vehicle.

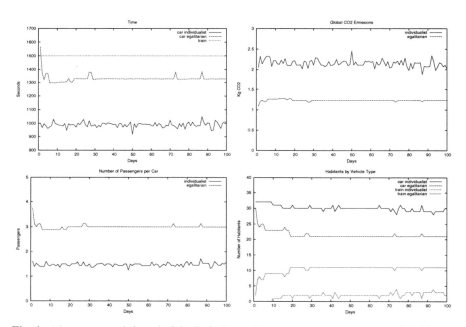

Fig. 1 a)Average travel time; b)CO_2 Emissions; c)Average passengers per car; d)Habitants per type of vehicle.

As shown in figure 1.a, individualist habitants are mainly interested in reducing the travel time. Hence, they usually prefer to travel by car (see the high number of agents using this type of vehicle in figure 1.d). Besides, they rarely share it with another partner as demonstrated by the low values in figure 1.c. Thus, this behavior is eventually reflected in the amount of CO_2 emitted by individualists which is higher than the pollution derived from the egalitarian society.

On the other hand, the average travel time of egalitarian citizens is logically higher (see figure 1.a) since they are also interested in balancing the CO_2 emissions and the monetary cost derived from their actions (see table 4). They manage to do this by increasing the degree of car sharing. For instance, when travelling by car, they normally share it with 2 other passengers (see figure 1.c). Additionally, a 30% of the habitants decides to travel by train (see figure 1.d). As a consequence of this behavior, the kilograms of CO_2 finally emitted by the egalitarian society is considerably reduced (see figure 1.b).

Although not included in this paper, we have also computed the delay incurred by both societies to verify that the simulation framework has been properly adjusted. The delay is calculated (for each day and agent) as the difference between the desired time to be at work and the arrival time coming from the simulator. The experiments carried out produce an average delay that converges quickly to a small negative value (around 1 minute), which indicates that the habitants are arriving just before they planned. This situation reveals that the scenario has been properly parameterized as different behavior emerge from the societies and both of them arrive on time.

5 Conclusions and Future Work

In this paper we have presented an urban mobility simulation framework developed over Jason that allows to model the mobility within a metropolitan area. The system uses the Jason Multi-modal Agent Decision Making (J-MADeM) library to model and implement the social decisions made by each habitant about how to get to work every day, e.g., by train, by car, sharing a car, etc. Therefore, the proposed approach focuses on the decision making aspects of this problem at a micro level, instead of focusing on the classical spatial or other macro level issues. The first results show the behavior of two societies of individualist and egalitarian citizens, which affect the average travel time, the use of the urban transportation and the amount of CO_2 emitted to the environment.

There is still work in progress to achieve the goal of developing a decision support platform for metropolitan planning. First, we are analyzing more complex scenarios that involve the use of new transports such as the bus or the bike. Second, we plan to extend the configuration files to include features such as the use of tolls or high-occupancy vehicle lanes. Regarding the infrastructure, we are currently studying the scalability of multi-agent systems in Jason so that we can run large-scale simulations.

Acknowledgements. This work has been jointly supported by the Spanish MEC and the European Commission FEDER funds, under grants Consolider-Ingenio 2010 CSD2006-00046 and TIN2009-14475-C04-04. The first author was also supported by the Spanish MEC under grant JC2009-00342. The last author is supported by Conacyt CB-2007-78910 fundings.

References

1. Bordini, R.H., Hübner, J.F., Wooldrige, M.: Programming Multi-Agent Systems in Agent Speak using Jason. Wiley, Chichester (2007)
2. Chevaleyre, Y., Dunne, P.E., Endriss, U., Lang, J., Lemaitre, M., Maudet, N., Padget, J., Phelps, S., Rodriguez-Aguilar, J.A., Sousa, P.: Issues in multiagent resource allocation. Informatica 30, 3–31 (2006)
3. E. Commission Towards a new culture for urban mobility (September 2007), http://ec.europa.eu/transport/urban/index_en.htm
4. Conte, R., Castelfranchi, C.: Cognitive and Social Action. UCL Press, London (1995)
5. Decker, K.S.: Task environment centered simulation. In: Simulating Organizations: Computational Models of Institutions and Groups, pp. 105–128. AAAI Press / MIT Press, Menlo Park (1998)
6. Dignum, V., Dignum, F.: Modelling agent societies: Co-ordination frameworks and institutions. In: Brazdil, P.B., Jorge, A.M. (eds.) EPIA 2001. LNCS (LNAI), vol. 2258, pp. 191–204. Springer, Heidelberg (2001)
7. Falcone, R., Pezzulo, G., Castelfranchi, C., Calvi, G.: Why a cognitive trustier performs better: Simulating trust-based contract nets. In: Proc. of AAMAS 2004: Autonomous Agents and Multi-Agent Systems, pp. 1392–1393. ACM, New York (2004)
8. Ferber, J., Gutknecht, O.: A meta-model for the analysis and design of organizations in multi-agents systems. In: Proc. of the 3rd International Conference on Multi-Agent Systems (ICMAS 1998), pp. 128–135. IEEE Press, Los Alamitos (1998)
9. Grimaldo, F., Lozano, M., Barber, F.: MADeM: a multi-modal decision making for social MAS. In: Proc. of AAMAS 2008: Autonomous Agents and Multi-Agent Systems, pp. 183–190. ACM, New York (2008)
10. Grimaldo, F., Lozano, M., Barber, F.: J-MADeM, an open-source library for social decision-making. In: Proc. of CCIA 2009: International Conference of the Catalan Association for Artificial Intelligence, pp. 207–214. IOS Press, Amsterdam (2009)
11. Grimm, V.: Ten years of individual-based modelling in ecology: what have we learned and what could we learn in the future? Ecological Modelling 115(2-3), 129–148 (1999)
12. Hexmoor, H.: From inter-agents to groups. In: Proc. of ISAI 2001: International Symposium on Artificial Intelligence (2001)
13. Hübner, J.F., Sichman, J.S., Boissier, O.: Developing organised multi-agent systems using the Moise+ model: Programming issues at the system and agent levels. International Journal of Agent-Oriented Software Engineering 1(3/4), 370–395 (2007)
14. Janssen, M.A., Ostrom, E.: Governing social-ecological systems. In: Tesfatsion, L., Judd, K.L. (eds.) Handbook of Computational Economics, 1st edn., vol. 2, ch. 30, pp. 1465–1509. Elsevier, Amsterdam (2006)
15. Kohler, T., Gumerman, G., Reynolds, R.: Simulating ancient societies. Scientific American 293(2), 76–83 (2005)
16. Prietula, M., Carley, K., Gasser, L. (eds.): Simulating Organizations: Computational Models of Institutions and Groups. AAAI Press / MIT press (1998)
17. Rao, A.S.: AgentSpeak(L): BDI agents speak out in a logical computable language. In: Perram, J., Van de Velde, W. (eds.) MAAMAW 1996. LNCS, vol. 1038, pp. 42–55. Springer, Heidelberg (1996)

18. Sichman, J., Demazeau, Y.: On social reasoning in multi-agent systems. Revista Ibero-Americana de Inteligencia Artificial 13, 68–84 (2001)
19. Tesfatsion, L.: Agent-based computational economics: A constructive approach to economic theory. In: Tesfatsion, L., Judd, K.L. (eds.) Handbook of Computational Economics, vol. 2, ch. 16, pp. 831–880. Elsevier, Amsterdam (2006)
20. Waddell, P.: Urbansim: Modeling urban development for land use, transportation and environmental planning. Journal of the American Planning Association 68, 297–314 (2002)

Diversity of the Knowledge Base in Organizations: Results of an Agent-Based Simulation

Friederike Wall

Abstract. The knowledge base used in organizations for decision-making usually is diverse due to various information systems or differing expertise of decision-makers. By an agent-based simulation the paper analyzes the question which level of diversity in the knowledge base in an organization is beneficial. The findings indicate that the preferable level of diversity subtly depends on the need and the mode to coordinate decisions in organizations. However, the findings provide support to rather unify the knowledge base on a medium level of quality than fragmenting and specializing the knowledge base. Furthermore, it appears that learning and adjustment capabilities in decentralized structures are more promising than the investment in a knowledge base as perfect as possible.

1 Introduction

An enduring challenge in organizations is to coordinate decisions of organizational units so that the organization's objective is achieved best possible. Among the problems is that decentral decision-makers usually dispose of diverse knowledge. This addresses the two sides of the same coin: Decisions are delegated to benefit from the expertise of specialists (e.g. [17, 3, 4]) while, at the same time, at least two problems emerge. Firstly, if knowledge differs among subunits there is no guarantee that these units estimate the same strategies to be preferable with respect to the organization's objective and, so, the decentral decisions might be unaligned. Secondly, asymmetrically allocated information may be used by self-interested decision-makers for their own best interest without that the headquarters or other decision-makers have the opportunity to detect this so-called opportunistic behavior. Powerful means to deal with these problems are, for example, *incentive systems* as well as *integrated information systems (IS)*. With incentive systems individual interests can be aligned to

Friederike Wall
Alpen-Adria-Universitaet Klagenfurt, Department of Controlling and Strategic Management
e-mail: friederike.wall@uni-klu.ac.at

Y. Demazeau et al. (Eds.): Adv. on Prac. Appl. of Agents and Mult. Sys., AISC 88, pp. 13–20.
springerlink.com   © Springer-Verlag Berlin Heidelberg 2011

the organization's objective and, so, opportunism is mitigated [13]. Integration of IS (e.g. [5, 15]) to a certain extent can unify the knowledge base of departmental decision-makers and contributes to mitigate opportunistic behavior as it most likely reduces information asymmetries. However, integration runs the risk to standardize what better should be distinct [6, 10]. Furthermore, different mental models and personal expertise are likely to exist regardless of the IS.

So, the very core of the problem remains: *Which level of diversity of the knowledge base in an organization is beneficial given the trade-off among specialization and coordination?* This represents the research question of the paper and can be put more into concrete for IS design in terms of *"What is the appropriate level of integration?"* [6, 5, 10] or might be related to the rather abstract issue of *"homogeneity or heterogeneity of resources"* in organizations (e.g. [7, 1]).

In order to investigate the research question an agent-based simulation is applied. Artificial organizations are sent to fitness landscapes and observed while searching for higher levels of performance. The agents of the model represent departments and the headquarters. Each agent is characterized by an information-level according to the particular expertise, decisional competencies, behavior and preferences. According to the three-dimensional framework of *agent-mediated knowledge management* elaborated by van Elst, Dignum and Abecker [2] our work

- relates to the stage of *analysis* in systems development
- allows to examine *homogeneous* as well as *heterogenous* multi-agent systems
- concentrates on the *utilization* of knowledge for decision-making.

So, we use an agent-based approach as a means to improve the understanding of the effects of diverse knowledge on organizational performance. For a more practical perspective the findings, for example, could contribute to identifying an appropriate level of IS integration and of personel expertise for various coordination needs.

2   Model

The interactions between the decisions that departmental managers are in charge of affects the need for coordination and, thus, reasonably represent an essential point for the effects of a diverse knowledge base. In order to deal with decisional interactions in a controllable way Kauffman's NK model [9, 8] is an appropriate framework. On this basis and similar to Siggelkow and Rivkin [19] we model organizations with self-interested departmental managers and coordination mechanisms. Though we use an advanced version of the NK model with adaptive walks on noisy fitness landscapes as introduced by Levitan and Kauffman [14] our model substantially is distinct as the knowledge base is diversified across the organization.

Our organizations face a ten-dimensional binary decision problem, i.e. have to make decisions $d_i \in \{0,1\}$ with $i = 1,..10$ (so $N = 10$). Each single state of decision d_i provides a uniformly distributed contribution C_i with $0 \leq C_i \leq 1$ to the overall performance $V(\mathbf{d})$ of the organization. In order to map interactions among decisions, C_i does not only depend on the decision d_i but also on K_i other decisions

$d_{j,j\in\{1,...K_i\},j\neq i}$ so that $C_i = f_i(d_i; d_{j=1},...,d_{j=K_i})$. The overall performance $V(\mathbf{d})$ of an organization is given as normalized sum of performance contributions C_i with

$$V(\mathbf{d}) = \frac{1}{N}\sum_{i=1}^{N}C_i = \frac{1}{N}\sum_{i=1}^{N}f_i(d_i; d_{j=1},...,d_{j=K_i}) \qquad (1)$$

where $j \neq i$. In our model the agents are the headquarters and $r \in \{1,2,3\}$ departments. Each department is responsible for a partial decision problem (i.e. some of the ten single decisions). In each period a department searches for the best configuration for its "private" partial decision problem and, while doing so, assumes that the other departments do not alter their prior choices.

The evaluation of configurations clearly depends on the incentives. For simplicity the departmental decision-makers are given firmwide incentives so that no conflicts of interest occur, i.e., departments seek to decide in the organization's best interest (thus, we can ignore issue 2 as mentioned in the beginning of the paper and concentrate on the diverse knowledge base). In particular, the compensation in a linear incentive scheme depends on two components, (1) the "own" performance P_r^{own} which is the sum of those C_i related to the subset of i decisions a department head is in charge of and (2) the "residual" performance P_r^{res}, i.e., the performance contributions of those decisions other departments are responsible for . As the head of department r seeks to maximize compensation that configuration $\mathbf{d}$ is preferred that maximizes the value base B_r for compensation. With cross-departmental interactions among decisions a departmental manager r might affect P_r^{res}, and also P_r^{own} might be affected by decisions of other departments.

Our organizations can operate under two alternative *coordination modes* [19]: either the departments propose configurations $\mathbf{d}$ according to their preferences to the headquarters which chooses that one with the highest overall performance ("central") or each department autonomously decides on the "own" partial decision problem ("decentral"). In the latter case the overall configuration $\mathbf{d}$ results as a combination of these departmental decisions without any central intervention.

Our agents use information for identifying superior solutions according to the known contribution to compensation or overall performance, respectively. To be more precise on the *diversity of knowledge base* the model distinguishes three dimensions of diversity:

- *Spread of specialized and general knowledge of a single decision-maker:* Each decision-maker is informed about the decisions he/she is responsible for and the resulting consequences for the own area of competence. Additionally, a decision-maker disposes of information related to the rest of the organization. In this dimension diversity depends on whether, for example, the decision-maker's knowledge is more specialized for the own area of competence or rather generalized with respect to the whole organization.
- *Number of different views within the organization:* According to the aforementioned dimension each decision-maker has a different view (or knowledge) of the organization's decisional problems due to specialization. So, the knowledge base that is applied is the more diverse the more the decision-problem is fragmented

and the higher the number of decision-makers (or departments) that these "partial" decisions are delegated to.

- *Diversity of information-processing capabilities among decision-makers:* Some decision-makers might be very well informed whereas others work on a very noisy basis of information; some might act on a specialized knowledge base according to the first dimension whereas others have no certain expertise.

We assume that departments decide on basis of the *perceived* value base B_r^* for compensation rather than the true value base. Therefore, we "distort" the true performance contributions of the single decisions d_i according to the dimensions of diversity as introduced above. In particular, the performances a department r *perceives* result as sum of true performances and an error term like

$$P_r^{*own} = P_r^{own} + e_r^{own} \tag{2}$$

and

$$P_r^{*res} = P_r^{res} + e_r^{res} \tag{3}$$

Accordingly, in case of coordination mode "central" the headquarters' makes the choice of the proposals on basis of the *perceived* overall performance $V^*(\mathbf{d})$ given as sum of the true overall performance $V(\mathbf{d})$ and an error term e_{head}. At least, with respect to accounting systems [12], it is reasonable that high (low) true values of performance come along with high (low) distortions. So, we reflect distortions as *relative* errors imputed to the true performance (for other functions [14]), and, for simplicity, the errors follow a Gaussian distribution $N(\mu;\sigma)$ with expected value $\mu = 0$ and standard deviation σ.

Furthermore, the organizations might have learning capabilities. To incorporate "learning" the errors related to a certain configuration $\mathbf{d}$ are reduced (with decreasing rates) whenever this configuration is realized (again) by the organization [21].

3   Results and Interpretation

The simulated organizations are thrown randomly somewhere in the fitness landscape in order to search for configurations $\mathbf{d}$ with superior levels of performance $V(\mathbf{d})$. As familiar for adaptive walks we use a hill-climbing algorithm: In the neighborhood of the status-quo-configuration our departments find two alternatives so that each department knows three options. Four scenarios of diversity in the knowledge base are simulated (table 1):

In scenario "Perfect" all decision-makers have perfect knowledge and, in consequence, there exists only one view of the decision-problem. This rather theoretical scenario serves as a kind of "benchmark". In scenario "Special" departmental managers are well informed about their own task but with quite vague knowledge about the cross-departmental effects. In scenario "General" the expertise is lower, but the organization-wide knowledge is better. Thus, among these two scenarios a "trade-off" between expertise and general knowledge is incorporated. In scenario "Nescient" one of the three departments has very vague knowledge. With "central"

Table 1 Scenarios of knowledge base in organizations

Dimension of diversity	Scenario "PERFECT"	Scenario "SPECIAL"	Scenario "GENERAL"	Scenario "NESCIENT"
Spread of knowledge on departments' site	$\sigma_r^{own} = 0$ $\sigma_r^{res} = 0$ $(\sigma_{head} = 0)$	$\sigma_r^{own} = 0.05$ $\sigma_r^{res} = 0.2$ $(\sigma_{head} = 0.12)$	$\sigma_r^{own} = 0.1$ $\sigma_r^{res} = 0.15$ $(\sigma_{head} = 0.12)$	like SPECIAL but dpmt. 3: $\sigma_3^{own} = 0.15$ $\sigma_3^{res} = 0.3$
No. diff. views	1	3 (4)	3 (4)	3 (4)
Differences in info.-processing capabilities	no	no	no	yes

coordination the headquarters' knowledge which is adjusted to a medium level of noise is involved in decision-making and, by that, a fourth perception is involved. The order of magnitude of the error terms was based on insights from organizational theory (e.g. [17, 3, 4, 7]) and calibrated due to findings on the information quality in organizations (e.g. [12, 16, 20]). With incorporating learning the diversity of the knowledge base diminishes in the course of the adaptive walk as the error terms are reduced according to the experience with the area in the fitness landscape.

The simulations were conducted for various interaction structures of decisions (i.e. coordination needs). Tables 2 and 3 report results for two exemplary *intensities of interactions* among decisions (for these and other interaction structures see [18]) which in a way represent two extremes: In the "low" case intra-departmental interactions among decisions are maximal intense while no cross-departmental interdependencies exist. In contrast, in the "high" case all decisions affect the performance contributions of all other decisions, i.e., the intensity of interactions and the coordination need is raised to maximum. While Speed 1 in the tables represents the performance improvement achieved in the first period, Speed 2 is able to reflect in the purest way possible the effect of learning because in period 2 for the first learning might affect decision-making. The average performance over 200 periods as well as the level of performance achieved in the end of the observation time might be regarded as condensed measures for the effectiveness of the search process.

In order to go more into the details and interpretation of the results, we begin with comparing the "Perfect" scenario to the others, then discuss the effect of learning and, subsequently, address the effect of the coordination mode.

So, comparing the "Perfect" scenario to the other scenarios the results provide broad, but no general support for conventional wisdom whereby decision-making based on perfect knowledge leads to higher speed and level of performance than achievable with imperfect knowledge. Two reasons support conventional wisdom [14]: (1) a *false positive* option appears favorable to decision-makers, whereas, in fact, it reduces performance compared to the status-quo level; (2) a *false negative* option appears not beneficial and is rejected while the current configuration of decisions **d** is perpetuated. Interestingly, the results show some exceptions to conventional wisdom and we return to that issue below.

Table 2 Performance improvements in decisions with low coordination need

Scenario	Learn (Y/N)	Speed 1 $(V_{t=1}$ $-V_{t=0})$	Speed 2 $(V_{t=2}$ $-V_{t=1})$	Average Perform. $(\overline{V}_{T^{200}})$	CI^a of $\overline{V}_{T^{200}}$	Perform. in $t=200$ $(V_{t=200})$	CI^a of $V_{t=200}$
Coordination mode "decentral"							
Perfect	-	0.18258	0.05122	0.98515	±0.00122	0.98726	±0.00123
Special	N	0.11481	0.01850	0.85214	±0.00247	0.85958	±0.00380
General	N	0.12914	0.02551	0.88003	±0.00245	0.88577	±0.00341
Nescient	N	0.10024	0.02101	0.83488	±0.00265	0.84192	±0.00424
Special	Y	0.11219	0.02876	0.90858	±0.00169	0.96314	±0.00293
General	Y	0.12813	0.03401	0.94707	±0.00133	0.98316	±0.00151
Nescient	Y	0.10028	0.02747	0.87241	±0.00183	0.92434	±0.00429
Coordination mode "central"							
Perfect	-	0.11973	0.07345	0.98488	±0.00117	0.98799	±0.00118
Special	N	0.08440	0.04328	0.87860	±0.00367	0.88062	±0.00373
General	N	0.08888	0.04558	0.88370	±0.00356	0.88573	±0.00362
Nescient	N	0.08273	0.04379	0.87357	±0.00369	0.87567	±0.00376
Special	Y	0.08288	0.05553	0.95359	±0.00183	0.96255	±0.00193
General	Y	0.08609	0.05799	0.95471	±0.00183	0.96260	±0.00192
Nescient	Y	0.08199	0.05313	0.94969	±0.00189	0.95936	±0.00200

^{a} Confidence intervals at a confidence level of 0.001

Each row represents 5000 adaptive walks, i.e. 5 walks per each of 1000 landscapes.

Not surprisingly learning capabilities apparently enhance speed of performance improvements and the average and end level of performance $V_{t=200}$ achieved within the observation period. With learning capabilities the knowledge base of all parties is improved - and inevitably more and more unified in direction of perfect knowledge.

In the *decentral* coordination mode and without learning the results indicate that organizations tend to achieve higher levels of performance with a more "generalist"-like knowledge base than with the "specialist" base. This provides support for the idea of rather integrating IS than having isolated specialized IS. However, with the *central* coordination mode in a way things seem to change: Firstly, the differences in performance level $V_{t=200}$ achieved between the scenarios with diverse knowledge base diminish. Obviously, involving the headquarters into decision-making equalizes differences of the various types of diversity of knowledge base. However, the performance levels achieved with central coordination are in some cases higher, in others lower than without involving the central information-processing power.

As mentioned above, contrary to intuition for higher levels of complexity decision-making on basis of diverse imperfect knowledge leads to *superior* overall performance $V_{t=200}$ than achieved with perfect knowledge. The reason might lie in the "false positive" decisions: With these an organization goes a "wrong way" for the short term, but with the chance to discover superior configurations. So, imperfect knowledge may afford the opportunity to leave a local peak in the fitness

Table 3 Performance improvements in decisions with high coordination need

Scenario	Learn (Y/N)	Speed 1 $(V_{t=1} -V_{t=0})$	Speed 2 $(V_{t=2} -V_{t=1})$	Average Perform. $(\overline{V}_{T^{200}})$	CI^a of $\overline{V}_{T^{200}}$	Perform. in $t=200$ $(V_{t=200})$	CI^a of $V_{t=200}$
Coordination mode "decentral"							
Perfect	-	0.04768	0.03538	0.87551	±0.00267	0.88092	±0.00273
Special	N	0.03519	0.01932	0.84131	±0.00320	0.86149	±0.00343
General	N	0.04026	0.02223	0.84912	±0.00309	0.86334	±0.00329
Nescient	N	0.03019	0.01865	0.82576	±0.00333	0.85036	±0.00361
Special	Y	0.03787	0.02423	0.87121	±0.00276	*0.89703*	±0.00277
General	Y	0.03973	0.02597	0.87547	±0.00265	*0.89196*	±0.00276
Nescient	Y	0.03328	0.01938	0.85681	±0.00298	0.89613	±0.00303
Coordination mode "central"							
Perfect	-	0.15903	0.03206	0.88159	±0.00271	0.88314	±0.00274
Special	N	0.13546	0.02534	0.83709	±0.00351	0.83826	±0.00355
General	N	0.13249	0.02816	0.83890	±0.00350	0.84010	±0.00355
Nescient	N	0.12953	0.02591	0.83200	±0.00360	0.83313	±0.00364
Special	Y	0.13185	0.03321	0.86789	±0.00297	0.87093	±0.00303
General	Y	0.13042	0.03256	0.86863	±0.00291	0.87139	±0.00296
Nescient	Y	0.12685	0.03167	0.86526	±0.00298	0.86842	±0.00304

^{a} Confidence intervals at a confidence level of 0.001 Notes: See table 2

landscape. Additionally, with higher interactions more local maxima exist and, so, the organization more likely might stick to a local maximum (e.g. [18, 11, 21]). Thus, the beneficial effect of imperfections might increase with the intensity of interactions. Interestingly, the "beneficial" effect in particular appears with learning capabilities: Possibly, feedback and adjustments as counterpart for making "false positive" decisions are supporting condition for this effect of imperfect knowledge.

4 Conclusion

The findings show that the effects of diversity in the knowledge base that an organization uses for decision-making subtly vary with the coordination needs and the coordination mode. If at all, it appears that a generalist-like knowledge base is superior to more specialized systems. This provides support for the idea of integrated information systems. However, involving central knowledge into decision-making obviously mitigates differences among the alternative knowledge scenarios. Furthermore, in complex and decentralized situations apparently the combination of diverse knowledge and effective adjustment mechanisms might be more appropriate than investments in a "perfect" knowledge base. Obviously, these insights put ambitions for perfect information systems in perspective.

References

1. DeSarbo, W.S., Di Benedetto, C.A., Song, M.: A heterogeneous resource based view for exploring relationships between firm performance and capabilities. Journal of Modelling in Management 2, 103–130 (2007)
2. van Elst, L., Dignum, V., Abecker, A.: Towards Agent-Mediated Knowledge Management. In: van Elst, L., Dignum, V., Abecker, A. (eds.) AMKM 2003. LNCS (LNAI), vol. 2926, pp. 1–30. Springer, Heidelberg (2004)
3. Galbraith, J.R.: Designing Complex Organizations. Addison-Wesley, Reading (1973)
4. Ginzberg, M.J.: An Organizational Contingencies View of Accounting and Information Systems Implementation. Account Org. Soc. 5, 369–382 (1980)
5. Giachetti, R.E.: A framework to review the information integration of the enterprise. Int. J. Prod. Res. 42, 1147–1166 (2004)
6. Goodhue, D.L., Wybo, M.D., Kirsch, L.J.: The Impact of Data Integration on the Costs and Benefits of Information Systems. MIS Quart. 16, 293–311 (1992)
7. Hitt, M.A., Ireland, R.D.: Relationships among corporate level distinctive competencies, diversification strategy, corporate structure and performance. Journal of Manage. Stud. 23, 401–416 (1986)
8. Kauffman, S.A.: The origins of order: Self-organization and selection in evolution. Oxford University Press, Oxford (1993)
9. Kauffman, S.A., Levin, S.: Towards a general theory of adaptive walks on rugged landscapes. J. Theor. Biol. 128, 11–45 (1987)
10. Kappos, A., Rivard, S.: A three-perspective model of culture, information systems and their development and use. MIS Quart. 32, 601–634 (2008)
11. Knudsen, T., Levinthal, D.A.: Two Faces of Search: Alternative Generation and Alternative Evaluation. Organ. Sci. 18, 39–54 (2007)
12. Labro, E., Vanhoucke, M.: A simulation analysis of interactions among errors in costing systems. Account Rev. 82, 939–962 (2007)
13. Lambert, R.A.: Contracting theory and accounting. J. Account Econ. 32, 3–87 (2001)
14. Levitan, B., Kauffman, S.A.: Adaptive walks with noisy fitness measurements. Mol. Divers. 1, 53–68 (1995)
15. Liu, X., Zhang, W.J., Radhakrishnan, R., Tu, Y.L.: Manufacturing perspective of enterprise application integration: the state of the art review. Int. J. Prod. Res. 46, 4567–4596 (2008)
16. Madnick, S., Wang, R.: Introduction to the TDQM Research Program. Sloan School of Management, Massachusetts Institute of Technology (1992), Available via DIALOG http://web.mit.edu/tdqm/papers/92/92-01.html (Cited 30 December 30, 2010)
17. Marschak, J.: Towards an Economic Theory of Organization and Information. In: Thrall, R.M., Coombs, C., Davis, R.L. (eds.) Decision Processes. Wiley, New York (1954)
18. Rivkin, R.W., Siggelkow, N.: Patterned interactions in complex systems: Implications for exploration. Manage Sci. 53, 1068–1085 (2007)
19. Siggelkow, N., Rivkin, J.W.: Speed and search: Designing organizations for turbulence and complexity. Organ Sci. 16, 101–122 (2005)
20. Tee, S.W., Bowen, P.L., Doyle, P., Rohde, F.H.: Factors influencing organizations to improve data quality in their information systems. Account Financ. 47, 335–355 (2007)
21. Wall, F.: The (Beneficial) Role of Informational Imperfections in Enhancing Organisational Performance. In: LiCalzi, M., Milone, L., Pellizzari, P. (eds.) Progress in Artificial Economics: Computational and Agent-Based Models. Lecture Notes in Economics and Mathematical Systems, vol. 645. Springer, Heidelberg (2010)

Weighing Communication Overhead against Travel Time Reduction in Advanced Traffic Information Systems

Rutger Claes and Tom Holvoet

Abstract. Advanced traffic information systems can assist drivers in reducing their travel times by making better use of available road capacity. In assessing their practical applicability, however, it is important to assess the overhead that various advanced traffic information systems bring. This paper evaluates the communication overhead for a decentralized, delegate multi-agent systems based advanced traffic information system, and for a centralized system. We document the relationship between the communication overhead and travel time reduction for both systems. This analysis can help weighing both factors when designing a practical traffic information system in a real-world scenario.

1 Introduction

The rising traffic demand continues to fuel development of advanced traffic information systems, or ATISs, that improve the performance of the traffic networks. Performance of traffic networks can be improved by making better use of the networks capacity and avoiding congestions. The implementation of ATISs result in a decrease of the total time spent by vehicles in the network.

The coordination of drivers in most ATISs depends on the sharing and spreading of information. Communication thus is key to the success of these systems. The overhead introduced by communication, is a cost that has to be balanced with the benefits caused by the ATISs. The tradeoff thus created is hard to explore. Measuring the cost of an ATIS in a real world setting is nearly impossible, as is measuring the benefits the system produces.

This paper explores the balance between communication overhead and decrease in vehicle travel times for two ATIS approaches. The first ATIS is decentralized and

Rutger Claes · Tom Holvoet
DistriNet Labs, Department of Computer Science, Katholieke Universiteit Leuven,
Celestijnenlaan 200A, 3001 Heverlee, Belgium
e-mail: rutger.claes@cs.kuleuven.be, tom.holvoet@cs.kuleuven.be

Y. Demazeau et al. (Eds.): Adv. on Prac. Appl. of Agents and Mult. Sys., AISC 88, pp. 21–31.
springerlink.com

uses link travel time predictions to guide vehicles, the second ATIS is centralized and based on real-time information. Link travel time predictions and the benefits they can introduce, even at low penetration rates, are described by Wunderlich et al. in [12]. Most link travel time prediction systems rely on information sent by the vehicles indicating their intentions and plans. The link travel time predictions used in the decentralized ATIS are generated using delegate multi-agent systems and online embedded simulations. These online simulations require information and thus communication. The link travel time predictions thus generated can lead to faster throughput times and shorter travel times for the vehicles participating in the system.

The second ATIS is a centralized system build around a central traffic information center. This approach aims to limit communication overhead by providing vehicles with real-time values instead of predictions for the link travel times. By avoiding forecasts for link travel times, the centralized ATIS eliminates the need for knowing the vehicles' intentions. Real time information is gathered from all roads in the network, but this causes less communication overhead. Both the decentralized and the centralized system are described more in-depth in respectively Section 2 and Section 3.

This paper uses simulation to evaluate both systems. The simulation includes a microscopic simulation of the traffic guided by the ATISs as well as the communication that is generated while guiding the traffic. Section 4 describes the experimental setup.

The benefits of both ATISs are evaluated by analyzing the sum and overall distribution of the travel time of all vehicles in the system. As the ATISs guide the vehicles more effectively, the travel time of the vehicles should decrease. By analyzing both the total travel time and the travel time distribution we include fairness in the evaluation. Section 5 presents and analyses the results of the experiments.

The costs of both ATISs are evaluated by comparing the number of messages sent in both approaches. As the ATISs require more information exchange, the communication overhead will be higher. This paper focusses on the balance between the benefits and costs of information exchange, and looks for an tradeoffs in which the benefits are significant while the overhead is manageable.

The paper concludes with an overview of related work in Section 6 and a conclusion in Section 7.

2 Delegate Multi-Agent Systems Based ATIS

The decentralized ATIS providing link travel time predictions is based on earlier work described in [2, 11] and uses the concept of *delegate Multi-Agent systems*, introduced in [4] combined with *online simulations*, as presented in [1, 5]. In this section we limit the description to an outline of the key principles of these techniques.

In this approach, vehicles are represented by *vehicle agents* and roadside infrastructure elements such as roads and junctions are represented by *infrastructure agents*. Both type of agents are positioned in a *virtual environment* that represents the actual real world environment: this virtual environment offers the agents means of communication and perception of the real world.

Coordination in this decentralized system is achieved using delegate multi-agent systems. Delegate multi-agent systems use lightweight agents, called *ants* to facilitate communication between agents. These ants move through the environment and can be considered smart messages that are forwarded from location to location.

Information about the vehicles' intentions is propagated by the vehicles using a delegate multi-agent system consisting of *intention ants*. These lightweight agents are dispatched by the vehicle agent at its current location. The intention ants then traverse the path the vehicle agent wants to use. Every infrastructure agent representing an infrastructure element on this path is notified of the pending visit. This notification includes the characteristics of the visiting vehicle and the time it intends to arrive. The notification is treated by the infrastructure agent as a pheromone, meaning the value of the information diminishes - or evaporate - over time. This evaporation has two consequences: (1) the vehicle agent has to resend intention ants at regular intervals; and (2) whenever a vehicle agent changes its intentions, it can disregard old notifications as they will evaporate automatically. The pheromone-like nature of the notifications thus ensures information about pending vehicle visits is kept up to date.

The information needed by the vehicle agents to decide on which route it will intends to follow is gathered from the infrastructure elements using a second delegate multi-agent systems. The lightweight agents in this second delegate multi-agent system are called *exploration ants* and explore possible routes through the traffic environment. During their exploration of a possible route, the exploration ants interact with the infrastructure agents they encounter. As exploration ants traverse the environment, they keep track of the time it would take the vehicle they represent to reach their location. They use this horizon to ask infrastructure agents link travel time predictions at the time the vehicle they represent would pass the infrastructure element, represented by the current horizon. Using this prediction, they can update their horizon and progress to the next infrastructure element. On reaching their destination, the ants report back to the vehicle about the expected travel times if the vehicle were to choose this route.

To constrain the exploration ants on their exploration journey, the vehicle agent limits the number of possible routes they explore. The vehicle agent uses an A^* based algorithm to select K feasible paths based on its own beliefs about the link travel times of the network. The exploration ants explore these K paths, and report their findings back to the vehicle agent which can then update its beliefs about these paths. From this description, it is clear that the choice of K greatly influences the communication overhead introduced by the exploration. The benefits of exploring more of the network as K rises, can lead to better performance. As vehicle agents have more information about potential routes, the options of avoiding congestion increase. Especially in traffic networks that offer many alternative routes to any destination.

3   A Central Advanced Traffic Information System

The distributed ATIS described in the previous section is evaluated by comparing it to a centralized alternative. The centralized ATIS emphasizes low communication over performance and uses real-time data instead of link travel time predictions. Developing a centralized ATIS that can provide link travel time predictions is possible, as discussed in section 6, but would require much more communication overhead.

The centralized ATIS implementation in this paper is a modification of the real world traffic guidance based on the Traffic Messaging Channel (TMC). In the implementation described here, information is provided to the vehicles by a central traffic information center (TIC). The information is broadcasted to the vehicles using the TMC. Where technical limitations limit the set of roads on which vehicles receive information in the real world system, the implementation used here disregards these limitations and allows vehicles to request real-time information on all roads.

Vehicles can request real-time information about a number of routes through the traffic network. The TIC responds with real-time link travel times for all links in those routes. This additional information can be used by the vehicles to make better choices when faced with congestion and makes the guidance offered by the central system more competitive with that offered by the decentralized system previously described.

This adaptation implies the use of two way communication between TIC and vehicle, a requirement the TMC based approach does not have. A second consequence of the adaptation is that the central TIC has to monitor all links in the network instead of just a limited set. The communication required by these adaptations, the two way communication between vehicles and TIC, and the communication required to inform the TIC about the current state of the road, is also included in the evaluation.

Using the information provided by the centralized ATIS described above, vehicle agents can make informed decisions on what route to take to reach their destination. Allowing the vehicles to request real-time information about all links in the network would put too much stress on the communication. Vehicles are allowed to request information about K routes. Such a request only counts for one message in the evaluation. The TIC replies to such a request with the real-time link travel times for all K routes. This reply also counts for just one message.

Based on local information from either a static network description or previously requested information, the vehicle agent uses an A^* based algorithm to construct K feasible routes to its destination. This process is very similar to that used by the vehicle agent using the decentralized ATIS when deciding how to send out the exploration ants. The vehicle agent using the centralized ATIS requests the link travel times from the central TIC and merge the response into its beliefs about the network. It then calculates the best route to its destination using the A^* algorithm and the newly updated information.

To avoid myopic behavior, the vehicle agents will hesitate before switching intentions. A newly discovered route has to be significantly better that the currently pursued route in order to switch intentions. Allowing myopic switching whenever a better route is discovered leads to instability and degrading performance.

Every minute, all road side elements inform the central TIC about the conditions on their links. This communication is essential for the TIC to be able to relay this information to the vehicles. The implementation assumes no delays. Information is sent by the road side elements to the central TIC and can be requested by the vehicle agents immediately. The updates sent by the road side elements are unidirectional. The TIC never contacts a road. Every update thus counts for only one message.

4 Experiment Setup

The experiments used to evaluate the ATISs described in the previous sections are based on simulations. This section will first describe the simulation used to obtain experimental results. Next, the experimental setup is described. This includes how the vehicles are generated and how the ATISs are deployed in the simulated environment.

The simulation environment models both the movement of the vehicles and the communication between agents deployed on vehicles or infrastructure elements.

The simulation environment only monitors the number of messages sent, it currently disregards the message size and the path the message traverses to reach its endpoint. When using delegate multi-agent systems to coordinate the vehicles' route choice, ants are considered to be smart messages and every movement of the ant counts as one message. Because ants always travel between two adjacent infrastructure elements, the discarded routing overhead will be minimal. The central system benefits most from the simplification. As all road side elements try to update the information at the central service.

Traffic in our evaluation is simulated on a microscopic level. Traffic is modeled in a multi-lane model. Traffic lights are not included in the simulation, all junctions have priority rules. This model allows a large scale simulation of traffic while maintaining a microscopic vehicle representation thus allowing an online agent based coordination of the vehicles. The simulated traffic environment is based on a real world city. The evaluation takes place in and around the city of Leuven. When choosing the simulated environment for our evaluations a balance is sought between the scale of the simulated environment and the intensity of the traffic. In order to assess the benefits of our approach, a saturated road network is simulated. The desired traffic intensity on the roads severely limits the size of the simulated area. Simulating a medium sized belgian city allows us to simulate saturated traffic on a reasonable scale.

Information about the simulated environment is extracted from OpenStreetMaps. OpenStreetMaps offers detailed information about road layout and characteristics. This allows the simulation to take into account the number of lanes, junction types, speed limits and other information that affects traffic. The resulting traffic network contains over 1600 links and 1250 junctions.

Low level driver behavior is modeled using a microscopic car following behavior [9], lane changing is done myopic. The drivers' behavior is independent of the routing approach chosen. Vehicles guided by both approaches will react identical in traffic.

All simulations start using the same OD matrix describing the origin and destination of an artificial population of drivers. This OD matrix is constructed by randomly picking origins and destinations on opposite sides of the city center. The selection algorithm thus favors trips across the city. It is important to note that all simulations, regardless of the traffic saturation, the OD matrices and thus the vehicles and their trips are identical.

Simulations use either the centralized or the decentralized ATIS to guide the vehicles to their destination. Four series of experiments are used: one using a centralized ATIS, referred to as *tic* and three other using a decentralized ATIS with different values for the K parameter. The K values are chosen to be 2, 5 and 10. Empirical results indicate that a value of $K < 5$ severely restricts the information obtained by vehicle agents, thus reducing the effectiveness of the ATIS. At values of $K > 10$, performance no longer improves as the increase in information is no longer valuable. These empirical observations are highly dependent on the characteristics and size of the traffic network used in the simulation.

The simulations using the centralized ATIS use a value of $K = 10$, the highest K value used in the decentralized ATIS simulations. This ensures the vehicles in the *tic* simulations have knowledge of 10 different routes, but receive this information by only sending and receiving one message. Increasing the K value would only increase the size of the message, something the analysis disregards.

All four series of experiments consist of 5 simulations. In every simulation the traffic saturation is increased. Within each series of experiments describing an ATIS, traffic saturations of 20, 40, 60, 80 and 100% are simulated. The maximum saturation is determined empirically, but is identical for all four series. The saturation is increased without altering the origin or destination of vehicles. The vehicles described in the OD matrix are simply introduced into the network at a faster rate. Every simulation will have 18000 vehicles.

5 Experiment Results

As previously described, the evaluation focusses on the number of messages exchanged between agents in both ATISs related to the performance obtained by reducing travel durations. First the travel durations for all previously described setups are shown and analyzed, next the messages needed to obtain these results are analyzed.

Figure 1a depicts the total travel duration, i.e. the sum of all time spent by vehicles to reach their destination. Because of the increasing saturation of traffic, the time it takes all vehicles to reach their destination increases in both the experiments using the decentralized ATIS as in those using the centralized ATIS. Because the decentralized ATIS allows the vehicles to take into account predictive link travel times instead of real-time travel times, the total travel duration is less for the decentralized ATIS. The decentralized ATIS using delegate Multi-Agent Systems with a parameter K of only 2 performs significantly worse than the other decentralized ATISs, especially at higher traffic saturations. As saturation increases, the limited amount of information received by the vehicle agents no longer allows them to choose suitable congestion free routes.

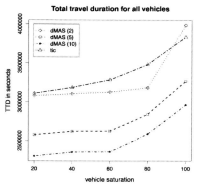 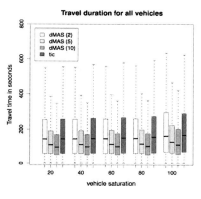

(a). Total travel duration   (b). Travel duration distribution

Fig. 1 Evolution of the travel duration for all vehicles as traffic saturation rises. *dMAS(K)* denotes the decentralized delegate multi-agent based ATIS with parameter K. *tic* stands for the centralized ATIS.

These observations are confirmed by looking at the distribution of travel times for all vehicles in the system. Figure 1b shows boxplots representing the entire population of vehicles in every simulation. These boxplots show how the travel times are distributed. Each boxplot illustrates the minimum travel time at the bottom and the maximum travel time at the top. The box in the middle of each boxplot starts at the fist quartile and ends at the third quartile. The line dividing the box into two shows the median or second quartile value.

The boxplots in 1b show that not only the mean, but also the third quartile and maximum value of the vehicle travel times is less when using decentralized ATISs with a parameter K of at least 5. The distribution of travel times is a measure of the fairness of the system, a criterium often neglected during evaluations.

The benefits of the decentralized ATIS and the high values for the K parameter come at a cost. Figure 2a shows the total amount of messages passed between agents as the ATISs try to guide the vehicles. The line showing the total message count of the centralized ATIS includes the messages between infrastructure agents and the TIC. Even with these messages included, the number of messages sent in the decentralized ATISs is higher.

It is worth noting that with the centralized ATIS, around 100.000 of the messages are sent between the road side infrastructure agents and the TIC. The rest of the messages are send between the TIC and the vehicle agents. All these messages are sent to or from one location, the TIC, thus creating a bottleneck.

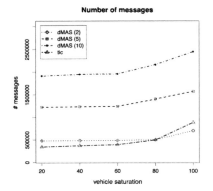
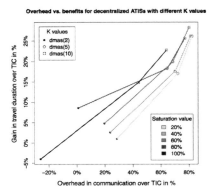

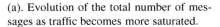

(a). Evolution of the total number of mes-   (b). Tradeoff between benefits and over-
sages as traffic becomes more saturated.   head for various values of K

Fig. 2

Figure 2b shows the tradeoff between the benefits and the overhead caused by decentralized ATISs for various values of the parameter K. As K increases, the points move to the top right corner, indicating that the performance improves and the overhead increases. The lines show for each traffic saturation value how both benefits and overhead evolve as K changes. If an acceptable communication overhead is known, Figure 2b can be used to estimate a reasonable K value.

The results presented in this section show that for all acceptable values of the exploration parameter K the overhead introduced by the decentralized system is higher. When comparing communication in decentralized and centralized systems this is often to be expected. This paper, nevertheless, argues that a decentralized ATIS offers benefits to scalability. Even though the total message count in the decentralized ATIS is higher than that of the centralized ATIS, it spreads the communication overhead more evenly over the system. The centralized system, on the contrary, focusses all messages - along with most computations - in one central location. For ATISs, scalability clearly is a requirement. This requirement suggests a decentralized approach. The bottlenecks introduced by the centralized system described in this paper threaten the scalability of the centralized approach.

Using delegate multi-agent systems to propagate and retrieve information from the distributed environment limits the communication overhead. If all distributed entities would communicate with each other directly, the communication overhead shown in Figure 2a would be higher. The use of delegate multi-agent systems allows the vehicles to share its intentions only with interested infrastructure agents and to request information only from relevant infrastructure agents. Managing the audience that receives the information and keeping this audience as small as possible reduces the communication overhead.

The use of delegate multi-agent systems has a second benefit over direct communication. Because the ants are smart messages, they are capable of keeping track of the time horizon as they travel from network link to network link. Updating this horizon is necessary to request the correct link travel time prediction at the next infrastructure element. With direct communication, the vehicle agent would request information from an infrastructure element, wait for the response, calculate the horizon and only then request information from the next infrastructure element. This would double the number of messages needed to obtain an estimate time of arrival for a route the vehicle agent is interested in. The lightweight ant agents take advantage of the path they have already traversed and the information they have accumulated when making the next request.

6 Related Work

The benefits of link travel time predictions, such as those provided by the delegate MultiAgent systems approach, are described by Park and Rilett in [6] where they state that,

> One of the major requirements of advance traveler information systems (ATISs) is a mechanism to estimate link travel times.

Work by Wunderlich et al. in [12] has shown the benefits of the presence of link travel time predictions, even when the penetration rate of the participating vehicles is limited.

Link travel time predictions can be generated using a variety of different techniques such as online simulation [5], artificial neural networks [6], swarm computing [8] and machine learning [7]. A common requirement for all these approaches is communication. All approaches require information from road side sensors or the intentions of vehicles to be aggregated and analyzed in order to predict link travel times.

Even though the benefits of link travel time forecasting are known and various link travel time predicting approaches are described in literature, research on the overhead these approaches introduce is scarce. Practical evaluations such as done by Van Der Horst and Nobel in [10], where they compare a distributed to a centralized task allocation, are needed.

The work of Wunderlich et al. [12] shows the benefits of link travel time forecasts using a central service. The reduction of travel time for participating and non-participating vehicles is shown using a number of simulations. The effect of the penetration rate on the reduction of travel time is evaluated very thoroughly. The communication overhead introduced by the approach is not discussed, the results are limited to an evaluation of the benefits of the approach.

Work by Hunter et al. [5] describes the link travel time prediction using online simulation situated in the vehicles. Much attention is given to accuracy of the online simulations. The need for synchronization between the in vehicle simulations and protocols to achieve this synchronization are described. The communication requirements of such a vehicle to vehicle synchronization would be interesting when evaluating the feasibility and scalability of this interesting approach.

In [8], Tatomir et al. describe a mechanism called *Ant Based Control* that is very similar to the decentralized ATISs described in this paper. Ant Based Control uses a combination of historical data and vehicle intentions to predict vehicle densities on the road. The relationship between speed and density is then used to predict the link travel time. The method of propagating and retrieving information used in [8] is very similar to that described in [11] and thus to the decentralized system described here, as both draw inspiration from *Ant System* described by Dorigo et al. in [3].

In their evaluation of decentralized and centralized task allocation systems in [10], the authors evaluate both systems by analyzing the robustness, performance and energy consumption of both approaches. Their results can be used when deciding between a centralized and decentralized deployment. The methodology used by Van Der Horst and Nobel can be applied to many self-organizing systems where the benefits of using a decentralized over a centralized systems are unclear.

7 Conclusion

The evaluation presented in this paper illustrates the tradeoff between cost and benefit when looking at decentralized services. The delegate multi-agent systems approach presented in this paper allows the vehicle agents to take into account link travel time predictions and thus allows for faster travel times. But the increased communication between vehicles and road side infrastructure comes at a cost.

Even though this paper evaluates a decentralized ATIS by comparing it with a centralized ATIS, the results indicate that the overhead produced by ATISs is also very sensitive to the implementation and configuration of the ATIS. Varying the exploration parameter K, greatly affects the communication overhead produced by the decentralized ATIS described in this paper. Besides analyzing the tradeoff between cost and benefits for centralized and decentralized approaches, it is also necessary to analyze the tradeoffs associated with parameters such as K.

Using delegate multi-agent systems to propagate and retrieve information is a way to control the audience when communicating. The use of smart messages behaving in an ant-like fashion ensures that information is only spread to relevant other parties. Because the ants can aggregate and analyze the information they obtain from the environment as they travel, less communication is required. Without the use of ants, vehicle agents would have to request information from the infrastructure agent, wait for a reply, calculate the estimated arrival time at the next link and then request a link travel time forecast at the next link. Ants can calculate the estimated time of arrival at the next link and continue on their way.

Acknowledgements. This research was funded by the IWT - SBO project 'MASE' (project no. 060823) and by the Interuniversity Attraction Poles Programme Belgian State, Belgian Science Policy, and by the Research Fund K.U.Leuven.

References

[1] Aydt, H., Turner, S., Cai, W., Low, M.: Symbiotic simulation systems: An extended definition motivated by symbiosis in biology. In: Proceedings of the 22nd Workshop on Principles of Advanced and Distributed Simulation, pp. 109–116. IEEE Computer Society, Los Alamitos (2008)

[2] Claes, R., Holvoet, T., Weyns, D.: A decentralized approach for anticipatory vehicle routing using delegate multi-agent systems. IEEE Transactions on Intelligent Transportation Systems (2011)

[3] Dorigo, M., Maniezzo, V., Colorni, A.: Ant system: optimization by a colony of cooperating agents. IEEE Transactions on Systems, Man and Cybernetics, Part B: Cybernetics 26(1), 29–41 (1996)

[4] Holvoet, T., Valckenaers, P.: Exploiting the environment for coordinating agent intentions. Environments for Multi-Agent Systems III, 51–66 (2007)

[5] Hunter, M., Sirichoke, J., Fujimoto, R., Huang, Y.L.: Embedded ad hoc distributed simulation for transportation system monitoring and control. In: Lee, L.H., Kuhl, M.E., Fowler, J.F., Robinson, S. (eds.) Proceedings of the 2009 INFORMS Simulation Society Research Workshop, vol. 1, pp. 13–17 (2009)

[6] Park, D., Rilett, L.R.: Forecasting freeway link travel times with a multilayer feedforward neural network. Computer-Aided Civil and Infrastructure Engineering 14(5), 357–367 (1999)

[7] Roozemond, D.: Forecasting travel times based on actuated and historic data. Environment 23, 25 (1997)

[8] Tatomir, B., Rothkrantz, L.J., Suson, A.C.: Travel time prediction for dynamic routing using ant based control. In: Rosetti, M., Hill, R.R., Johansson, B., Dunkin, A., Ingalls, R.G. (eds.) Proceedings of the 2009 Winter Simulation Conference, pp. 1069–1078. IEEE Computer Society, Los Alamitos (2009)

[9] Treiber, M., Hennecke, A., Helbing, D.: Congested traffic states in empirical observations and microscopic simulations. Phys. Rev. E 62(2), 1805–1824 (2000)

[10] Van Der Horst, J., Noble, J.: Distributed and centralized task allocation: When and where to use them. In: IEEE International Conference on Self-Adaptive and Self-Organizing Systems Workshops (2010)

[11] Weyns, D., Holvoet, T., Helleboogh, A.: Anticipatory vehicle routing using delegate multi-agent systems. In: Intelligent Transportation Systems Conference, ITSC 2007, pp. 87–93. IEEE, Los Alamitos (2007)

[12] Wunderlich, K.E., Kaufman, D., Smith, R.: Link travel time prediction for decentralized route guidancearchitectures. IEEE Transactions on Intelligent Transportation Systems 1(1), 4–14 (2000)

Information Extraction System Using Indoor Location and Activity Plan

Bjørn Grønbæk, Pedro Valente, and Kasper Hallenborg

Abstract. In this paper we present an agent based system for extracting live and historic position data on multiple persons in a real environment, using a commercial off the shelves Real Time Location System (RTLS). We present a context model for representing the location data which allows composition of location data with preexisting knowledge on the environment and activities taking place at the locations. We describe our experiences using the RTLS deployed in non-laboratory environment with hundreds of participants.

1 Introduction

Since Schilit and Theimer introduced the notion of context to distributed computing in 1994 [8] location has always been among the most profound information used for adaptation in pervasive and ubiquitous computing. In decades indoor locations systems have been a challenging task in research and ranged in base technologies from infrared light, ultrasound, camera, UWB, WiFi, cameras, RFID, etc, in systems like [2, 7, 9]. In smart environments for ambient assisted living (AAL) indoor locations systems starts to play an important role in coordinating and providing intelligent services to the users, including activity recognition, task allocation, information services, monitoring, etc. However, in most of the commercial setups the systems are often limited to the last location registered, which could come from pressure sensors, passive tags (RFID) entering zones, or doors being opened. More advanced systems, such as [4], use algorithms to better estimate a user's current position based on pattern recognition or to predict a user's next location. Another problem common to such simple locations systems is identification of persons and

Bjørn Grønbæk · Pedro Valente · Kasper Hallenborg
The Maersk Mc-Kinney Moller Institute - University of Southern Denmark,
Odense, Denmark
e-mail: {bjgr,prnv,hallenborg}@mmmi.sdu.dk

Y. Demazeau et al. (Eds.): Adv. on Prac. Appl. of Agents and Mult. Sys., AISC 88, pp. 33–38.

activities, as described by [6]. Pressure mats or movement sensors do not distinguish efficiently between different persons, and may provide uncertain sensor data.

The choice of using a multi agent base approach it motivated by the natural mapping of having several hundreds of autonomous participants acting in the same system, each with an individual set of preferences. The reasoning and planning aspects normally considered by participants of a conference map naturally to similar components in an agent architecture.

In this paper we describe a practical experiment using an indoor location system, tracking hundreds of persons simultaneously in real time at a large conference venue with multiple rooms, hallways and floors. We will present and discuss implementation of a multi agent system , the AAL Butler system, that provides both live data access as well as historical data about participants. In addition the system provides contextual information about participants activities by combining knowledge about their location with predefined knowledge on scheduled activities taking place in rooms.

2   Architecture Overview

Figure 1 show the significant agents in the AAL Butler system. The layered organisation of agents is based on the type of services and functionality provided by each agent, following the design principles described by [3]. We do not enforce a strictly layered communication in the system, since each agent provides its own interface that matches the parts of the context model that is relevant, but use the layers to provide a logical separation of responsibilities between the agents.

The **Sensor** layer holds logical and physical sources of data. In the AAL Butler system this layer is made up by programs providing API access to two databases and the RTLS server used for the experiment. The AAL database provides information from the conference venue, most importantly the time schedules, room assignments, and participant information. Additionally this database holds information that relates the identity of the physical rooms with the location ontology's room concept. The RTLS database stores information from the RTLS server for historical use. Finally, the RTLS provides us with an API for accessing the live RTLS data.

The agents in the **Data Retrieval** layer extracts data from the sensors and matches the data to the context model used in the system. They

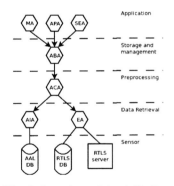

Fig. 1 Overview of the AAL Butler agent architecture. Agents are shown layered according to functionality or services they provide.

provide an ontology based access to the data sources in the sensor layer. The **Ekahau Agent (EA)** is responsible for providing access to the RTLS data. The agent provides two primary services, as well as a number of utility services. The first service is a subscription based notification service that allow other agent to receive information on new positions and events captured by the RTLS server. The second service provides access to the historical database of RTLS data. The **AAL Information-tion Agent (AIA)** retrieves the context data in its raw form from the conference session database, matches the data with the relevant concepts of the context model, like `Participant` or `Session`. The main services provided by the AIA is getting information on the schedule of the conference, information on particular sessions and getting participant profiles.

The **AAL Context Agent (ACA)** in the **Preprocessing** layer provide contextual information by aggregating multiple data sources into a single context source. This is done by collecting the data from the context atoms in the data retrieval layer and providing a higher level of context. The agent's responsibilities include providing a service for agents to retrieve information on the location of a participant in the form of a room. This is done by extracting relevant information about the participants id, querying the EA for the area of that participant, and then matching the area to a room by querying the AIA. The agent is also capable of providing a information on a given participants session history, by comparing the participants location history with location of sessions. The method used for providing this service is described further in section 3.1.

The **AAL Butler Agent (ABA)** in the **Storage and Management** provides a logical interface to applications build on top of the system by exposing the functionality of the AAL Butler infrastructure for the purposes of doing information extraction and providing context information on participants at the conference.

The **Application** layer is meant for agents that are end-users, either in terms of software agents or human agents. Application specific context management is also handled in this layer. The **AAL Participant Agent (APA)** agent serves two roles in the system. First, in future implementations, it is intended to serve as an interface between users of the system and the system's agents. This includes routing information to the users most convenient interface, Ekahau tag, PC or mobile device, and capturing user input for the system. Additionally, it currently implements a simple user model by storing a participants user preferences along with the history of the participant to enable the agent to perform simple recommendations to the user. The **MapperAgent (MA)** agent provides an user interface for examining live and historic position data. The **SessionEvaluationAgent (SEA)** agent provides an user interface for accessing the session history of a participant. It instructs the ABA agent to perform the relevant data processing, and present the context information for the user in a human readable way.

3   Context Modelling

The AAL Butler system expresses its content using an ontology based model, which according to [3] fulfil a few basic requirements that we wished to achieve for the context model: Simplicity, extensibility and expressiveness. By embracing these properties we seek to ensure that we can extract and extend elements of the presented ontology to future projects.

The domain specific AAL Butler ontology supports the task of providing both live and historical information on participants, and also supports reporting on a given person's current or past activity. The location specific part of the ontology supports the representation of indoor environments in a simple form. The focus is on representing locations in the form of either coordinates, rooms or special points-of-interests (POI). The primary

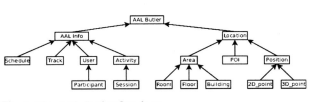

Fig. 2 The AAL Butler Ontology

concepts in the ontology needed for the system are described in figure 2, which shows the taxonomy of the relevant elements of the ontology used in the AAL Butler system. These concepts provides enough expressiveness to implement the AAL Butler system when combined with a number of predicates to describe the status of the system.

3.1   Determining Session Participation

To detect which session a participant has spent time in, the gathered positions are clustered into groups. While k-means clustering analysis is simple to use, we don't know the number of locations a participant will visit in a given time interval, and therefor can not know the number of clusters to use, which is a requirement when using k-means. Instead the AAL Butler system utilizes the DBSCAN algorithm [5], in an approach similar to [1]. The DBSCAN algorithm create clusters based on the density of the spatial data, and does not require knowledge about the number of clusters to find. To handle the time dimension of the gathered position we analyze the data not as the complete set for a participant during the whole conference, but instead for one session time slot at a time. This effectively eliminates the need for considering the time-line of the positions, since we can then assume that every position in a given location will be connected to a particular session currently going on in that room, if the cluster is found to be in that room. After the clustering has been performed, the most significant of the clusters are found. The location of the cluster is then matched against all sessions that takes place in the time slot being investigated, and if any session has a room that matches the location, the participant is considered to have been at the session. For simplicity we do not consider the duration of the stay at the session, or the possibility of attending more than one session in a time slot.

4 Experiment

The system implementation was evaluated in a practical experiment, collecting location data and extracting predefined knowledge, in the context of the Intellicare[1] research project at the AAL Forum 2010 conference on Ambient Assisted Living. The conference venue included three plenary rooms, ten session rooms, a lunch and break area, some additional areas, and all the connecting corridors etc. between those rooms. The conference was placed in a single building with two floors and the dimensions of the conference area was more than 7000 m^2. The RTLS system employed [2] is based on Wi-Fi technology and provides 2D real-time position, with one meter precision and accuracy errors of three meters. The system use tags as it's tracking asset, with the tags providing a two line screen and three buttons, making the system capable of two-way communication.

In terms of performance we tracked 268 of the 736 participants simultaneously, resulting in 2.6 million position records, with each user represented as an agent in the system, and observed satisfactory performance on both the RTLS server and the AAL Butler system. Each agent was able to keep track of the location of it's related RTLS tag, observe the participant's session history and perform a simple prediction on the next session of the participant. Table 1 shows a comparison of the session participation found by the AAL Butler system for a particular participant and the actual session history of that participant. We have data on six slots, out of which the AAL Butler system correctly found four, missed one session completely and found one session incorrectly, e.g. found a different session than the participant actually attended. After investigation we have determined these inaccuracies to be the result of low RTLS coverage and bad calibration, rather than a problem with the AAL Butler systems clustering technique. We therefore consider the AAL Butler system to have determined the participants session history in satisfactory way.

Table 1 Comparison of a participant's session history with results from the AAL Butler system. For each time slot the name of the session found by the system and the name indicated by the participant is shown. Only time slots for which we have information is show.

time slot	16.30-18.00	09.00-10.30	10.30-11.00	12.00-13.00	13.00-14.30	12.30-15.00
known history	R1	F2	Break	F3	Lunch	Closing
AAL Butler	R1	F2	Break	C6	-	Closing

5 Conclusion

The work presented in this paper is a initial attempt at creating a system for providing contextual information on persons, or agents, based on their current and previous location data by aggregating that information with predefined knowledge on

---

[1] http://www.intellicare.dk

[2] http://www.ekahau.com

time schedules and locations of activities. The current implementation was focused on a specific use case, described in the experiment section of this paper, and on providing functionalities for that use case. We have described an approach to do mapping between the information presented as predefined knowledge and the location data of a person using an ontology based context model. We have designed an agent-based system that consists of a number of cooperating agents, that can utilise the context model to provide data access to the location of a person as well as context information on that person's activities. In our future work will utilise the MAS approach of the AAL Butler system by implementing an user agent with a Belief-Desire-Intention model, which will allow us to focus on prediction and recommendation functionality. We are currently working with partners on adapting the system for use in elder care environments.

Acknowledgements. The presented work has been conducted under the IntelliCare project, which is supported by The Ministry of Science, Technology and Innovation in Denmark.

References

1. Adams, B., Phung, D., Venkatesh, S.: Extraction of Social Context and Application to Personal Multimedia Exploration. ACM Press, New York (2006)
2. Addlesee, M., Curwen, R., Hodges, S., Hodges, S., Newman, J., Newman, J., Hopper, A., Steggles, P., Steggles, P., Ward, A., Ward, A.: Implementing a sentient computing system. IEEE Computer 34(8), 50–56 (2001)
3. Baldauf, M., Dustdar, S., Rosenberg, F.: A survey on context-aware systems. International Journal of Ad Hoc and Ubiquitous Computing 2(4), 263 (2007)
4. Das, S.K., Cook, D.J., Bhattacharya, A., Heierman, E.O., Lin, T.Y.: The role of prediction algorithms in the MavHome smart home architecture. IEEE Wireless Communications 9(6), 77–84 (2002)
5. Ester, M., Kriegel, H.-P., Sander, J., Xu, X.: A density-based algorithm for discovering clusters in large spatial databases with noise. In: Published in Proceedings of 2nd International Conference on Knowledge Discovery and Data Mining, vol. 96, pp. 226–231. AAAI Press, Menlo Park (1996)
6. Hong, X., Nugent, C., Mulvenna, M., McClean, S., Scotney, B., Devlin, S.: Evidential fusion of sensor data for activity recognition in smart homes. Pervasive and Mobile Computing 5(3), 236–252 (2009)
7. Andy Hopper, R.W., Falcao, V., Gibbons, J.: The active badge location system. ACM Transactions on Information Systems 10(1), 91–102 (1992)
8. Schilit, B., Theimer, M.: Disseminating active map information to mobile hosts. IEEE Network 8(5), 22–32 (1994)
9. http://www.ubisense.net (2010)

Using Multi-Agent Systems to Visualize Text Descriptions

Edgar Bolaño-Rodríguez, Juan C. González-Moreno,
David Ramos-Valcarcel, and Luiz Vázquez-López

Abstract. Usually people make a visual representation of what they read. This is a fact inherent to their nature. Many efforts have been done in the last years to try to find a direct method to translate a text wrote on a natural language to a graphical representation. Some solutions try to represent mainly a static scene with the objects appearing in the text, also there are few approaches that try to build an animation from the action described in text. But in any case all the approaches found on literature are not reciprocal approaches; that is, if the visual result is not satisfactory another text must be wrote in order to get a better one. Software Engineering and in particular Agent Oriented Software Engineering treat this kind of problematic using model driven techniques over different intermediates models. In this paper such kind of solution is proposed to get not only a visual model of the text wrote, but also to get an intermediate model that could be used to modify the visual model without modify the text and also to propose an alternative narration if necessary. In order to validate the visual model obtained, a previous multi-agent system has been adapted to provide a representation to be run on the Alice System.

Keywords: Agent-based Animation, Storytelling, Avatars, Multi-Agent Systems, Agent Oriented Software Engineering, Model Driven Techniques.

1 Introduction

Natural language could be understood as a vital and ubiquitous part of human intelligence and their social relations. Therefore, it is natural to believe that technologies

Edgar Bolaño-Rodríguez · Juan C. González-Moreno · David Ramos-Valcarcel ·
Luiz Vázquez-López
Departamento de Informática, Universidad de Vigo
e-mail: `jcmoreno@uvigo.es, david@uvigo.es, luisvazquez@uvigo.es`

Y. Demazeau et al. (Eds.): Adv. on Prac. Appl. of Agents and Mult. Sys., AISC 88, pp. 39–45.
springerlink.com   © Springer-Verlag Berlin Heidelberg 2011

Fig. 1 An example on Alice representing the characters and objects from the text description: "Merliño gives the old woman a box". The visual model presents a virtual environment with a men, a women, and a box. Despite its simplicity a qualified designer must consume several hours of hard work.

which can automatically process natural languages (NLP technologies) could facilitate the everyday lives of ordinary people. Another vital part of human intelligence is their ability to visualize abstractions, more exactly their ability to visualize what people describe when speaking or writing. This ability is basic in many domains in which people needs to visualize a description in order to understand what happens. Moreover, nowadays computer animation is a significant part of our society.

Currently, the creation of computer animation is mostly done manually. This manual process can be tedious, time consuming and labor intensive, as an example consider the scene that is shown in Figure 1. This scene usually represent only the started point for a particular animation. If someone wish to tell a story taking place on this virtual environment, he/she will need to somehow manipulate the entities within this environment in a particular way so that the story could be properly visualized. Usually this will be done accordingly to a stablished script wrote on a natural language. In the last years some works [10, 8, 7] have been proposed to automatize this work based on the use of natural language instructions.

As pointed in [10]: *"generating computer animation from natural language instructions is a complex task that encompasses several key aspects of artificial intelligence including natural language understanding, computer graphics and knowledge representation"*. Moreover, traditionally this task has been approached by the use of rule based systems that are highly successful on their respective domains, but that are not easy to generalize to other domains, and of course without provide a well stablished mechanism to come back from the visual representation to the original phrase. That is, they don't allow feed back between the visual model and the text description they used to generate it.

In contrast this paper proposes a complementary kind of approach, based on the use of an intermediate model based on the agent paradigm. Basically the idea is to use similar techniques to thus described in the literature to process the text description on natural language, but generating not only the visual model but also a software agent model. This model will include abstract elements (related with agents concepts) that refers both to grammar elements on the text and to visual elements on the final model.

The main advantage of this approach is that it will allow to work, if necessary, not only, at a different level of abstraction in order to modify the visual model without modify the original text description; but also to change the text description without modify the intermediate model or their visual model. For example, in the Figure 1 one would expect *a smaller box* or that the starting point consists in *Man taking with him the box* and not *taking it from the floor*.

The organization of the rest of the paper is as follows: In section 2 some relevant and related works are summarized. Section 3 summarizes the NLP4INGENIAS MAS and presents the relation identified between the agent and visual model. And to conclude section 4 picks the conclusions and future works.

2   State of the Art of Visualizing Text Descriptions

Over the last two decades, many researchers in the NLP community jointly with the computer graphics community have been developing techniques to enable computers to understand the human natural language. These techniques could aid artists to create virtual reality for storytelling or help researchers to understand some social and human behavior. It is usual for NLP systems the use of knowledge bases containing linguistic information to analyze sentences and the production of data structures representing the syntactic structure, jointly with the semantic dependency of such information. In the computer games and animation industry, computer artists create virtual characters, props, and whole scenes of stories. The construction of these scenes is a labour so intensive, that consumes much more time than expected, and that is hardly reusable. To solve this gap in the last decade several works have tried to get a system to automatically generate virtual reality from stories in natural language.

A first related work is CONFUCIUS [7], an intelligent storytelling system, which converts natural language into 3D animation and audios. CONFUCIUS is implemented using VRML, Java and Javascript, and it integrates existing tools for language parsing and text-to-speech synthesis. The input of CONFUCIUS is natural language text, but it only deals with single sentences with non-metaphor word senses, and output animations of rigid objects and human bodies. Althought CONFUCIUS' multimodal output includes 3D animation and a presentation agent (Merlin the narrator), it doesn't provide feed back with user, nor between the animation and the text.

In [10] the problem is foreseen as a black box paradigm, in which the final system has two inputs and one output, namely: As Inputs: "S", a natural language description of a list of visualizable actions and "V", an accessible and manipulatable virtual environment in which all the actions in S take place. And as output a "*computer animation*" that faithfully and coherently visualizes S within V. The main contribution os this approach is to include novel methods for performing semantic role labeling on prepositional phrases, and for performing verb sense disambiguation, and most importantly, a domain-independent empirical approach for mapping verb semantics to animation procedures based on training data. Nevertheless as the previous approach no feed back is proposed.

Recently [8] proposed an animation system based on a framework for a motion database that stores numerous motion clips for various characters. In the proposal the motion synthesis generates an animation by smoothly connecting given motion clips. The interactions between characters and between a character and objects are handled by this module based on the information that the motion clips have. The scene information contains characters and objects and their initial states, including postures, positions, and orientation. Each object has certain object information including names, a default contact point, alternative contact points and their names. For example, a desk object has *"desk"*, *"table"*, etc. as its names. An object also has sets of pairs consisting of a part name and its position (e.g., *"above"*, *"under"*, *"side"*, etc.). This information is used to search for appropriate motions and determine appropriate contact positions according to an adjective that is used in the input text. In addition, an object has a default position which is used when no adjective is specified. Of course the system assumes that the scene information is provided in advance by the user.

3   Building Alice Programs from Scratch

[1] presents a 3D Electronic Institutions methodology, which supports human integration into MAS-mediated environments. One of the basis of the proposal is the equivalence established between the meta-model elements needed to specify an electronic institution and those needed to model a 3D virtual world. Taking into account the characteristics that presents the Alice system [6, 3] jointly with those of the INGENIAS meta-model in this section it is presented a two phase process that could be used as an alternative to the systems described in section 2. The first one will consist on the use of the NLP4INGENIAS Multi-Agent System to get the agent modeling of the text sentences; the second one will consist in the translation of the intermediate model into a valid specification for the Alice System.

Recently some works [5, 4] have shown the viability in order to get an initial agent model (using INGENIAS [9]) by using NLP techniques. Although the main goal of these works is to obtain the initial agent model for a software application and they took as starting point a text description of the requirements, the framework proposed could be use also to establish an abstract specification based on agents for any text description. Briefly the underlying idea is similar as used in [7] to associate single sentences with output animations.

It should be noted that the process of generating the model is not fully automatic. This is mainly due because of the ambiguity of Natural Language and because it is not usual that users give a fully and precise description of what they really needs. In order to manage this problem a Multi-Agent System (NLP4INGENIAS [5]) has been proposed as a virtual assistant for the process. This solution is different from proposed in [10] in which ambiguity is solved by performing semantic role labeling on prepositional phrases. The main advantage in the use of the NLP4INGENIAS approach is the active involvement of user in the disambiguation process.

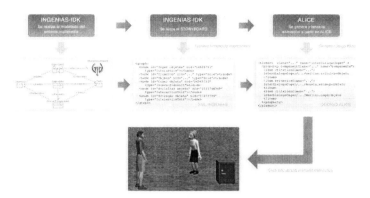

Fig. 2 The process of translation from INGENIAS model to a visual model running on Alice.

An important point to take into account in the solution adopted is that Alice system [6, 3] is not a traditional programming environment, moreover the Alice program structure (on the version 2.2) consists in a compressed file which contains a directory structure consisting of binary files and *"xml"* files that collect the behavior and the structure of the animation. Binary files refer to the visual elements while the xml structure describes the behavior and the positioning of such elements on the virtual world.

As the MAS system NLP4INGENIAS provides an initial model on INGENIAS that specify the description text. The easier way to get the final visual model would consist in translating the INGENIAS model into a representation on Alice of such model. INGENIAS [9] uses a set of meta-models (*The Agent Metamodel. The Interaction Metamodel. The Organization Metamodel. The Environment Metamodel. The Tasks and Goals Metamodel.*) that describe the elements that define the MAS using the following five viewpoints: *The definition, control and management of each agent mental state. The agent interactions. The MAS organization. The environment. The tasks and goals assigned to each agent.*

Considering the specification of an Alice program from the point of view of a Multi-Agent system. The organization model will specify each scene needed on the animation with the elements and characters appearing on it. It has been taken into consideration in this part of the specification that not any character play the same role, nor have the same kind of participation. For this reason the following restrictions have been fixed over the original INGENIAS organization model: Only one organization diagram will be used to specify an Alice scene. The diagram will contain at most three predefined groups of agents: *Dynamic, Static* and *Tools*. A Dynamic agent (i.e. main characters) refers to a character that has an own dynamic behavior or it is managed dynamically by users. A Static agent (i.e. secundary characters) refers to a character whose behavior (may be none) is static and is partially present in the full scene. And a tool agent (i.e. helping characters) is a character that could appear or not (the CONFUCIUS presentation agent) in the scene depending on the behavior of main characters, with a predefined functionality supplementing the knowledge or the behavior of the rest. Using only this model the static view presented on Figure 1 could be obtained. In order to provide dynamic behavior to the characters each character must have an agent model that specify the roles, tasks and goals that the character will do on the scene modeled. The Interactions between characters and between characters and objects are specified by means of the interaction diagrams. The task and goal model express the relations between goals, the tasks that satisfy each goal and what changes will apply to the state of each character.

4 Conclusion and Future Works

This paper presents a new approach to the problem to get a visual animation from a text description in natural language. The approach differs from those presented in the literature on the establishment of an intermediate agent model (an INGENIAS model). The main advantage of this approach could be appreciate on the Figure 2 that shows how a change on the intermediate model could be done to clarify the visual behavior without change the syntax, nor the semantic of the original text. In this case a modification on the task and goal model is introduced to state that the box must be on the floor. Another possible advantages that could be introduced are to have several visual animation versions for one text description, and also that changes on the visual model could propagate to the text description by the use of the intermediate agent model by modifying the MAS used to get the agent model.

As future works it is planned to use the learned lessons from this experiment to develop a new methodology based on the use of agents to specify and build multimedia applications.

References

[1] Bogdanovych, A., Esteva, M., Simoff, S.J., Sierra, C., Berger, H.: A methodology for developing multiagent systems as 3d electronic institutions. In: Luck, M., Padgham, L. (eds.) Agent-Oriented Software Engineering VIII. LNCS, vol. 4951, pp. 103–117. Springer, Heidelberg (2008)

[2] Corchado, J.M., Rodríguez, S., Llinas, J., Molina, J.M. (eds.): International Symposium on Distributed Computing and Artificial Intelligence, DCAI 2008, University of Salamanca, Spain, October 22-24. Advances in Soft Computing, vol. 50. Springer, Heidelberg (2009)

[3] Dann, W.P., Cooper, S., Pausch, R.: Learning To Program with Alice, 2nd edn. Prentice Hall Press, Upper Saddle River (2008)

[4] González-Moreno, J.C., Vázquez-López, L.: Design of multiagent system architecture. In: COMPSAC, pp. 565–568. IEEE Computer Society, Los Alamitos (2008)

[5] González-Moreno, J.C., Vázquez-López, L.: Using techniques based on natural language in the development process of multiagent systems. In: Corchado et al. [2], pp. 269–273

[6] Kelleher, C., Pausch, R.: Using storytelling to motivate programming. Commun. ACM 50, 58–64 (2007)

[7] Ma, M.: Automatic Conversion of Natural Language to 3D Animation. PhD thesis, University of Ulster. Faculty of Engineering (2006)

[8] Oshita, M.: Generating animation from natural language texts and semantic analysis for motion search and scheduling. The Visual Computer 26(5), 339–352 (2010)

[9] Pavón, J., Gómez-Sanz, J.J., Fuentes-Fernández, R.: The INGENIAS Methodology and Tools, article. IX, pp. 236–276. Idea Group Publishing, USA (2005)

[10] Ye, P.: Natural Language Understanding in Controlled Virtual Environments. PhD thesis, University of Melbourne. Department of Computer Science and Software Engineering (2009)

Empowering Adaptive Manufacturing with Interactive Diagnostics: A Multi-Agent Approach

Thomas M. Hubauer, Christoph Legat, and Christian Seitz

Abstract. This paper presents a novel approach towards proactive manufacturing control that integrates automated diagnostics with human interaction, resulting in a flexible adaptation of machine capabilities which helps to avoid damage in case of abnormalities. The model-based interpretation process supports predictive diagnostics using abductive reasoning, relying on plausibility thresholds and human intervention to resolve the resulting ambiguity between competing solutions. This enables the system to detect and avoid potential failure states before they actually occur. The proposed architecture additionally integrates intelligent products as mobile sensors, improving robustness and dependability of the production system.

1 Introduction

A growing demand for improved reactivity to market trends and towards provision of individualized products results in an ongoing trend from supplier-driven to customer-driven manufacturing. This results in higher flexibility requirements for manufacturing systems, namely (i) provision of flexible processes to enable small lot sizes and (ii) robustness to failures within the technical system. Especially the customization of products challenges traditional engineering approaches unattainably. For this reason, a number of projects explore new ways of addressing these requirements on different levels in accordance to the IEC 62264 [6] standard.

The application of agent-technologies within production systems is a lively research topic, recent examples include the ILIPT project [13] which focuses on enabling adaptable supply chains using market-based approaches, the EUPASS project [12] focusing on reorganizing automation systems dynamically using modular assembly components, and the PABADIS'PROMISE project [8] which aims at

Thomas M. Hubauer · Christoph Legat · Christian Seitz
Siemens AG - Corporate Technology - Intelligent Systems & Control - Agent Technologies
e-mail:{thomas.hubauer.ext,christoph.legat.ext,
    ch.seitz}@siemens.com

Y. Demazeau et al. (Eds.): Adv. on Prac. Appl. of Agents and Mult. Sys., AISC 88, pp. 47–56.
springerlink.com   © Springer-Verlag Berlin Heidelberg 2011

realizing a distributed manufacturing execution system including reconfiguration, but without using diagnostic information. Agent-based approaches have also been applied to real-time control of production systems [17], the capabilities of these agents are however very limited due to the hard real-time constraints that have to be addressed. Several other concepts from computer science have recently been adopted in the context of production system control, e. g. the development of modular material flow systems [3] based on the internet of things [7], and the use of diagnostics to enhance the detection of system failures which is explored in the MAGIC project [1]. In contrast to the approach presented here, the latter does not consider a direct interaction with the automation control system leading to an automated reaction and does not integrate intelligent products. Thus, the proposed approach provides a benefit over existing systems due to a tighter interaction between diagnostics, adaptation and product interaction facilitated by generic service description.

Product-driven manufacturing systems supported by intelligent products are a promising way to overcome limitations with respect to small lot sizes. An intelligent product enhances the currently used Auto-ID technologies [10], in which a computing entity is attached to (part of) a product during the whole product life cycle, collecting information and cooperating with other devices in its ambient environment. We go one step further and beyond existing approaches surveyed in [11] by (i) enabling the intelligent product to interact and cooperate with the production control system and (ii) alleviating the lack of integration between the shop floor control system and higher level supervisory operations accomplished by human personnel. Although no real-world production system operates completely unsupervised, there is to our best knowledge no other approach investigating the interaction of automated diagnostics and human supervision within an autonomous flexible production environment. The result of our approach is a highly flexible and robust framework for product-driven manufacturing based on cooperation of a model-based diagnostic system and human operators.

We demonstrate the concept in the context of a flexible production scenario enhanced with autonomous products. More specifically we consider a T-style component of a (directed) inner logistic system composed of one incoming conveyor, a switch and two distinguishable outgoing conveyors (typically leading to different machines). Each conveyor section is impelled by a live axle equipped with a sensor measuring axle motion. As some products may be sensitive to vibrations at certain time points during production (e. g. due to a freshly glued joint), products are equipped with a digital product memory and an acceleration sensor during production, storing acceleration measurements for quality assurance. However, this information can also be combined with other measurements to diagnose the conveying system: If the axle of a conveyor turns but a product located on the belt does not move, this indicates a transmission problem which renders this section of the conveyor unusable until repair. Irregular movement with sudden accelerations or vibrations may indicate bearing problems, which only excludes vibration-sensitive products from using this conveyor. Depending on whether the affected section is the shared incoming or one of the two available outgoing conveyors, the usability restrictions either hold for the complete conveying system or just for one of the two possible routes. These restrictions have to be incorporated into production planning to optimize machine usage and throughput while guaranteeing product quality.

The remainder of this paper is structured as follows. In Sect. 2 we present the conceptual architecture of the proposed multi-agent system and introduce the semantic models employed. Some selected details on the realization of the components are presented in Sect. 3, before we conclude the presentation of our ideas in Sect. 4.

2 Conceptual Architecture

This section presents the general architecture of the proposed multi-agent system. The functionality of the system is realized by agents implementing four roles which are detailed in the upcoming subsections: symptom provider, situation analyzer, production controller, and supervision mediator. The upper half of Fig. 1 depicts these roles and their associated meta-models. The realization of these roles by agents (which is focus of the successive Sect. 3) and exemplary concrete models are illustrated in the lower section of the same figure. Please note that in favor of conciseness we only depict concepts referenced in the remainder of this paper, rather than the complete models. The framework additionally provides a discovery service enabling agents to find interaction partners, implementing yellow pages functionality.

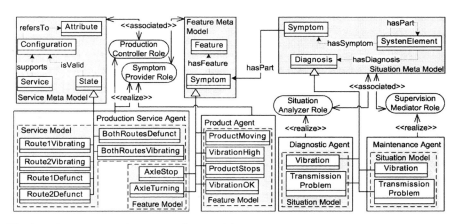

Fig. 1 Roles, Agents and corresponding Models

The meta-models incorporate the relevant domain knowledge ranging from process information to definitions of diagnoses, whereas the specific models encapsulate the knowledge about the concrete products and processes. The models are used for diagnostics, but also as a common vocabulary for service discovery bringing together information providers and consumers based on the type of information offered and requested. To facilitate automated information processing and to increase reusability and interchangeability of the model, it is employed with formal semantics based on the Web Ontology Language (OWL). More specifically, the situation meta model is based on $\mathcal{EL}$, a subset of lightweight profile OWL 2 EL [16] allowing for polynomial-time query answering, and the feature meta model uses the profile OWL 2 RL [16] to facilitate interpretation by a rule engine; both profiles were chosen to optimize the tradeoff between expressivity and performance w. r. t. the intended evaluation method.

2.1 Symptom Provider

A symptom provider (SP) realizes the gathering of Features within a given field of responsibility defined by a set of SystemElements of the proposed metamodel. To offer its functionality to the agent community, a SP registers with the discovery service by announcing the ids of the SystemElements it is responsible for. Consumers can register with a SP for certain event classes denoted by subconcepts of Symptom. Upon receiving data (a Feature) from an associated information source or preprocessing component, e. g. a piece of machinery, the SP automatically infers the corresponding Symptom based on its feature model. Subsequently, all consumers having subscribed for this event or any of its superconcepts are notified by the SP.

2.2 Situation Analyzer

The situation analyzer role (SA) encapsulates the process of interpreting symptoms based on the diagnosis meta model depicted in Fig. 1. To receive the necessary input data, the SA subscribes for the Symptoms it wants to be notified about as follows: First, it queries the discovery service for symptom providers responsible for the SystemElements allocated to this agent for surveillance. Then, the SA subscribes with each of them for the concrete symptoms (i. e. subconcepts of Symptom) of the associated component it is interested in.

Upon receiving a Symptom the SA triggers the interpretation process which determines a set of defects that might possibly cause the observations. By allowing to assume the presence of symptoms that have not been detected yet, the SA is even able to detect faults which have not yet fully manifested. Obviously a Diagnosis not requiring the assumption of additional symptoms will more likely apply than another one based on assumptions, the SA therefore attributes its interpretations with a measure of their plausibility. Due to a potentially high number of calls to its services, performance of a SA is a vital aspect. To ensure reactivity and performance, the proposed realization of the situation analyzer presented in Sect. 3.3 uses an anytime approach and a lower bound $pl_\downarrow$ on the quality of solutions.

Dependent on configurable thresholds pl_μ and pl_Δ the SA triggers both automated reactions and operator interaction: Firstly, the SM is informed about the complete set of diagnoses, their necessary assumptions and plausibility values. Then, if and only if the plausibility of the best Diagnosis determined by the SA exceeds

pl_μ and in the second-best alternative is at least by an amount of pl_Δ less plausible than the best one, the SA additionally commands the production controllers in charge for the components in question to automatically initiate protective steps, sending a notification about this decision to the supervision mediator (SM).

2.3 Production Controller

The production controller (PC) is responsible for the interaction with the automation system. It realizes higher-level management and control functionality of the associated SystemElements, which in turn provide specific capacities within the production system. The functionality realized by the PC is offered to higher level control systems (as e. g. production planning) as a service. In order to be discoverable, the PC registers with the discovery service (DS) as described in [9].

In case of a reliably (w. r. t. pl_μ and pl_Δ) identified malfunction of the system, the PC is informed about the Diagnosis by a situation analyzer (SA). This information enables the PC to restrict the production services it provides via the discovery service in order to reduce the probability of machine and product damage. The semantic model and the detailed failure state identified by the SA allow the PC to restrict as few machine capabilities as necessary, as compared to the complete temporal decommissioning of a machine in case of an emergency stop in reaction to an unspecific general failure state.

2.4 Supervision Mediator

The supervision mediator (SM) provides an interface between the agent-based automated control system and the human operator, integrating her into the diagnostics and control cycle: Using the SM, the operator assigns situation analyzers (SA) to selected components of the system and configures the plausibility threshold values pl_μ, pl_Δ (used by the production controller) and $pl_\downarrow$ (used by the situation analyzer). Upon detection of a set of plausible diagnoses by a SA, the SM enables the operator to inspect the alternatives, manually change interpretations, and revoke automated reactions if necessary. Additionally, the SM supports the operator in acquiring additional information not available by automated sensors and feeding it into the diagnostic process. The SM is therefore central to the proposed interactive approach to diagnostics in production.

3 Realization

The prototypal realization of the proposed framework in a real-world industrial production environment is depicted in Fig. 2. As to not disturb the real time communication of the automation system based PROFINET/Profibus, we decided to implement the communication of the multi-agent system using installed enterprise links based on TCP/IP. We use a specialized in-house multi-agent platform; the basic system could however be ported straightforwardly to other, public platforms since we only rely on services available in most available systems. The following subsections present selected aspects of the prototype system.

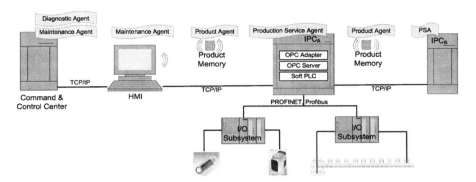

Fig. 2 Architectural overview of the realized system

3.1 Production Service Agent

A production service agent (PSA) acts both as a symptom provider and a production controller. In this latter role, the PSA allows the operator to configure the production control system, and it controls the production process to ensure that configurations which might damage the machinery of the product are avoided. In the symptom provider role, the PSA interprets data provided by machinery-mounted sensors yielding Features, which it combines subsequently using the feature model to derive Symptoms. This behavior of the PSA is realized straightforwardly by a rule engine, as the feature meta model and the feature models are restricted to OWL 2 RL. The necessary sensor and parametrization data can be accessed as process variables of the real-time control kernel the upcoming OPC UA Standard [5] based on a locally executed client-server protocol. All components including the OPC UA infrastructure and the agent itself are realized on an industrial PC (IPC) equipped with a TPC/IP communication link and a software programmable logic controller (PLC) that exercises the real-time control as shown in Fig. 2.

In the running example, the PSA allotted to the exemplary transportation system uses the axle motion sensors to derive one of the symptoms AxleStop or AxleTurning depicted in Fig. 1, which is then sent to the diagnostic agent (DA) registered for the respective symptom. If the PSA is notified by a DA about a plausible Diagnosis, the actual State is derived in order to to restrict the set of Configurations of the production service which are announced to the higher-level planning systems.

3.2 Product Agent

A product memory is an embedded device attached to a product during its complete life-cycle and equipped with sensors to autonomously observe the ambient environment of the product. A product agent (PA) implementing the symptom provider role controls the product memory and makes its measurements available to the production control system as `Symptoms`, which can be mapped to `SystemElements` based on product location. `Symptoms` are derived from `Features` based on the feature model using a rule engine installed on the product memory; details on this embedded reasoning approach can be found in [15]. Figure 3(a) depicts the conceptual architecture of a product memory, a photograph of a prototype is shown to its right in Fig. 3(b).

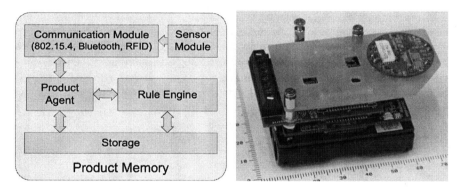

Fig. 3 Architecture of a digital product memory (a) and photograph of a prototype (b)

Regarding the exemplary inner logistics component, the PA uses the local feature model to interpret the measurements of its onboard 3-axis motion sensor, determining whether or not the product is moving (`ProductMoving` or `ProductStops`), and whether the motion trajectories indicate excessive vibrations that might damage the product (`VibrationHigh`). The `Symptoms` are sent to subscribed diagnostic agents as they are detected.

3.3 Diagnostic Agent

The main task of a diagnostic agent (DA) implementing the situation analyzer role is knowledge-based diagnostic reasoning which has been a vivid research topic over years, leading to a multitude of approaches (see [2] for a survey). The current implementation uses logic-based abduction [14] for this task as abductive reasoning naturally handles incomplete information, enabling predictive diagnostics. More specifically, we use a method extending the ideas presented in [4] to determine a set of plausible diagnoses along with the assumptions they require and the resulting plausibility score. Omitting most technical detail due to space limitations, diagnosis generation can be seen as finding optimal paths in a hypergraph, where each path represents a valid derivation. The structure of the graph is determined by the models, the observed and assumed Symptoms, and the set of possible diagnoses; its size is polynomial in the size of the situation model due to the restrictions on the representation language. The plausibility of a path depends on two factors, namely the observations it explains and the assumptions it requires to be made, inducing a partial order. As the number of paths can be exponential, we use an incremental anytime algorithm to determine them one by one in order of decreasing plausibility, stopping as new information arrives or the lower plausibility bound $pl_\downarrow$ is reached. As delineated in Sect. 2.2, the parameters pl_μ and pl_Δ determine whether an instant automated reaction is taken via the production service agent, additionally the complete set of competing diagnoses is provided to the maintenance agent along with information on required assumptions and plausibility.

In the running example, assume that a product located on outgoing belt 1 of the logistics component detects increased vibration values (Vibration) while the respective component signals a turning axle (AxleTurning). The most straightforward explanation for both observations requiring no assumptions then states that only this route of the logistic system vibrates (Route1Vibrating). Alternatively, assuming for example that the shared incoming belt does not run smoothly either gives rise to the diagnosis BothRoutesVibrating.

3.4 Maintenance Agent

Maintenance agents (MA) realize user interaction as defined by the supervision mediator role. To this end, a MA offers two separate graphical user interfaces, the supervision view integrated into the command and control center of the factory, and the maintenance view built into the SCADA system WinCC. The former allows the operator to assign diagnostic agents (DA) to components, set the thresholds pl_μ, pl_Δ and $pl_\downarrow$, introspect diagnoses derived by the DA and manually pick one of them, in which case the new diagnosis is signaled back to the DA where it is forwarded to the responsible production service agent. The maintenance gui located at the machine control panel supports the operator or technician in sharpening the result of the analysis by pointing out missing relevant data (determined from the assumptions created during reasoning), and feeding measurements made by the operator back into the diagnostic process.

In context of the running example, the operator might for example decide to make sure the other belts to not exceed vibration limits by visual inspection to reduce the risk of products being damaged, using the maintenance gui.

4   Conclusions and Future Work

We have presented an architecture for an agent-based flexible production system which integrates a service-based approach to production with intelligent products and interactive diagnostics of the production machinery. We extend previous work on using diagnostic reasoning in automated production control by integrating mobile intelligent products and by enabling a human operator to directly interact with the diagnostic agent. Here, digital product memories serve as additional mobile sensor units within the production system providing both product- and machine-centered data. Using a Human-in-the-Loop approach to interactive diagnostic reasoning solves several prominent problems of automated diagnostics: Firstly, it alleviates the problem of decision making in the context of a huge amount of possibly imperfect data by automated pre-processing and default decisions which can be modified due to human experience. Furthermore, lack of information is handled by enabling the diagnostic system to draw hypothetical conclusions which are then validated or refuted on the basis of additional data provided by the operator. Moreover, the proposed system is based on real industrial automation systems and hardware, thus providing a direct integration path into existing products.

We are currently building a prototype system. Major challenges include optimizing memory consumption of the embedded reasoner (e. g. by forgetting facts), fast sensor data provision from real-time control systems using OPC, and addressing real-time requirements in reconfiguration. For the future, we intend to integrate model-based planning algorithms into the production service agent to realize optimized configuration scheduling and distributed production planning, and market-based mechanisms for the coordination of production service agents and product agents. Another interesting topic of research is the (semi-)automated extraction of the required models from the huge amounts of engineering information produced during factory planning and available e. g. in plant lifecycle management systems, which could significantly reduce the effort of implementing such a model-based approach. Another idea is to extend the approach towards more expressive models and complex structures such as groups of factories and supply chains. Using a hierarchical approach, diagnoses determined for a factory can serve as symptoms for the whole supply chain, allowing to infer even more information which can be employed to reduce production loss. Finally, we intend to evaluate system performance as well as quality of diagnoses and automated reactions.

References

1. Albert, M., Längle, T., Wörn, H., Capobianco, M., Brighenti, A.: Multi-agent systems for industrial diagnostics. In: Staroswiecki, M., Wu, N. (eds.) Proceedings of the 5th Symposium on Fault Detection, Supervision and Safety of Technical Processes 2003, vol. 1, pp. 459–464. IFAC, Elsevier Science Ltd., Washington D.C., USA (2003)
2. Dressler, O., Puppe, F.: Knowledge-based diagnosis - survey and future directions. In: Puppe, F. (ed.) XPS 1999. LNCS (LNAI), vol. 1570, pp. 24–46. Springer, Heidelberg (1999)
3. Günthner, W., ten Hompel, M. (eds.): Internet der Dinge in der Intralogistik. Springer, Heidelberg (2010) (in german)

4. Hubauer, T.M., Lamparter, S., Pirker, M.: Automata-based abduction for tractable diagnosis. In: Haarslev, V., Toman, D., Weddell, G.E. (eds.) Proceedings of the DL Home 23rd International Workshop on Description Logics (DL 2010), CEUR Workshop Proceedings, pp. 360–371. CEUR-WS.org., Waterloo, CA (2010)
5. International Electrotechnical Commission. IEC 62541 - OPC Unified Architecture Specification
6. International Electrotechnical Commission. IEC 62264 - Enterprise-control system integration (2003)
7. ITU Strategy and Policy Unit: ITU Internet Reports 2005. The Internet of Things. Itu internet reports, ITU (2005)
8. Kühnle, H. (ed.): Distributed Manufacturing - Paradigm, Concepts, Solutions and Examples. Springer, Heidelberg (2010)
9. Legat, C., Lamparter, S., Seitz, C.: Service-oriented product-driven manufacturing. In: Proceedings of the 10th IFAC Workshop on Intelligent Manufacturing Systems (IMS 2010), pp. 161–166 (2010)
10. McFarlane, D.: Auto-ID based control - an overview. Whitepaper, Auto-ID Centre. University of Cambridge, UK (2002)
11. Meyer, G.G., Främling, K., Holmström, J.: Intelligent products: A survey. Computer in Industry 60(3), 137–148 (2009)
12. Onori, M., Barata, J., Frei, R.: Evolvable assembly systems basic principles. In: Proceedings of the 7th IFIP BASYS 2006, Ontario, Canada, vol. 220, pp. 317–328 (2006)
13. Parry, G., Graves, A. (eds.): Build To Order - The Road to the 5-Day Car. Springer, Heidelberg (2008)
14. Paul, G.: Approaches to abductive reasoning: An overview. Artificial Intelligence Review 7(2), 109–152 (1993)
15. Seitz, C., Lamparter, S., Schoeler, T., Pirker, M.: Embedded rule-based reasoning for digital product memories. In: Proceedings of the 2010 AAAI Spring Symposium. AAAI Press, Palo Alto (2010)
16. W3C OWL Working Group: OWL 2 Web Ontology Language: Profiles. W3C Recommendation (2009), http://www.w3.org/TR/owl2-profiles/
17. Wannagat, A., Vogel-Heuser, B.: Agent oriented software-development for networked embedded systems with real time and dependability requirements in the domain of automation. In: Chung, M.J., Misra, P. (eds.) Proceedings of the 17th International Federation of Autonomic Comtrol World Congress, 1. International Federation of Automation Control (IFAC). Elsevier Science Ltd., Seoul (2008), http://www.ifac2008.org/

Multi-Agent Cooperation for Advanced Teleoperation of an Industrial Forklift in Real-Time Environment

F.J. Serrano Rodríguez, J.F. Rodríguez-Aragón, B. Curto Diego, and V. Moreno Rodilla

Abstract. The creation of a multi-agent system that allows an advanced teleoperation of the movement of an industrial forklift, its fork and the control of an on-board pan-tilt-zoom camera is described in this paper. The developed agents, that fall into an AuRA architecture, have been integrated into the MissionLab platform. Thus, they take advantage of the communication and agents cooperation capabilities which are provided by this framework.

1 Introduction

The integration of a teleoperation system in a robot, mainly autonomous, offers great advantages. The operator can take control when the robot must perform complicated actions in order to carry out some specific work manually within a preset automatic mission, or simply to override the default behavior when there is a hardware or software malfunction. This feature can be implemented by using a control architecture. Currently, there are three different kinds of control architectures [9]: deliberative, reactive and hybrid. Hybrid architectures bring the best of deliberative and reactive ones together. They recover the capacity for planning from deliberative architectures and take advantage of the simplicity and speed of the reactive behaviors. An example of an hybrid architecture can be AuRa [1], the architecture implemented in the MissionLab [2] platform.

A study case could be an automated industrial forklift like the one used in this work, which has a mission consisting on loading pallets and move them form one place to another. Loading and unloading pallets is a task that requires quite accuracy, so it could be done manually, while the transfer of the pallet from one point to another can be more easily made automatically. It is also usual for an industrial

F.J. Serrano Rodríguez · J.F. Rodríguez-Aragón · B. Curto Diego · V. Moreno Rodilla
Robotics Group, University of Salamanca
e-mail: fjaviersr@usal.es, jraragon@usal.es, bcurto@usal.es, vmoreno@usal.es

Y. Demazeau et al. (Eds.): Adv. on Prac. Appl. of Agents and Mult. Sys., AISC 88, pp. 57–62.
springerlink.com © Springer-Verlag Berlin Heidelberg 2011

vehicle like a forklift to be used in some monotonous tasks such as transporting materials from one place to another. However, during the day schedule, it may be required to perform some small and unplanned work such as removing some material that blocks the way to other machines, or doing a transfer slightly different from the rest. An operator can then take the control of the vehicle remotely, do the special work and then let it continue working in the mission that has been scheduled.

By default, MissionLab has a very basic teleoperation behavior that is integrated into most of elemental behaviors. Thus, the control can be taken at any time during a mission based on these basic behaviors. However, the teleoperation system included in MissionLab only allows to control the movement direction and the speed. This is useful for robots that only perform low complexity tasks, but when the problem is to deal with industrial vehicles with a larger number of actuators, this such basic control is insufficient. For example, in the Robotics Group at the University of Salamanca (GROUSAL) we have a fully automated industrial forklift [10]. In this forklift, we can control the height and tilt of the fork where materials are loaded and also includes an onboard pan-tilt-zoom (PTZ) camera. In the processes of loading and unloading hazardous materials, it should be desirable to teleoperate all the actuators of the vehicle and to have that teleoperation integrated in MissionLab, so that we can take the most of its advantages.

In this paper, we present an extension to the MissionLab teleoperation behavior, using a multiagent architecture, that allows us to control, using a joystick, additional analog and digital actuators. In the next section (Section 2) MissionLab, a reactive behavior-based multiagent architecture, is introduced. Section 3 presents a new multiagent-based behavior which is used in the navigation of an industrial forklift, and it has the capability of being teleoperated at any time during the navigation. Section 4 presents a study case of the new behavior used in a real system. Finally, the summary and conclusions in Section 5 complete this paper.

2   MissionLab: Multi-agent Control Architecture

MissionLab is an AuRA architecture implementation that allows mission creation and management in multirobot platforms. The deliberative part is a state machine (FSA), using the methods of the Temporal Sequence [3], which is an agent itself that maintains the control of all the agents and it activates and disables them according to the current state. There are two types of agents: atomic and assemblage. Atomic agents are the basic primitive bahaviors while assamblage agents are coordinated societies of agents with a common goal. There are several ways of coordination [6]: subsumption style [5] and potential or cooperative style [4]. In AuRA, the coordination is mainly cooperative.

MissionLab provides many software components: Mlab, CfgEdit, HServer and the Mission itself. *Mlab* allows the user to monitor the progress of the robots during the execution of the mission, showing the current location, the map and the obstacles detected. *CfgEdit* is the mission configuration editor. Missions can be built like a state machine where every state is a behavior or group of behaviors. *HServer* is the hardware server that provides the robot and sensors state. Besides, it is also responsible for the direct control of the hardware and its configuration. The mission defines the behavior of the robot, which is controlled through *HServer*, and it sends data to *Mlab*, that monitors the robot state and receives teleoperation orders. Communications between all components are made via a communications server called *IptServer*.

MissionLab provides an agent-based language that allows the creation of new agents as a collection of already known agents, that are coordinated following the Societal Agent Theory [8]. The Configuration Description Language (CDL) specifies the data structure used by a behavior, the agents involved in that behavior and the coordination among them. However, CDL does not provide the implementation of the agents functionality, but it is a generic description.

Thus, when a configuration, specified by CDL, is deployed into a hardware device within an AuRA architecture, the CDL compiler generates a Configuration Network Language (CNL) specification of the configuration. CNL is a hybrid language [7] based on C++ that provides itself the implementation of the communications and the agents instantiation. The CNL compiler result is the final robot executable.

3   Implementation of the Advanced Teleoperation

MissionLab has been chosen for the implementation because, unlike other frameworks like ROS or CARMEN, it's an implementation of an hybrid architecture (AuRA), provides a complete agent infrastructure and has graphical tools to create complex missions out of the box. It fits perfectly into our project.

MissionLab has already implemented an agent that controls the robots movement using a joystick (*Telop*). This agent can interact cooperatively with other agents so the robot can go to a certain place in an autonomous way, avoid obstacles and be teleoperated at any moment. This agent gets the joystick state from the database that the mission maintains internally and indicates, in its return value, the direction and magnitude of the desired movement. *Mlab* component is the responsible for the communication with the joystick and the one that updates its state in the mission database, while the *HServer* component is the one that finally obtains the desired movement and transfers it to the robot. (Fig. 1) In our application, the forklift movement is needed, but to control an onboard PTZ camera and the fork height and tilt is needed as well. We have developed a new agent, called *GROUSAL_Telop*, following the idea of the basic *Telop* agent implemented in MissionLab. We have integrated that agent into a new behavior called *GROUSAL_GoTo* (Fig. 2) where the new functionality of advanced teleoperation interacts with the rest of navigation and obstacle avoidance agents implemented in MissionLab. The cooperative agent *GROUSAL_GoTo* has been described using the CDL language in the same way as the previous basic *GoTo*. In this new behavior, the agent responsible for the teleoperation process is our new advanced agent: *GROUSAL_Telop*.

At the same time, the agent *GROUSAL_Telop* is described using the CDL language (Fig. 3). Our teleoperation parameters are based on the joystick state

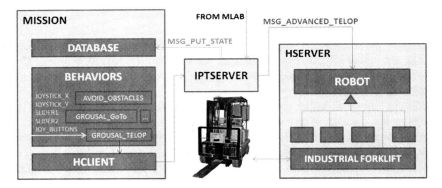

Fig. 1 Overall View of the System

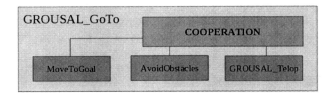

Fig. 2 GROUSAL_GoTo: Assemblage Agent

```
GROUSAL_TELOP(
  telop_mode = DATABASE_INT(key = {"telop_mode"}, initial = {0} ),
  robot_heading = GET_HEADING(cur_pos = $RobotLocation),
  joystick_mag = DATABASE_DOUBLE(key = {"joystick_magnitude"}, initial = {0.0}),
  joystick_x   = DATABASE_DOUBLE(key = {"joystick_x"},  initial = {0.0}),
  joystick_y   = DATABASE_DOUBLE(key = {"joystick_y"},  initial = {0.0}),
  joystick2_x  = DATABASE_DOUBLE(key = {"joystick2_x"}, initial = {0.0}),
  joystick2_y  = DATABASE_DOUBLE(key = {"joystick2_y"}, initial = {0.0}),
  slider_1     = DATABASE_DOUBLE(key = {"slider_1"},    initial = {0.0}),
  slider_2     = DATABASE_DOUBLE(key = {"slider_2"},    initial = {0.0}),
  joy_buttons  = DATABASE_INT   (key = {"joy_buttons"}, initial = {0}))
```

Fig. 3 GROUSAL_Telop: CDL Description File

provided by our modified *Mlab*: 3 analog axis, the movement magnitude and a bitmap with the state of up to 32 toggle buttons. Several different agents, called *DATABASE_DOUBLE* and *DATABASE_INT* are responsible for obtaining the needed values from the mission database and transfer them to the agent *GROUSAL_Telop*. The functionality of *GROUSAL_Telop* has been implemented, as the rest of MissionLab agents, using the CNL language. The implementation receives the input parameters, calculates the desired movement according to the joystick state and the robot heading, and sends the correct orders to *HServer*. This hardware server has been also modified so it can receive these parameters and carry out the teleoperation of the rest of the new elements: PTZ camera and the fork.

4 Results

We have developed a mission using CfgEdit. The mission goal was to transport pallets from a warehouse to another crossing a street which contains obstacles. Outdoor transport was performed in an autonomous way, always monitored. The loading and unloading operations were performed in a teleoperated way.

As expected, several executions of this mission have shown that all teleoperation features work well without any appreciable lag. The teleoperator can take and release the control at any time and can drive the forklift with the help of the PTZ camera as if he were inside it, but just using a joystick instead of a steering wheel and a gear stick.

In Fig. 4a we change the values of several sensors in order to get a simulated malfunction of the forklift. The operator, who is monitoring the mission, can take the control of the forklift remotely so that it can be redirected avoiding any type of incident. After solving sensor malfunctions, the forklift may go on with the desired mission. Fig. 4b shows the teleoperated unloading operation of a pallet inside a warehouse. Fig. 4c shows the performance of the forklift while executing the outdoor part of the mission (between both warehouses). Therefore, it makes an autonomous navigation based on the data provided by its sensors.

Fig. 4 Application of the New Agents in the Real Industrial Forklift Indoors and Outdoors

5 Conclusions

A new agent called *GROUSAL_Telop* that allows to teleoperate the movement of an industrial forklift, its fork and an onboard PTZ camera has been implemented with success. This software component interacts with the rest of agents already implemented into MissionLab according to the specifications of the AuRA architecture. We have taken advantage of the multiagent capabilities of MissionLab by creating a new additional agent called *GROUSAL_GoTo* that associates our advanced teleoperation agent with the automatic navigation and avoidance of collisions MissionLab agents.

Thanks to this development, to take control of the forklift, to move the camera or to handle the fork can be done at any moment, even if the forklift is moving in an autonomous mode. Besides, our teleoperation is supervised by the obstacle avoidance agent, that will prevent us from crashing with any object detected by the sensors, even if we are in teleoperated mode.

The choice of a multiagent approach and the use of the MissionLab platform for this project has proven to be successful. It has made the communication tasks among the different software modules easier and it has allowed us to develop new agents with a low coupling with the rest of the system.

Acknowledgements. The work has been carried out within projects financed by the Junta de Castilla y León SA030A-07 and the Spanish Ministry of Science and Innovation DPI2007-62267. F.J.Serrano Rodríguez has worked under the support of a University of Salamanca fellowship. J.F.Rodríguez-Aragón has worked under the support of a Junta de Castilla y León fellowship.

References

1. Arkin, R.C., Balch, T.: AuRA: principles and practice in review. Journal of Experimental & Theoretical Artificial Intelligence 9(2-3), 175–189 (1997)
2. Arkin, R.C., et al.: MissionLab (2006),
   http://www.cc.gatech.edu/aimosaic/robot-lab/research/MissionLab
3. Arkin, R.C., MacKenzie, D.C.: Temporal coordination of perceptual algorithms for mobile robot navigation. IEEE Transactions on Robotics and Automation 10(3), 276–286 (1994)
4. Arkin, R.C.: Towards Cosmopolitan Robots: Intelligent Navigation of a Mobile Robot in Extenderd Manmade Environments. PhD. Dissertation, University of Massachusetts, Department of Computer and Information Science (1987)
5. Brooks, R.A.: A robust layered control system for a mobile robot. IEEE Journal of Robotics and Automation 2(1), 14–23 (1986)
6. Connell, J.: A colony architecture for an artificial creature. AI Tech. Report 1151. MIT, Cambridge (1989)
7. Lee, B., Hurson, A.R.: Dataflow architectures and multithreading. IEEE Computer (1994)
8. Mackenzie, D.C., Arkin, R.C., Cameron, J.M.: Multiagent Mission Specification and Execution. Autonomous Robots (1997)
9. Murphy, R.: Introduction to AI Robotics. MIT Press, Cambridge (2000)
10. USAL Robotics Group. Diseño y desarrollo de una arquitectura para el guiado de un vehculo en condiciones adversas (2010),
    http://gro.usal.es/proyectocarretilla

Map Partitioning to Approximate an Exploration Strategy in Mobile Robotics*

Guillaume Lozenguez, Lounis Adouane, Aurélie Beynier, Philippe Martinet, and Abdel-Illah Mouaddib

Abstract. In this paper, we present an approach to automatically allocate a set of exploration tasks between a fleet of mobile robots. Our approach combines a *RoadMap* technique and Markovian Decision Processes (MDPs). We are interested in the problem of exploring an area where several robots need to visit a set of points of interest. This problem induces a long term horizon motion planning with a combinatorial explosion. The *RoadMap* allows us to represent spatial knowledge as a graph of paths. It can be modified during the exploration mission requiring the robots to use on-line computation. By decomposing the *RoadMap* into regions, an MDP allows the leader robot to evaluate the interest of each robot in every single region. Using those values, the leader can assign the exploration tasks to the robots.

1 Introduction

The problem of exploring an environment with a fleet of robots is a persistent topic in mobile robotics [1]. It is difficult to consider the global problematic in a long term horizon but several studies have contributed in different orientations. Indeed, the topic can be decomposed to 2 main parts: planning and controling of the robot's

Guillaume Lozenguez · Abdel-Illah Mouaddib
GREYC, Campus Côte de Nacre, Bd Marchal Juin, BP 5186, 14032 Caen Cedex France
e-mail: firstname.lastname@info.unicaen.fr

Guillaume Lozenguez · Lounis Adouane · Philippe Martinet
LASMEA, Campus des Cézeaux, 24 Avenue des Landais, 63177 Aubiere Cedex France
e-mail: firstname.lastname@lasmea.univ-bpclermont.fr

Aurélie Beynier
LIP6, University Pierre & Marie Curie, Boîte courrier 169, 4 place Jussieu,
75005 Paris France
e-mail: firstname.lastname@lip6.fr

* Supported by the Nationnal Reserch Agency of France (ANR) through the *R-Discover* project.

Y. Demazeau et al. (Eds.): Adv. on Prac. Appl. of Agents and Mult. Sys., AISC 88, pp. 63–72.
springerlink.com   © Springer-Verlag Berlin Heidelberg 2011

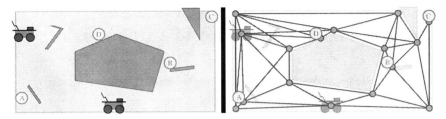

Fig. 1 Exploration problem with 4 points of interest $\{A, B, C, D\}$ and the attached *RoadMap*.

motion [2][3] with localization [4] ; communicating and computing the collaborative exploration strategies [1][5][6]. The notion of *RoadMap* (as a topological map increased by metric informations [7]) appears as an interesting tool to connect the sensors control and decision making [8]. The approach in this paper proposes the use of the *RoadMap* in on-line decision making considering a long-term horizon.

Our research is employed in a project named R-Discover that aims at exploring an external area using a fleet of robots. An Unmanned Aerial Vehicle (UAV) takes several pictures of an area and then a fleet of ground robots are set to refine knowledge about this area. Indeed, the pictures allow us to build a first map regarding detected obstacles. Next, an operator defines key positions which must be visited to increase the map definition as a set of points of interest in the map (Fig. 1).

The main concern of the paper is based on the robots' capacities to cooperate in order to calculate and adapt exploration strategies along the mission. Robots are equipped with communication devices efficient in a given radius. In this paper, we focus on the particular steps of the mission where two or more robots of the fleet can communicate. We suppose that robots are able to share their knowledge. The present robot with the higher level in a defined hierarchy is set to be the current leader. We are interested in the capacity of the current leader to re-allocate the local set of points of interest $I = \{A, B, C, \ldots\}$ between the present robots in few seconds.

Allocating the set of points of interest is computed in a way that maximizes the sum of individual expected gains. But the complexity of evaluating the interest of a robot in visiting a sub-set of points of interest does not permit a computation of an optimal solution for real size problems $|I| > 20$ (Section 2.2). To bypass the complexity of finding a solution on-line, heuristics are used to partition the *RoadMap* which allows the robots to plan and reason over regions instead of the set of points of interest. The approach permits to solve the problem of exploring an area with a fleet of robots with planning under uncertainty over long term horizon using methods to minimize the on-line computation time.

2 Working Context

The used architecture presented in the next section allows us to separate the robot locomotion problem from the deliberative aspect. This way, we concentrate on the problem of computing the exploration policies using several levels of abstraction in order to allocate the points of interest between the present's robots.

2.1 The Robot's Control Architecture

Each robot is composed of two modules. The first is reactive and permits the robot to move between two positions. The second is deliberative and aims to organize the tasks and gives the decision to the reactive module (Fig. 2).

The reactive module controls movements according to events of low importance such as little discovered obstacles to avoid. A hybrid multi-controller [3] is used for the navigation of the mobile robot in cluttered environments. This architecture is based on a flexible switching between different atomic controllers as attraction to a target or obstacle avoidance. That allows the robot to reach a position while avoiding obstacles.

In this paper, we are not interested in developing the reactive module. In fact, we are interested in describing the deliberative module. A task (Fig. 2) matches a target position to reach by avoiding obstacles. A Probabilistic *RoadMap* [2] is defined as a graph $< W, P >$ where W is the set of way-points (nodes) and P is the set of paths (edges). The way-points set W is composed of the points of interest I in addition to the environment way-points (points around the known obstacles)(Fig. 1). The *RoadMap* is assumed fully connected.

$$RoadMap = \{W, P\}, \quad where: \quad W = \{w_1, ..., w_k\}$$
$$P = \{(w_1, w_2, \overrightarrow{v}, c, u) \mid w_1, w_2 \in W, \quad \overrightarrow{v} \in \mathbb{R}^2, \quad c \in \mathbb{R}, \quad u \in [0, 1]\}$$

Each path p is defined by the current w_1 and the targeted w_2 way-points. A vector $\overrightarrow{v}$ gives the relative position of w_2 from w_1. The attribute c is the associated cost; it depends on the distance and the quality of the path. The attribute u is the probability to reach a target position without crossing important obstacles.

Structuring knowledge of collision-free connectivity in a *RoadMap* permits to use graph algorithms like $A*$ to plan the movements of an agent or a fleet of agents [6]. Graph theory does not directly fit to the stochastic aspect to find an optimal policy in uncertain environments. However it is possible to combine the *RoadMap* and Markovian Decision Processes. It was used to improve path finding by minimizing the movement cost and considering collision safe paths [8]. Due to non-deterministic actions, we propose to compute robot's policies using Markov Decision Processes (*MDP*) [9], where the MDP model is built from the *Roadmap* elements.

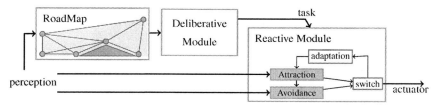

Fig. 2 The Robot Architecture Schema.

2.2   The Markovian Decision Processes

An MDP is defined as a tuple $< S, A, t, r >$ with S and A respectively, the state and the action sets that define the system and its control possibilities. t is the transition function defined as $t : S \times A \times S \rightarrow [0, 1]$ that gives the probability $t(s, a, s')$ to reach the state s' from s by doing action $a \in A$. The reward function r is defined as $r : S \times A \rightarrow \mathbb{R}$, $r(s, a)$ gives the reward obtained by executing a from s.

A policy function $\pi : S \rightarrow A$ assigns an action to each system state. Optimally solving an MDP consists in searching an optimal policy π^* that maximizes the expected gain. π^* maximizes the value function of Bellman equation [10] defined on each state. For a policy π:

$$V^\pi(s) = r(s, a) + \gamma \sum_{s' \in S} t(s, a, s') V^\pi(s'), \quad a = \pi(s)$$
$$V^{\pi^*}(s) = \underset{a \in A}{argmax}(\; r(s, a) + \gamma \sum_{s' \in S} t(s, a, s') V^{\pi^*}(s') \;)$$

The γ parameter in $[0, 1]$ balances the importance between future and immediate rewards. In case of finite horizon problems, as visiting a set of points, γ is set to 1.

Several studies [4][5] use MDPs in mobile robotics. In routing problem, an action is added for each path p ; transitions and rewards are done considering the values u and c linked to p (Section 2.1). To visit a set of points, an MDP state needs to include the set of already visited points of interest I' and the number of states increases exponentially regarding the total number of points of interest ($|S| > 2^{|I|}$). We can not consider a problem with more than 20 points of interest to solve it on-line.

2.3   MDPs Decomposition

Decomposition permits us to decrease the complexity of the policy computation by building a hierarchy between local problems and a global solution. It is particularly efficient in spatial problems as it is based on the topological aspect of transitions. The idea of Decomposed MDPs is to aggregate strongly connected states together in sub-MDPs to compute the policy in a distributed way [11][12]. Several policies are computed for each sub-MDP depending on neighboring parameters values.

The problem of computing an optimal graph partition is known to be NP-Complete [13][14]. In case of Decomposed MDPs, the optimal partition on S is the partition that allows to compute the policy the most quickly possible. Independent sub-MDPs are desired in order to decrease the number of policies computation in each sub-MDP. Generally, the criteria is to built partitions as balanced as possible by minimizing connexions between sub-MDPs [15][16].

The presented approach is based on a greedy decomposition (Section 3.2) of the *RoadMap* in order to build a global abstract MDP on the regions set. This global MDP allows robots to evaluate their interest of exploring each of the regions in order to allocate the mission between them (Section 3.3).

3 The Deliberative Module

During the mission, robots meet each other. At this moment, present robots are able to communicate in order to merge their knowledge. We are interested in how the current leader can build, on-line, a new partition of the updated *RoadMap* and allocate the exploration tasks to the present robots. All points of interest need to be allocated while minimizing the sum of the expected movement costs of all the robots. The solution must be able to handle 3 communicating robots, up to 120 targets to visit and an exploration area including around a thousand way-points.

3.1 The RoadMap Partition

After updating its *RoadMap* regarding transmitted information from present robots, the current leader partitions the *RoadMap* into regions in order to re-allocate the set of regions of interest between the present robots.

We search the k-partition that maximizes the ratio between the number of paths contained in each region over the number of paths which connect two regions. We denote a partition as $\Phi_k = \{R_1 \ldots R_k\}$, defined on k regions of the *RoadMap*. A partition covers the way-points set with no intersection between two regions:

$$\bigcup_{R_i = \Phi_k} R_i = W, \qquad \forall R_i, R_j \in \Phi_k, \qquad (R_i \neq R_j) \Rightarrow (R_i \cap R_j = \emptyset)$$

We define $Input(\Phi_k) \subset P$ as the set of all internal paths contained in regions. In contrast, the set $Output(\Phi_k) \subset P$ contains all intersected paths by the current partition.

$$Input(\Phi_k) = \{p(w_1, w_2, \vec{v}, c, u) \mid \exists R_i \in \Phi_k, \quad (w_1, w_2) \in R_i \times R_i\}$$
$$Output(\Phi_k) = \{p(w_1, w_2, \vec{v}, c, u) \mid \exists R_i, R_j \in \Phi_k, \quad R_i \neq R_j, \quad (w_1, w_2) \in R_i \times R_j\}$$

Finally we search the optimal partition Φ_k^* defined on the *RoadMap* which maximizes the criteria $\frac{|Input(\Phi_k)|}{|Output(\Phi_k)|}$. Graph partitioning is known to be NP-complete and we need to compute a solution on-line during the mission. Therefore, we use a greedy heuristic. We choose to build regions incrementally by adding the way-points which maximize the criteria for a current region.

3.2 The Greedy Heuristic

Similarly to $Input(\Phi_k)$ and $Output(\Phi_k)$, during the construction of the partition Φ_k we define $Input(R_i, w)$ and $Output(\Phi_k, R_i, w)$ the sets of paths that connect a region R_i to a way-point w and that connect w to way-points not contained yet in the current Φ_k. Starting with a given way-point w_0, Algorithm 1 builds the region R_i by selecting the way-points which maximize $criteria(\Phi_k, R_i, w) = \frac{|Input(R_i, w)|}{|Output(\Phi_k, R_i, w)|}$. In case of no way-point w has a criteria value $criteria(\Phi_k, R_i, w)$ up to a bound b (fixed to 1 in our study), a new region begins with the closest free way-point to w_0.

Algorithm 1. Greedy Partitioning

Require: *RoadMap* $M = \{W, P\}$, $\quad w_0 \in W$, $\quad b \in \mathbb{R}^+$, $\quad \Phi_k \leftarrow \emptyset$

$\quad$ **while** $\bigcup_{R_i \in \Phi_k} R_i \neq W$ **do**

$\qquad R' \leftarrow \emptyset$

$\qquad$ choose w' the closest way-point to w_0 in term of distance where $w' \in W - \bigcup_{R_i \in \Phi_k} R_i$

$\qquad$ **repeat**

$\qquad\qquad R'.add(w')$

$\qquad\qquad$ choose w' that maximizes $M.criteria(\Phi_k, R', w')$ and $w' \in W - \bigcup_{R_i \in \Phi_k} R_i$

$\qquad$ **until** $M.criteria(\Phi_k, R', w') > b$

$\qquad \Phi_k.add(R')$, $\quad (k \leftarrow k+1)$

$\quad$ **end while**

$\quad$ **return** Φ_k

Furthermore, we control the expected size of regions by balancing $Output(\Phi_k, R_i, w)$ with a parameter $e(R_i)$ that simulates future paths inside R_i. The effect of $e(R_i)$ is bounded by the number of $Output(\Phi_k, R_i, w)$. We define the criteria as:

$$criteria(\Phi_k, R_i, w) = \frac{|Input(R_i, w)|}{|Output(\Phi_k, R_i, w)| - \min(\,|Output(\Phi_k, R_i, w)|,\, e(R_i)\,) + \varepsilon}$$

This way, negative values of $e(R_i)$ force the selection to close the region by increasing the number of cut paths. At each step, $e(R_i)$ is defined inversely proportional to the number of points of interest added to the region. We denote by d the desired number of points of interest contained in a region. Furthermore, k^* denotes the expected number of regions ($k^* = |I|/d$). In fact, the greedy partitioning does not guaranty that $k = k^*$.

$$e(R_i) = \frac{d - |R_i \cap I|}{d} \cdot \frac{|P|}{|W|}, \quad d \in \mathbb{N}^+$$

3.3 The Global MDP and Region Allocation

We will explain how MDPs are used to model a global exploration problem. The global MDP allows the leader to valuate each robot interest in exploring a sub-set of regions. From a partition $\Phi_k = \{R_1 \dots R_k\}$, a *RoadMap* and a set I of points of interest to allocate, we define a set of regions of interest $J \subset \Phi_k$ as the set of all regions with at least one point of interest. A state $s = (R_s, J_s)$ of the global MDP includes the region R_s where the robot is positioned and $J_s \subset J$ the set of explored regions. A specific state called *block* is added to represent the situation where an unknown obstacle prevents the robot from reaching a task. Falling in the *block* state means that the robot needs to recalculate the global computed policy. This situation comes with important updates of the *RoadMap* that modifies the region configuration. In this model, when a robot is in a state (R_s, J_s), its possible actions are moving to an adjacent region R'_s or exploring the current region R_s.

$$S = \{(R_s, J_s) \mid R_s \in \Phi_k, \quad J_s \subseteq J\} \cup \{block\}$$
$$A = \{goto_{R'_s} \mid R'_s \in \Phi_k\} \cup \{explo_{R_s} \mid R_s \in J\}$$

In the first case the robot ends up in state (R'_s, J_s), and in the second case it ends up in state $(R_s, J_s \cup \{R_s\})$. We denote $suc(s, a)$ the reached state if the execution of the action a is successful. The transition function $t(s, a, s')$ gives the probability to reach the state $suc(s, a)$ or to get *blocked*.

Optimal transition and reward evaluation depend on the shape of each region. It depends on the local region policies and the crossed way-points. The computation time does not permit us to compute all local policies so we chose to approximate the transition and the reward functions. The approximation is done proportionally to the values of uncertainty u and the cost c linked to each path p of the *RoadMap* and the number of included points of interest $|R_i \cap J|$ of the explored region R_i.

Using the *ValueIteration* algorithm [9] on the global MDP allows the robots to compute an abstract policy as moving between regions and exploring them. We search to automatically allocate a set of regions to explore to each robot of the n present robots at a communication step of the mission.

We want to find the best allocation $\mathbb{J}_n^* = \{J_0, \ldots, J_n\}$ where each $J_i \in \mathbb{J}_n$ matches the set of regions allocated to the robot Ag_i. The optimal allocation maximizes the sum of the robots expected gains in the $n^{|J|}$ possible allocation. Knowing the set R_{Ag_i} of the actual regions of the robot Ag_i, the computed policy π^* of the global MDP and the value function of bellman $V^\pi(s)$, we search:

$$\underset{\mathbb{J}_n^* \in \{\mathbb{J}_n^0, \ldots, \mathbb{J}_n^{n^{|J|}}\}}{argmax} \left(\sum_{i=0}^{n} V^{\pi^*}(R_{Ag_i}, J_i) \right)$$

By considering up to 3 robots and 12 regions ($n \leq 3$ and $k* \leq 12$) it is possible to test all the set of n-allocations. The leader, for each possible allocation $\mathbb{J}_n^j$, computes the sum of expected gain for each robot Ag_i ($V^{\pi^*}(R_{Ag_i}, J_i^j)$) and holds $\mathbb{J}_n^*$ with the maximum sum. The architecture of a *RoadMap* with multi-level MDPs permits to consider more than 3 robots for on-line computations. However, the calculation of $\mathbb{J}^*$ by an exhaustive way limits us to consider only few robots in this paper.

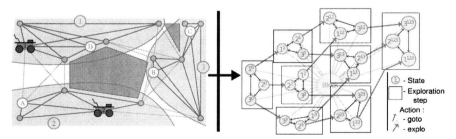

Fig. 3 A partitioned *RoadMap* and the attached global MDP. For example, a state 2^{13} means that the robot is in the region 2 and the regions 1 and 3 are explored

4 Experiments

A visual representation of the built partitions (Fig. 4) allows us to conclude: when the environment is more structured with a coherent expected region size, the algorithm builds a partition closer to the expected one. Otherwise, we observe intersected regions in free space environments. It is due to the non-consideration of the path cost in the used criteria. This phenomenon is reduced in cluttered environment.

We are interested in evaluating this approach in regard to the number of regions, their sizes and the time needed to compute the global policy and allocate the set of regions of interest. The considered problem involved up to 120 points of interest, 3 present robots and a desired number of 10 points of interest by region ($|I| \leq 120$, $n = 3$ and $d = 10$). Our experimentation generates targets randomly (Fig. 5) (500 random generations for different numbers of points of interest $|I|$ growing by 5). We use an Intel Core2 Quad CPU Q9650 at 3.00GHz to compute the partition, the global policies and the region allocation.

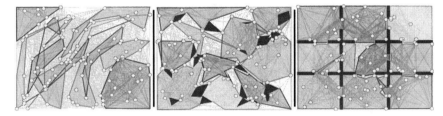

Fig. 4 Examples of partitions built from differently structured environments (RoadMaps)

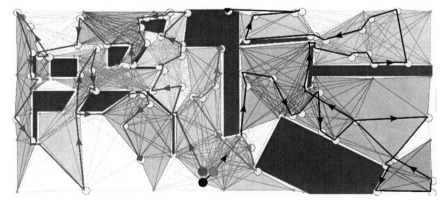

Fig. 5 The 3 likely trajectories following the computed policy ($n = 3$, $|I| = 80$ and $d = 10$). Each robot computes its local policies for each crossed region in a similar way to Section 3.3. Local policies are oriented by the current interest of neighbors regions evaluated in the Global MDP.

Table 1 This table shows, for a selected set of experiments (function of the number of points of interest $|I|$), the average number of built regions (k), the average bounds of regions sizes (in regard of the number of contain points of interest), the average computing time and the average allocated number of regions.

$|I|$	k	R_i sizes bounds	time (s)	allocation
5	1.13	$4.382 - 4.968$	0.058822	$1 - 0 - 0$
20	3.414	$3.428 - 8.126$	0.047248	$2 - 1 - 0$
40	5.6	$4.12 - 10.144$	0.0526601	$3 - 2 - 1$
60	7.708	$4.292 - 11.44$	0.11589	$4 - 2 - 1$
80	9.426	$4.728 - 12.742$	0.339208	$5 - 3 - 1$
100	10.634	$4.356 - 14.192$	0.897122	$6 - 3 - 1$
120	12.422	$3.744 - 14.81$	4.9466	$6 - 4 - 2$

Table 2 A presentation of experiments regarding the numbers of built regions k in function of the expected number k^* (a) and the number of contained points of interest in the worst region (b). The percentages are calculated from experiments with between 5 and 120 random points of interest.

(a)

number of region k	$k^* - 3$	$k^* - 2$	$k^* - 1$	k^*	$k^* + 1$	$k^* + 2$	$k^* + 3$	$k^* + 4$	$k^* + 5$	$k^* + 6$
experiment (%)	0.0	0.13	3.76	21.87	48.44	23.15	2.56	0.09	0.01	0.0

(b)

worst region size	8^-	9	10	11	12	13	14	15	16	17	18	19	20
experiment (%)	7.78	11.0	12.1	11.6	12.5	13.3	13.6	9.21	5.13	2.42	0.88	0.26	0.12

Table 1 and Table 2 present some experimentation results that validate the control of the expected number of regions, the numbers of points of interest in a region and the associated computation time. MDP sizes and computation time grow exponentially with the number of regions k or, with the number of points of interest in a region (in local MDPs). We notice that the allocations are unbalanced, it is normal in regard to the placement and shapes of obstacles. Indeed, the intention was to minimize the sum of robots movement cost.

Experiments with few points of interest in regard to the number and the structure of obstacles lead to a partition with more regions than expected ($k > k^*$) and worst regions which have a lower size than desired size d (Table 2). That denotes the capacity of using the shapes of existing obstacles to built different regions.

5 Conclusion and Future Work

This paper presents a decision making architecture for mobile robots sharing an exploration mission. It is difficult for a leader to evaluate the allocations of many tasks between robots. The study is based on knowledge organized in a *RoadMap* and a solver based on a abstract MDP. Indeed the problem size and the constraint of on-line computation impose to decompose the input data. A greedy decomposition is instantaneous regarding the computation of decision policies. We have demonstrated how a greedy decomposition permits to perform on-line a multi-agent long horizon problem in a stochastic domain by controlling the number of tasks in regions and the number of regions.

In future work, the partitioning quality will be improved by using algorithms based on finding minimal cuts. The idea is to converge to a local optimum from the first greedy decomposition and assume other characteristics as disjunction between regions. Furthermore, adding cooperation in the local MDPs solving will permit several robots to explore a region together. But, that will impact the Global MDP where few robots can explore parts of a same region.

References

1. Burgard, W., Moors, M., Stachniss, C., Schneider, F.: Coordinated multi-robot exploration. IEEE Transactions on Robotics 21, 376–386 (2005)
2. Kavraki, L., Svestka, P., claude Latombe, J., Overmars, M.: Probabilistic roadmaps for path planning in high-dimensional configuration spaces. In: IEEE International Conference on Robotics and Automation, pp. 566–580 (1996)
3. Adouane, L.: Hybrid and safe control architecture for mobile robot navigation. In: 9th Conference on Autonomous Robot Systems and Competitions, Portugal (May 2009)
4. Foka, A.F., Trahanias, P.E.: Real-time hierarchical pomdps for autonomous robot navigation. Robotics and Autonomous Systems 55(7), 561–571 (2007)
5. Teichteil-Königsbuch, F., Fabiani, P.: Autonomous search and rescue rotorcraft mission stochastic planning with generic dbns. In: IFIP AI, pp. 483–492 (2006)
6. Bayazit, O., Lien, J., Amato, N.: Swarming behavior using probabilistic roadmap techniques. In: Swarm robotics: SAB International Workshop, Santa Monica, USA (July 2004)
7. Kuipers, B., tai Byun, Y.: A robot exploration and mapping strategy based on a semantic hierarchy of spatial representations. Journal of Robotics and Autonomous Systems 8, 47–63 (1991)
8. Alterovitz, R., Siméon, T., Goldberg, K.: The stochastic motion roadmap: A sampling framework for planning with markov motion uncertainty. In: Robotics: Science and Systems (2007)
9. Puterman, M.L.: Markov Decision Processes: Discrete Stochastic Dynamic Programming. John Wiley & Sons, Inc., Chichester (1994)
10. Bellman, R.: A markovian decision process. Journal of Mathematics and Mechanics 6, 679–684 (1957)
11. Dean, T., hong Lin, S., hong Lin, S.: Decomposition techniques for planning in stochastic domains. In: 14th International Joint Conference on Artificial Intelligence (1995)
12. Boutilier, C., Dean, T., Hanks, S.: Decision-theoretic planning: Structural assumptions and computational leverage. Journal of Artificial Intelligence Research 11, 1–94 (1999)
13. Garey, M.R., Johnson, D.S., Stockmeyer, L.: Some simplified np-complete problems. In: 6h Symposium on Theory of Computing, pp. 47–63. ACM, New York (1974)
14. Bichot, C.-E., Siarry, P.: Graph Partitioning. Wiley-ISTE (2011)
15. Parr, R.: Flexible decomposition algorithms for weakly coupled markov decision problems. In: 14th Conference on Uncertainty in Artificial Intelligence, pp. 422–430 (1998)
16. Sabbadin, R.: Graph partitioning techniques for markov decision processes decomposition. In: 15th Eureopean Conference on Artificial Intelligence, pp. 670–674 (2002)

ROAR: Resource Oriented Agent Architecture for the Autonomy of Robots

Arnaud Degroote and Simon Lacroix*

Abstract. This paper presents a multi-agent system to organize the various processes that endow a robot with autonomy. The main objectives are to allow the achievement of a variety of missions without an explicit writing of control schemes by the developer, and the possibility to augment the robot capacities without any major rewriting. The proposed architecture relies on a partition of the *decisional layer* in separate *resources*, each one managed by a specific agent. The architecture of *resource* agents and their interactions to guarantee a coherent system are depicted.

1 Introduction

Besides progresses in the robotic functional layer, whether on algorithms or in architecture with some well-understood component architecture like GeNoM or ROS, it is the *assembly* of theses components that leads to autonomy. This assembly, often referred to as "decisional architecture", is in charge of configuring, scheduling, triggering and monitoring the execution of the various processes. It should be designed in order to endow the robot with *(i)* the capacity to achieve a *variety* of high level missions, without manual configuration; and *(ii)* the capacity to cope with a variety of events which are not necessarily a priori known, in a mostly unpredictable world – these two capacities being essential characteristics of autonomy.

Related work. The most popular architectural paradigm in the roboticists community is probably the three layered architecture. In [5], E. Gatt argues that the consideration of the internal state naturally yields the definition of three layers: an

Arnaud Degroote · Simon Lacroix
CNRS; LAAS; 7 avenue du colonel Roche, F-31077 Toulouse,
Université de Toulouse; UPS, INSA, INP, ISAE; LAAS; F-31077 Toulouse, France
e-mail: {arnaud.degroote, simon.lacroix}@laas.fr

* This work has been partially supported by the DGA founded Action project – action.onera.fr.

Y. Demazeau et al. (Eds.): Adv. on Prac. Appl. of Agents and Mult. Sys., AISC 88, pp. 73–78.

intermediate layer is necessary to tie the functional layer, that has no or ephemeral internal state, with the decisional layer, a symbolic planner that strongly relies on a long lasting internal state and plan. Several languages have been proposed in the literature, to ease the implementation of this layer, as PRS [6] for the LAAS architecture [1], or TDL [10] for the *Remote Agent* system [3].

Even if there are some differences between these approaches, they all rely on the main idea that an intermediate layer is required to fill the gap between the functional and symbolic worlds. This leads to different representations of plans, models and information that coexist in the different layers. This discrepancy of representations makes the diagnostics of plan failures difficult, because the planner does not have relevant information about the failure causes, and it hinders the efficiency of the plan executions, because the executive layer does not have a global view of the plan. A first step to solve these issues has been done by the CLARAty system [4]: even if there are still two different tools and representations for the decisional layer (CASPER) and the executive layer (TDL), the system has some way to reflect changes from one representation to another, and exploits heuristics to decide which subsystem will handle the faults. IDEA [9] and T-REX [8] define a two-layer architecture: the problem is partitioned into several agents relying on the *same plan model*, each one composed of a planner and an execution layer. In this way, the planning and the execution phase are consistently interleaved. Moreover, during execution the different agents are synchronised to maintain a consistency of the global plan. T-REX goes further by proposing some systematic formulation to synchronously exchange states between these agents.

Another issues of a three-level architecture are their lack of modularity (decisional and execution layers are two separate "monolithic" blocks, and changes in their model often leads to heavy side effects) and scalability (as the deliberation time increases exponentially with the number of robot functionalities). Mc Gann *and al* [7] state that having one big plan and one execution layer is not scalable on the long run, and conclude that the problem needs to be portioned to be efficiently handled: the use of different planning agents, with different timing constraints, partially solves the scalability issue. However, their partition is constructed by the programmer, based on the mission needs. If the nature of the mission changes, the whole partition must be reorganized: this kind of construction does not scale well over a large variety of missions, missing the objective of a versatile architecture.

Requirements. The principle of *partitioning* the robot functionalities into a network of components is essential to simplify the overall system control. This partition must be carefully designed: in particular, it must allow the addition or removal of some components without breaking the whole system. In other words, each component and its interactions with the other components must be defined by an abstract formal description, thus yielding to a *composability* property of the whole system.

A key feature of autonomy is the ability to properly react to unpredicted events or situations (though handling correctly *any* situation remains a challenge): such events should be asynchronously treated as they occur, and each agent must select the most suitable strategy, on basis of its model and its knowledge of this event.

Finally, the architecture must be *robust to failures*: in case of an agent failure due to a logic or programming error, or to a physical failure, the framework must pursue its operation if possible, using alternative strategies to handle the mission.

Overview. We propose in this paper the definition of ROAR[1], an architecture based on a partition scheme that aims at fulfilling these *composability, reactivity* and *robustness* requirements.

We follow the decomposition principle proposed in IDEA or T-REX , but contrary to these architectures in which the decomposition is defined according to a set of tasks, our proposal is to decompose the robot abilities into a set of separate *resources*. The term "resource" has to be understood in its most general sense here: a resource can be a physical resource, an information resource or a planification resource. Each resource is embedded within a separate agent, a *resource agent*, which is responsible of the consistency and the good use of the resource. The decomposition in resources still must be done by a domain's expert but depends only on the domain, and not a specific robot or functional layer.

Resource agents do not expose directly the resource they embed, but a list of different points of control, which are called *free variables*. Other resource agents ask for specific behaviour of one resource by constraining these *free variables* (*i.e.* binding these variables with specific values). The constraints between agents form a dynamic directed acyclic graph, where vertices are the agents, and the edges are the constraints between agents at a time t. The ROAR framework is in charge of maintaining this graph, ensuring that the required relations are satisfied. For this purpose, each agent is endowed with a solver that locally enforces the constraints set on it. In case of impossibility, the failure goes back through the graph until it is solved by a defined policy – *e.g.* by a call to a planner or (in last resort) to a human operator. In this way, the framework can handle a variety of problems without changing the definition of each agent: the system adapts the information graph to handle the problem at hand, and the different logic solvers locally schedule the access to the resources.

2   Resource Agents

Each agent only exposes a set of *free variables* to specify modifications on the state of the resource it embeds, and is structured along a 2-tier approach (figure 1): upon reception of a new constraint, the *logic layer* states if it can be enforced on the basis of the current *agent context*. If yes, it selects a set of tasks to achieve the transition from one logic state to another one. For each task, the *execution layer* selects a recipe to achieve it on the basis of current *task context*.

Agent Logic layer. On reception of a new constraint message, a resource agent needs to decide if the requested constraint is compatible with the current logic state of the resource. This can be done with a Finite State Machine for simple resources

---

[1] ROAR stands for "Resource Oriented Agent architecture for Robots".

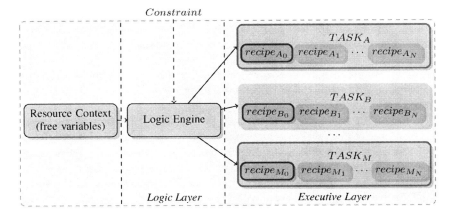

Fig. 1 Global mechanism for a resource agent. The surrounded boxes are the tasks / recipes currently selected by the logic layer/execution layer.

with few *free variables*, but the complexity of the state machine exponentially grows with the number of variables, making it difficult for the developer to guarantee its consistency or to update it after modifications of the *free variables* interface. The programming language community has proposed several languages to solve this kind of problem according to a *logic paradigm*. We propose to use such a paradigm to declare and decide if, in the current context, a constraint is enforceable. The part of the system which handles this is called *Logic Engine*.

The logic layer of an agent is the set of the resource specific legal transitions, described by their name and their pre- and post-conditions. The post-conditions are predicates that are true at the end of the transition, and the pre-conditions are predicates which must be true before the beginning of the transition. Considering the current facts and the required constraints, the *Logic Engine* chooses the combination of transitions to fulfill the constraints or reject them if it can not find any transition. This transition between two states is called a *task*.

Since the notion of resource is robot-agnostic, the notion of transition between two states for a resource is also robot-agnostic. This layer is therefore generic, and reusable for different robots. This layer is also essential to guarantee the consistency of the resource – as long as the developer does not make errors in its specification.

Agent Execution layer. *Tasks* describe the possible transitions between two logic states, they do not describe how to handle them: their execution is handled by some *recipes*. The objective of the decomposition in *tasks* and *recipes* is twofold: *(i)* reduce the complexity to select the action to perform in an agent, and so reduce the time to make this choice, and *(ii)* provide multiple strategies to achieve a transition from one state to another state – in particular, one can stack different strategies to handle different robots capabilities. In other words, this decomposition improves the *reactivity* of the system, and its portability over different robotic platforms.

Recipes are described as a set of pre- and post-condition, similarly to *tasks*, and a body, which contains the real description of the new behaviour of the agent. The body is mainly constructed on top of two constructions, *make* and *ensure* which asynchronously send a constraint message, respectively for discrete and continuous constraints. For conjunction or disjunction of predicates, they dispatch in parallel the constraints to the different agents, and will evaluate lazily the result. The layer evaluates the pre-conditions of each recipe, and execute the most adequate one (the one with the largest number of satisfied pre-conditions).

Error handling. Errors or unexpected events often occur during robotic missions, and it is essential to correctly handle them to successfully achieve the autonomous mission. In the three layer architectures, there are several strategies to handle errors:

- exception handlers, as proposed by Simmons in [10], are piece of application code responsible to handle a specific error. Our framework proposes this approach too. As said previously, recipes are selected on the basis of their pre-conditions. To handle specific error E, we just need to add some recipes with pre-condition check for this specific error.
- the plan has computed several strategies for one task. In ROAR , each task can be handled by multiples recipes. If one recipe fails because of specific agent A, the system can choose another recipe, not using A, to achieve the task.
- the planner is able to repair its plan. In our architecture, it means that the agent can try to choose another combination of tasks to handle a constraint.

All these solutions are local to an agent, and are the preferred method to handle a problem. However, if the agent cannot find a solution by itself, the failure goes up in the agent graph, with a full *error context* (*i.e.* the list of agents that fails, with the associated constraints set on them during the failure). At each step, the system tries to find an alternate strategy using the methodology described previously. When executing this new strategy, the *error context* is passed to each agent, so they can decide of their local strategy, knowing the history of the global task. This improves the behaviour of the whole system, avoiding to use a path which leads to failure.

3   Discussion

We have presented the design of a framework to control the runtime configuration of complex robotic systems. Even if the decomposition in several layers in each agent is similar to the one presented in three-level layer architecture, the whole system is decomposed in several agents: in this way, the system is more reliable (no single point of failure), and scales better, as computations are split by each agent and not handled by the complete system (reducing the complexity at the expense of the optimum). It is modular, and the linked model for executive and planning allows better error handling. In comparison of the T-REX architecture, the use of an asynchronous model yields more reactivity, and it eases network transparency, *i.e.* the seamless use of agent on remote machines or robots. The strict decomposition in resources also make the system *more composable*.

Various language based solutions to control autonomous robots have been proposed, *e.g.* PRS [6], via TDL [10], a C++ extension with (parallel) task semantic. All these languages can express some high-level tasks, but we think that they lack the possibility to precisely deal with resource conflict. PRS may be the better alternative with respect to this, but lacks modularity and robustness. Our work is grounded on some robust and concurrent language like Erlang [2] to provide these features, and provides some methodology to ensure the composability of the framework.

The overall implementation of the framework is still partial, but is however already integrated within two of our robots. We now focus to make the implementation more robust. Our compiler generates some high-level C++ from the specification language. One of the motivation to targeting a high level language instead of a custom runtime is, apart from development time, the possibility to easily integrate with third party libraries without the need to define dedicated interfaces.

Ongoing work include putting efforts on making it applicable to multi-robots scenarios. Another of our goals is to make the system valid by design, *i.e.* to have some guarantees about execution of each agent and about their interaction, in particular avoiding deadlocks situations between different agents.

References

1. Alami, R., Chatila, R., Fleury, S., Ghallab, M., Ingrand, F.: An architecture for autonomy. The International Journal of Robotics Research 17 (1998)
2. Armstrong, J., Virding, R., Wikstrom, C., Williams, M.: Concurrent Programming in Erlang, 2nd edn. (1996)
3. Bernard, D., Dorais, G., Fry, C., Gamble Jr., E., Kanefsky, B., Kurien, J., Millar, W., Muscettola, N., Pandurang Nayak, P., Pell, B., Rajan, K., Rouquette, N.: Design of the Remote Agent experiment for spacecraft autonomy. In: IEEE Aerospace Conference (1998)
4. Estlin, T., Volpe, R., Nesnas, I., Mutz, D., Fisher, F., Engelhardt, B., Chien, S.: Decision-making in a robotic architecture for autonomy. In: Proceedings of the International Symposium on Artificial Intelligence, Robotics and Automation in Space (2001)
5. Gat, E.: On three-layer architectures. In: Artificial Intelligence and Mobile Robots, pp. 195–210. AAAI Press, Menlo Park (1997)
6. Ingrand, F.F., Chatila, R., Alami, R., Robert, F.: PRS: A high level supervision and control language for autonomous mobile robots. In: IEEE International Conference on Robotics and Automation, Mineapolis (1996)
7. McGann, C., Py, F., Rajan, K., Olaya, A.G.: Integrated Planning and Execution for Robotic Exploration. In: International Workshop on Hybrid Control of Autonomous Systems (2009)
8. McGann, C., Py, F., Rajan, K., Thomas, H., Henthorn, R., Mcewen, R.: A Deliberative Architecture for AUV Control. In: IEEE International Conf. on Robotics and Automation (2008)
9. Muscettola, N., Dorais, G., Levinson, C., Plaunt, C.: IDEA: Planning at the Core of Autonomous Reactive Agents. In: International NASA Workshop on Planning and Scheduling for Space (2002)
10. Simmons, R., Apfelbaum, D.: A task description language for robot control. In: Proceedings of the Conference on Intelligent Robots and Systems, IROS (1998)

Engineering Agent Frameworks:
An Application in Multi-Robot Systems

Jérôme Lacouture, Victor Noël, Jean-Paul Arcangeli, and Marie-Pierre Gleizes

Abstract. In this paper, we present a novel development process called SPEARAF (Species to Engineer Architectures for Agent Frameworks) and evaluate its relevance to ease the implementation of Multi-Agent Systems in the context of a multi-robot project for crisis management. SPEARAF allows to build component-based architectures for agents and their infrastructure. We show the advantages of using an architecture-based process to realise an application-specific agent framework adapted to the requirements of such a system. SPEARAF gives guidelines to enables the use of architecture-oriented practices for agent implementation.

1 Introduction

The Rosace (Robots and Embedded Self-Adaptive Communicating Systems) project[1] aims at developing means to specify, design, implement and deploy a set of mobile autonomous communicating and cooperating robots and personal devices. The designed system has to be safe, to enable self-healing, to achieve a set of missions and to self-adapt in a dynamic environment. The main case study considers a crisis management situation with a control center, Autonomous Ground Vehicles (AGVs), Autonomous Aerial Vehicles (AAVs) and human actors all carrying mobile communicating devices. Rosace promoted solution is the design of a Multi-Agent System (MAS) to manage the self-adaptation/self-management of the robots collective activities, where each actor (AAV, AGV, control center, human actor...) is represented by a software agent. The project intends to implement and compare the following multi-agent strategies for several scenarios: Adaptive Multi-Agent System (AMAS) [4], Bonnet-Torres and Tessier approach [4], and Gascuena *et al.* model [4]. In this paper, we focus on the AMAS strategy.

Jérôme Lacouture · Victor Noël · Jean-Paul Arcangeli · Marie-Pierre Gleizes
Université de Toulouse, Institut de Recherche en Informatique de Toulouse,
118, route de Narbonne, 31 062 Toulouse Cedex, France
e-mail: firstname.lastname@irit.fr

[1] http://www.irit.fr/Rosace,737

Y. Demazeau et al. (Eds.): Adv. on Prac. Appl. of Agents and Mult. Sys., AISC 88, pp. 79–85.
springerlink.com

Because the development of such systems is complex and time-taking, we propose SPEARAF, a development process, to design the different entities (AGVs, AAVs...) using component-based software architectures. It enables us to ease the development, produce reusable artifacts and easily reuse them, as well as clearly design architectures supporting evolution while taking into account the requirements from the different stakeholders involved in the project.

In this paper, we focus on a simplified scenario in order to illustrate SPEARAF: AGVs with local perception have to rescue victims scattered in a forest on fire. Information about the existence and location of victims arrives dynamically and the rescue tasks must be self-allocated in a distributed manner by the fleet of AGVs to adapt to the disturbances (fire spreading, evolution of AGV tasks priorities, rescue team reorganisation, communication or material breakdowns, etc.). Based on these functional requirements, our objective is to build a system to implement and compare the strategies to solve this problem. We want to run simulations and take measures w.r.t. a set of metrics to test the feasibility, the consistency and the performance of the strategies in a virtual world before deploying it in real conditions (*i.e.* with real AGVs). A demo of the final system illustrating this scenario with 5 AGVs and 18 victims can be found on the website of the project.[2]

A constraint of the project is to use the provided simulator, called Morse[3], also built in the context of the Rosace project. It provides means to connect a program to a simulated AGV and control all its functions. We also have identified, with the stakeholders of the project, the following non-functional requirements for this system: 1. extensibility and reuse: build a prototype for one strategy that can be improved and extended with new strategies; 2. portability: minimise the effort to port the different strategies from the Morse simulator to real AGVs; 3. abstraction: provide abstract and high-level mechanisms to enable the developers of the strategies to focus on their expertise domain (functional concerns).

In the following, we discuss related works before presenting the process and applying it for our use case. Then we evaluate the process and conclude.

2   Related Works

In this section, we focus on existing approaches that can help the design of MAS with similar properties than the studied use case. Existing design methods [4] do provide guidance to develop MAS by identifying agents and environment as well as designing their behaviours and interactions. Their objectives are to design the functionality of the system and to realise the system itself. Other works such as [2] or Malaca [1] proposes to use component-based and aspect-oriented software engineering to build agent architectures and their crosscutting concerns. In the robotic community, several works tackle the building of robot architectures using components, such as YARP, OROCOS, Orca or more recently GenoM3.

---

[2] http://www.irit.fr/Rosace,1196

[3] http://morse.openrobots.org

In most of these approaches, we are missing a way to build specific types of agents for the specific problem and application we are building, without resting on a generic agent model such as the FIPA one. Moreover, we want to be able to build specific architectures depending on the chosen strategy instead of using a predefined one such as goal-oriented or subsumption architecture. The other approaches that enable reuse and extensibility doesn't give the possibility of abstracting specific types of agents: they at most focus on the component and mechanisms level, if not only the behaviour level.

3 Engineering Architectures for Species of Agents: SPEARAF

In this section, we present SPEARAF (Species to Engineer Architectures for Agent Frameworks), a development process based on [5].

To complete existing design methods that focus on designing the functionality of a MAS, SPEARAF promotes the engineering of application-specific frameworks for the development of multi-agent applications. It enables to take into account the non-functional requirements expressed by the developers of a MAS in order to help them to focus on the functionality of the system. Such frameworks provide what we call "species of agents": **species define sets of agents with common structural characteristics**. In the context of Rosace, there can be a species of agents representing AGVs and another one for those representing AAVs. Also, agents in the different strategies presented Sect. 1 differ from their individual behaviours but share the same structural elements such as GPS, camera or radio that makes them member of the species of AGVs. **A subspecies is a species that derives from a parent species after evolution** (*i.e.* refinement and possibly modification). By defining species, the idea is to provide specific types of agents that fit functional requirements. Developers can rely on species both when designing and implementing a MAS, they don't need to deal with operational concerns and can focus on the agent's functional behaviours.

Species of agents are realised by component-based software architectures and building them happens in two steps: 1. identifying a species of agent for the application and 2. assembling and reusing software components in architectures to create a framework for the species. Moreover, we differentiate two roles in the process: a) creation of the framework by the **framework developer** and b) use of this framework to develop the MAS by the **framework user**. Indeed, when programming the MAS, *hotspots* in the frameworks can be instantiated (possibly with sub-architectures) by the framework user to specify the behaviour of the agents using a set of agent-oriented and application-specific programming primitives defined by the framework developer. In practice, the architectures are defined using the MAKE AGENTS YOURSELF[4] tool that supports SPEARAF using model-driven engineering and editors while the frameworks are implemented with JAVA.

On top of the classical objectives of architecture and component-based software engineering such as modifiability, abstraction, testability or reuse of produced

---

[4] http://www.irit.fr/MAY

artifacts (architectures or components), SPEARAF focuses on concerns specific to MAS development by providing guidelines to build the frameworks. Moreover, it enables the use of software architectures for the development of the behaviours of a produced framework, thus giving all the advantages of such practices in all the development of the whole application.

4 Application: Engineering Architecture for Species of AGVs

In this section, we detail the SPEARAF development process by applying it to the Rosace use case and we show how it enables us to architecture, design and implement a solution to the requirements expressed previously. As we will show, SPEARAF proposes to build the architecture of the species of agents but it also enables to take care of the infrastructure to execute them, however, we will not address this second point since the infrastructure is managed by the Morse simulator.

4.1 Identification of the Species

Identifying the species of agents is twofold and is based on: 1) defining the agents of the species dynamics such as their lifecycle, the way they process information and take decisions; 2) defining the high-level abstract constructs the application developer will use to define the behaviour of the agents. Here, we are interested in the species of AGVs concerned with AGV dynamics and mechanisms.

Based on the species of AGVs, we differentiate the subspecies of AMAS AGVs, concerned with decision and interaction behaviour, by defining how it uses the mechanisms available from its parent species. In the scenario, the objective of the AMAS AGVs is to plan and allocate tasks (to move to a destination to "rescue" a victim) among them. For that, each AMAS AGV follows a perceive-decide-act lifecycle and exchange messages with other AMAS AGVs by following a cooperative task allocation protocol defined using the AMAS theory (see [3]). Tweaks of the behaviour are focused on the evaluation of tasks criticality and their acceptance.

4.2 Architecture of the Species of AGVs

Based on the species presented previously, Fig. 1 shows the architecture of AGVs. In particular, the `Behaviour` component is in charge of orchestrating the internal components functions to achieve a coherent global behaviour. It gathers elaborated information from the rest of the components, makes choices, orders execution of actions, monitors results, and sends control information to relevant components when necessary. `Behaviour` is considered as an abstract component where will be implemented the specific strategy we want to test (AMAS in this paper). Other components are considered generic for the species of AGVs (and thus for all strategies).

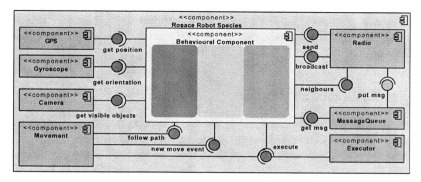

Fig. 1 Component view of the AGVs architecture (UML 2)

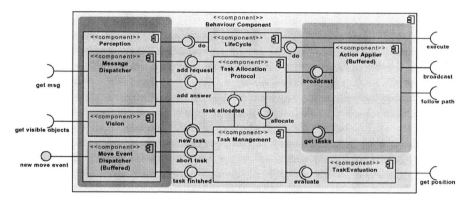

Fig. 2 Component view of the AMAS AGVs architecture (UML 2)

4.3 Architecture of the Species of AMAS AGVs

To match with the AMAS theory, the `Behaviour` component of the species of AGVs architecture is implemented with a set of components categorised into three main activities of the agent: perception, decision, action. A specification has been elicited from a description of the AMAS strategy for the victim rescue task allocation. From this specification, we were able to design the layered architecture of the subspecies of AMAS AGVs depicted Fig. 2.

At the perception level, AMAS AGVs use the components defined in the architecture of the species of AGVs in order to interpret messages and perception. It dispatches information to the decision level depending on their content. At the decision level, AMAS AGVs have to individually evaluate tasks and collectively decide to (re)allocate them. At the action level, AMAS AGVs execute tasks, broadcast messages (decisions) and forward victims location.

With `Behaviour` is seen as an architecture, the high-level abstract constructs for defining the behaviour of this subspecies are `Task Evaluation`, `Perception` or `Task Management`. This last point highlights the differences between user and developer of the framework: building the AMAS AGV subspecies is to use the AGV framework, but at the same time is to build a new specific framework.

5   Feedback and Evaluation

In this section, we briefly evaluate the produced species (architectures and framework) and SPEARAF in regards to the requirements detailed in Sect. 1.

From the AMAS specialist point of view, SPEARAF provides an easy way to implement the subspecies of AMAS AGVs. The developer doesn't need to know implementation details of the species of AGVs: indeed, specification of the interfaces and the dynamics is enough to use a specific species. In this sense, SPEARAF enable to implement MAS strategies with a common framework in order to factorise the effort and compare the different solutions. Moreover, responsibilities of components of the subspecies are clear enough to focus on improving the parameters (task evaluation and acceptation) of the task allocation protocol by a less expert developer.

Then, reuse and extensibility of species and components are clear advantages of SPEARAF. From the species of AGVs, it is easy to provide the other subspecies to implement the strategies presented Sect. 1. Consequently, the development itself produces reusable artifacts such as components (*e.g.* task allocation, vision) or species (by extension) for the development of other species, other scenarios and even other applications in robotics. An interesting example is reusing the species of AMAS AGVs to produce an species of AMAS AAVs with same goals: it doesn't need an important effort and would only consist in developing the AAVs specific components to build a species of AAVs. Moreover, using species enables to first prototype the implementation of an AGV then create a real subspecies by building a derived architecture from the parent species.

From the point of view of the framework developer, SPEARAF provides guidelines dedicated to architectures for agents by explicitly making the developer define precisely the agent dynamics, behaviour and mechanisms. Other advantages of SPEARAF for this role are not pointed up by this paper, in particular at the infrastructure level with the separation of agent concerns from environment concerns (MAS and runtime).

6   Conclusion

By relying on the SPEARAF development process, we built a system enabling the comparison in a simulator of different multi-agent strategies for multi-robot task allocation in a dynamic environment. As the evaluation shows, we were able to answer non-functional requirements needed by such a system. SPEARAF provides helps and guidelines for the development of MASs, in particular at the agent level and its programming. The concept of "species" encourages to explicitly build an architecture realising the dynamics of the agents, its interactions with its environment and enabling the definition of its behaviour through high-level abstract constructs.

Using such concepts enables the species user to build subspecies and profit from the advantages of software architecture practices till the end of the development.

References

1. Amor, M., Fuentes, L.: Malaca: A Component and Aspect-Oriented Agent Architecture. Information & Software Technology 51(6), 1052–1065 (2009)
2. Garcia, A., Lucena, C.: Taming Heterogeneous Agent Architectures with Aspects. Communications of the ACM 51(5), 75–81 (2008)
3. Georgé, J.P., Gleizes, M.P., Garijo, F., Noel, V., Arcangeli, J.P.: Self-adaptive Coordination for Robot Teams Accomplishing Critical Activities. In: Demazeau, Y., Dignum, F., Corchado, J.M., Bajo, J. (eds.) Advances in PAAMS. Advances in Intelligent and Soft Computing, vol. 70, pp. 145–150. Springer, Heidelberg (2010)
4. Gleizes, M.P., Camps, V., Georgé, J.P., Capera, D.: Engineering systems which generate emergent functionalities. In: Weyns, D., Brueckner, S.A., Demazeau, Y. (eds.) EEMMAS 2007. LNCS (LNAI), vol. 5049, pp. 58–75. Springer, Heidelberg (2008)
5. Noël, V., Arcangeli, J.P., Gleizes, M.P.: Between Design and Implementation of MAS: A Component-Based Two-Step Process. In: EUMAS 2010 (2010)

Lightweight Trusted Routing for Wireless Sensor Networks

Laurent Vercouter and Jean-Paul Jamont

Abstract. Communication in *ad hoc* network, such as Wireless Sensor Networks (WSN), needs the use of decentralised routing algorithms requiring that several sensors behave in an expected way. This introduces a vulnerability as the global issue of decentralized tasks depends on local behaviors and is compromised in case of failures or malicious intrusions. We propose here an adaptation of a routing protocol for WSN, the MWAC model, that introduces trust decisions to detect and avoid sensors that exhibit an incorrect behavior. The proposed trusted routing algorithms takes into account the low energy, communication and memory capacity of sensors to provide a realistic improvement of the routing robustness.

1 Introduction

Communication in Wireless Sensor Networks (WSN) is usually supported by the creation of an *ad hoc* network connecting each sensor to the ones that are within its communication range. The decentralized nature of such networks implies the use of a multi-hop communication protocol in which several nodes are involved in the routing tasks. A drawback of relying on a collective activity for such a global task is that it increases the system vulnerability faced to a failure or a malicious intrusion. If a sensor does not behave as expected, it will influence the issue of the global task. Moreover, another specificity of WSN

Laurent Vercouter
École Nationale Supérieure des Mines de Saint-Étienne, ISCOD/LSTI group,
158 cours Fauriel, 42023 Saint-Étienne cedex 02, France
e-mail: `laurent.vercouter@emse.fr`

Jean-Paul Jamont
University of Grenoble, LCIS Labs, 51 rue Barthlmy de Laffemas,
26000 Valence, France
e-mail: `jean-paul.jamont@iut-valence.fr`

Y. Demazeau et al. (Eds.): Adv. on Prac. Appl. of Agents and Mult. Sys., AISC 88, pp. 87–96.
springerlink.com   © Springer-Verlag Berlin Heidelberg 2011

is that the sensors have low energy, communication and memory capacities. Lightweight mechanisms are required and that represents an obstacle to the use of classical trust management techniques to protect the system against incorrect local behaviors.

We propose in this paper a lightweight trust model to improve the robustness of routing in WSN. The main originality of this trust model is that it is designed to work in a network where authentication cannot be ensured. As a node cannot authenticate its neighboors, our approach is based on trust estimation of the nodes' overall neighborhood. If it is untrustworthy, a node switches to a backup mode so that its neighborhood becomes in quarantine. Our proposal has been integrated to the MWAC model [3] that allows a low cost routing in WSN.

The second section describes the MWAC model. Then, the trust model we propose is described as well as its integration in MWAC algorithms. Section 4 shows a practical implementation of our proposal in the MWAC simulator and evaluates experimentally the benefits of using trust for routing.

2 The MWAC Model

MWAC[1] has been proposed in previous works [3] to handle communication in WSN. The specificity of WSN is that each sensor has a limited communication range and low energy, memory and computing capacity. The MWAC model proposes a multi-hop routing mechanism that decreases the energy expense compared to flooding techniques. It relies on an organisational structure based on agent roles and groups. Each group is composed by:

- one and only one *group representative agent* (r) managing the communication in its own group;
- some *connection agents* (c) belonging to more than one group and that are connected to the representative agent of each of their groups;
- some *simple members* (s) that do not have any routing task to ensure (unless they are the final sender or receiver of a message).

This section gives a general view of the routing and self-organisazing mechanisms performed in MWAC. The detailled mechanisms are described in the MWAC reference paper [3]. The last subsection presents the problem of vulnerability against failures and malicious intrusions that appears in most of the decentralized routing mechanisms and that we tackle in the section by proposing a robust variant of the MWAC model.

Message routing in MWAC. A message in MWAC follows a path between a source (a) and a receiver (b) corresponding to the definition of the following equation : $((a, r),^* [(r, c), (c, r)], (r, b))$

---

[1] *Multi Wireless Agent Communication model.*

Representative agents use local routing tables to choose some connection agents to which the message should be sent. Then, these connection agents propagate the message to the representatives of the groups they belong to. Each of these representative agents will continue this propagation with other connection agents unless the final receiver is in its own group. The local routing tables are updated by the way of eavesdropping. If a routing table does not allow a representative to build a message path, it uses a flooding protocol.

The energy saving comes thanks to the fact that the propagation is only directed to the representative agent of the groups and to some connection agents, instead of all the neighbors of a node as in flooding algorithms.

MWAC self-organisation process. The efficiency of the routing protocols depends on the allocation of roles to agents and to the maintenance of a consistent organisation. A self-organisation process is used to allocate dynamically the agent roles and to build an efficient organisation. The general idea of the algorithm is that an agent checks the role of its neighbors. If there is no representative in its neighborhood, it creates a group and adopts the representative role. If there is more than one representative in its neighborhood, it becomes a connection agent. In the other cases, it is a simple member. In order to communicate with their neighbors, agents follow an introduction protocol in which they send periodically a message to describe themselves. This message contains the id, role and the groups of its sender.

Conflicts occur when two or more representative agents communicate. In this case, the representatives exchange another message containing a score calculated from the sender's amount of energy and its number of neighbors. The representative with the highest score keeps this role while those with lower scores drop it to become simple members or connection agents.

Vulnerability against intrusions and failures. MWAC is especially suited to deal with high scale WSN composed of nodes having very limited communication and computing capacities. However, as it is often the case when one considers decentralized algorithms, it is assumed that the working network nodes interact as specified in the local algorithms. If a sensor does not behave as expected (because of a failure or a malicious intrusion) when interacting with other sensors, it will impact the decisions taken by other nodes and it may corrupt the overall functionning of the system. In this paper, we will use the term "lie" to refer to incorrect interactions even if they are not intentional and due to failures.

Lies can disturb both the routing and the self-organising processes. The work described in this paper is only focused on the impact of lies on the issue of self-organisation. Increasing the robustness of MWAC against lies during the routing process will be considered in our future works. Here, we consider essentially lies that occur in the introduction message in which an agent declares its id, its role and its groups.

We propose in this paper a variant of the MWAC protocol that includes a trust model in the decision process. Trust is calculated from the detection of the occurence of lies and is used to adapt the role allocation process.

3   A Robust Variant of the MWAC Model

Weaknesses against failures and malicious intrusions are commonly encountered when dealing with networks that rely on decentralized algorithms. As these algorithms are executed by several agents, if one of them does not execute correctly its portion of code, it can make the global algorithm fail. Trust has been proposed in existing works to tackle this problem. However, some specificities of WSN, especially the limitation of energy and communication costs makes it difficult to apply existing trust models.

The first subsection shows why existing approaches of decentralized trust management cannot be used for WSN. The second subsection presents the local trust assessment processes that we propose for MWAC. An adaptation of the role allocation algorithm using trust is then described.

3.1   *Decentralized Trust Management for WSN*

Trust management aims at protecting a system against bad behaviors of some of its entities. The idea is to observe other agents' behaviors, to compute and assign them trust values and to avoid agents that are not trustworthy. Decentralized trust management mechanisms have also been proposed to increase the reliability of routing in large scale networks such as *ad hoc* [1] or peer-to-peer [5] networks. A global overview of existing trust systems for multi-agent systems can be found in this survey [4] and a review.

However, all these existing systems share the mutual assumption that there exists an authentication system. It is indeed essential that an id is assigned to each agent without any doubt, so that agents can attach trust estimation to identities. Service infrastructures, such as Public Key Infrastructure, are frequently used to provide authentication. Yet, in WSN, sensors have very low capacities for communication or data storage, especially in large scale networks where the cost of each sensor should be as low as possible. It is therefore not realistic to consider that each sensor stores a public key for every other sensor, nor that each one can communicate with a central repository storing all these keys.

The work proposed in this paper follows a new approach to decentralized trust management that can be used when authentication is not possible. We suggest to use trust to assess the reliability of the neighborhood as a whole rather than a separate assessment for each neighbor. An untrustworthy neighborhood would then mean that there should be at least one malicious or defective agent in it.

When a node believes that its neighborhood is not trustworthy, it changes its behavior to work in a backup mode, performing only the minimal required tasks. This backup mode should involve as less as possible the node's neighbors. The global aim of this approach is that all the neighbors of a malicious or failing sensor will progressivly detect a wrong behavior in their neighborhood and switch to the backup mode. The part of the network composed of the malicious agent and its neighbors will then be in quarantine and will no longer have the possibility to influence self-organisation with lies. The messages used to assess trust in the neighborhood can be all the messages exchanged in this neighborhood if eavesdropping can be used.

3.2 Trust Assessment

Trust is estimated locally by each agent by supervising the messages sent in its neighborhood. Even if authentication is not possible, nodes have to use an id (real or fake) when sending a message. We propose to use trust in an id (rather than trust in an agent authentified with an id) in order to represent the way an id has been used in the past.

Trust initialisation and updates. A new trust value is created each time a new id value is used in the sender's neighborhood. A trust value takes a value in the range $[0; 1]$. The initial value of an id id is the highest $(Trust(id) = 1)$.

In the proposed trust model, trust values are decreased when a lie is detected or suspected but it is not increased when correct messages are perceived. As an id is first fully trusted, there is no need to reward good behaviors in using this id, whereas a lie should imply a drastic decrease.

Depending on its nature, a lie can be either detected without any doubt or only suspected. In the first case, trust in the corresponding id is decreased at the lowest value (0). In the second case, the occurrence of a lie is only suspected as an unusual event occured but it may not be caused by a lie. For instance, a node may send a false information if it has temporarily a false belief. If a lie is suspected, trust should be decreased with an importance proportional to the likelihood that it is a lie.

Table 1 summarizes the lies that can occur during the introduction protocol.The first and the third cases refer respectively to the case where a node perceives a message using its own id and where a representative see that one of its neighbors hide the group it represents. Trust is here set to the minimal value.

The second case corresponds to a new neighboor which declares to be a workstation. Workstations are used in WSN to be the final recipient of messages and are generally not mobile. It is therefore unlikely that a new one appears in a neighborhood. If this happens the sensor should already have some routes to reach the workstation as it is a final recipient of messages. These old routes are used to check that it is really the claimed workstation.

Table 1 Lie in the introduction message

Lie on	Detection by node n	Trust decrease
id	if $usedid = id$ of n	$Trust(id) = 0$
	if $usedid = id$ of a new worsktation	$Trust(id) = Trust(id) - \eta$ and call $check\_old\_route\_procedure$
$group$	if n is a representative and his group is not included in the set of groups of the message	$Trust(id) = 0$
	if a new group is presented and if it is the only agent that connects it *see below the* new group checking *process*	*see below*
$role$	if the node claims he has a connection role, it is the same situation than the presentation of a new group (see previous case)	*see below*

The fourth and fifth cases correspond to a node claiming to be a connection node creating a link with a new group. The appearance of a new group is rare and it is unlikely that a node is the only connection to another group. However, this may happen and a lie can only be suspected. The node suspecting such a lie should then follow a *new group checking* process consisting trying to reach the representative of the new group with a query asking the ids of its group members. Due to a lack of space, the detailled mechanism to send the query is not detailed here. An important characteristic is that this specific query should be sent by flooding in paths trying to avoid the suspected id. Several replies may then be received. Depending on these replies, trust in the suspected id is updated as shown in table 2.

If no reply is received or if all the received replies state that the node is inside the representative's group, there is probably no lie and trust should not be decreased. If all the replies state that the suspected node is not in the representative's group, it may be a lie or a false belief of the suspected node. Trust should be decreased by a value α but not at the minimum value as there are still some doubts.

The last case occurs when some different replies from the representative arrive, some stating that the node is in the group and some stating that it

Table 2 Result of the new group checking process

Replies received	Trust decrease
no reply	no decrease
All replies stating that the node is in the group	no decrease
All replies stating that the node is not in the group	$Trust(id) = Trust(id) - \alpha$
Inconsistent replies received	$Trust(id) = Trust(id) - \beta$

is not. The most likely here is that fake replies are sent by a liar. But it can also happen without any wrong behavior, for instance if the neighborhood changes between the sending of two replies. As these situations will be quite rare, the sanction β on trust should be stronger. We propose then to set these sanctions in the interval $0 < \beta < \alpha < 1$. Trust values are set to 0 if they are decreased to negative values.

Trust recovery. It is important that trust in the neighborhood can be recovered, especially if the neighborhood of an agent may change (and maybe the malicious node that caused trust decreases has left). Trust recovery is done by forgiveness with a slow trust increase as time goes by. Algorithm 1 shows this recovery.

Algorithm 1. Trust recovery algorithm

---

    for all *id* in *neighborhood.getUsedIds()* **do**
      $Trust(id) = (1 - \lambda) * Trust(id) + \lambda$
    end for

Two parameters are used to configure the speed of trust recovery: (i) the frequency ν of invocation of the trust recovery function; (ii) the evaporation rate λ, with $0 \leq \lambda < 1$ setting the amount of trust recovered. The value of these parameters should be set according to the expected mobility of nodes. If the nodes are very mobile, the neighborhoods will frequently change and it may be interesting to have high evaporation and frequency. Otherwise, if the network is very static, these values should be quite low.

Trust decision algorithm. Trust in the ids is used to estimate the trust-worthiness of the whole neighborhood of a node. The neighborhood of a node is considered untrustworthy if there is at least one id that is distrusted (algorithm 2). The threshold θ_1 represents the limit of the trust value under which an id is distruted. It takes a value in the range: $0 < \theta_1 < 1$.

Algorithm 2. Neighborhood trust decision algorithm

---

    for all *id* in *neighborhood.getUsedIds()* **do**
      if $Trust(id) < \theta_1$ **then**
         return *distrusted*
      end if
    end for
    return *trusted*

3.3 Adaptation of the MWAC Model

The original MWAC model has been modified to integrate trust in a node's neighborhood. If a node distruts its neighborhood, it should place itself in quarantine and run in a backup mode in which he does not participate anymore to routing. In MWAC, the backup mode consists in adopting systematically a role of *Simple Member*.

The proposed adaptation of the MWAC role attribution process is shown in algorithm 3.

Algorithm 3. Adaptation of the role attribution algorithm

---

if *neighborhood.isEmpty()* then
   // *No possible organizational structure*
else if *neighborhood.trust() = distrusted* then
   myRole ← SIMPLEMEMBER
else if *myRole = REPRESENTATIVE* and *neighborhood.nbOfRepresentative()* > 0 then
   conflictResolutionProcedure()
else if *neighborhood.nbOfRepresentative() = 0* then
   myRole ← REPRESENTATIVE
else if *neighborhood.nbOfRepresentative() = 1* then
   myRole ← SIMPLEMEMBER
else {*neighborhood.nbOfRepresentative() > 1*}
   myRole ← CONNECTION
end if

---

4 Experimental Evaluation

The robustness improvement provided by trust integration in MWAC has been experimentally evaluated in the MASH simulator [2]. In this section, we describe this experimental protocol and then give an insight to the results of this evaluation.

Experimental protocol. Instrumentation applications of WSN involve two type of entities are evolved: sensors and a collect workstation. Sensors are spatially distributed measurement nodes which monitor environments. The collect workstation is the final destination node that should acquire the data.

A popular attack of this type of WSN consists in usurping the collect workstation id. This is a way to spy the data collected or to prevent the real collect workstation to obtain them. We introduced in the MWAC simulation some sensors with this malicious behavior.

A scenario in which sensors send a measure every 5 seconds has been implemented. The network contains 600 sensor agents and 1 to 10 malicious agents. The network topology is partially visible on the background of the figure 1. Malicious agents are uniformly distributed on this topology.

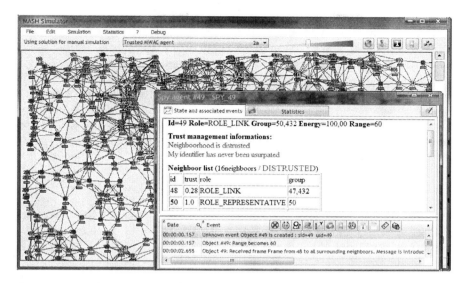

Fig. 1 The MASH simulator (simulated WSN in background, internal state of an agent in foreground i.e. role, available energy, neighboor list...)

Results. Figure 2 gives an insight of the performances of the trusted version of MWAC in comparison its original version. The charts represent the number of missing measure, *i.e.* sensor data sent that are not received by the workstation. $m1$, $m5$ and $m10$ are the results obtained with the original version of MWAC deploying respectively 1, 5 and 10 malicious agents. $t1$, $t2$ and $t10$ are the results obtained with the trusted version deploying respectively 1, 5 and 10 malicious agents.

Before $t = 0cs$ the WSN has stabilized its self-organization and sensor agents have exchanged measures. Some representative agents have learn some path to route messages to the workstation.

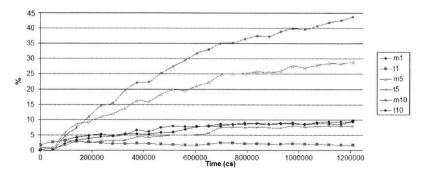

Fig. 2 Percentage of missing measures

At $t = 0cs$, some sensors agent are replaced by malicious agents. The experiments show the high impact that a few malicious behaviors have in the original version of MWAC. The percentage of missing measures is close to respectively 45% and 30% with only 10 and 5 malicious agents.

With the trusted version of MWAC, the percentage of missing measures is significantly lower. It reaches 9%, 7% and 2% with respectively 10, 5 and 1 malicious agents. It may be surprising that there are still missing measures whereas all the malicious agents are in quarantine. In fact, the undelivered messages are not intercepted by malicious nodes but corresponds to the measures directly emitted from quarantine zones.

5 Conclusion

We describe here an extension of the MWAC model to increase its robustness against failures and malicious intrusion. MWAC is used to enable message routing in WSN and this kind of application raises an original problem for trust modelling which is to take into account the absence of authentication. We propose here to use a trust model in an agent's neighborhood, rather than in its neighbors, to move in quarantine untrusted parts of the networks.

This is an ongoing work that has for the moment be applied to the self-organisation in MWAC. Our proposal has been experimentally evaluated against attacks consisting in stealing the identity of a workstation. Experiments show an important decrease of the messages lost when trust is used. Our current work is to extend the usage of trust to the other processes used in self-organisation and to message routing.

References

1. Griffiths, N., Jhumka, A., Dawson, A., Myers, R.: A simple trust model for on-demand routing in mobile ad-hoc networks. In: Proceedings of the 2nd International Symposium on Intelligent Distributed Computing - IDC 2008, Catania, Italy. SCI, vol. 162, pp. 105–114. Springer, Heidelberg (2008)
2. Jamont, J.P., Occello, M.: A multiagent tool to simulate hybrid real/virtual embedded agent societies. In: Proceedings of the 2009 IEEE/WIC/ACM International Conference on Intelligent Agent Technology, pp. 501–504. IEEE Computer Society, Los Alamitos (2009)
3. Jamont, J.P., Occello, M., Lagrèze, A.: A multiagent approach to manage communication in wireless instrumentation systems. Measurement 43(4), 489–503 (2010)
4. Sabater, J., Sierra, C.: Review on computational trust and reputation models. Artif. Intell. Rev. 24(1), 33–60 (2005)
5. Vercouter, L., Muller, G.: L.I.A.R.: achieving social control in open and decentralized multiagent systems. Appl. Artif. Intell. 24, 723–768 (2010)

Privacy-Preserving Strategy for Negotiating Stable, Equitable and Optimal Matchings

Maxime Morge and Gauthier Picard

Abstract. The assignment problem has a wide variety of applications and in particular, it can be applied to any two-sided market. In this paper, we propose a multi-agent framework to distributively solve this kind of assignment problems, by providing agents which negotiate with respect to their preferences. We present here a realisation of the minimal concession strategy. Our realisation of the minimal concession strategy has useful properties: it preserves the privacy and improves the optimality of the solution and the equity amongst the partners.

1 Introduction

Negotiation over the assignments of agents is a new challenging area [10]. This problem has the potential for attracting interests, as resource allocation [5], from microeconomics and social choice theory on the one hand and computer science and AI on the other. The assignment problem has a wide variety of practical applications and in particular, it can be applied to any two-sided market: students/projects, carpool, home swapping, service provider/requesters, etc.

A particular instantiation of the assignment problem consists of the *stable marriage problem* (SMP) which it is commonly stated as mapping between two communities (e.g. men and women). In this paper, we propose a multi-agent framework to distributively solve this kind of problems, by providing agents which negotiate with respect to their preferences. Here assignments are viewed as emergent phenomena resulting from local agent negotiations. The objective of such procedure is to find an assignment that is *optimal*. For this purpose, we can consider different notions

Gauthier Picard
ISCOD team, Ecole des Mines de Saint-Etienne
e-mail: `picard@emse.fr`

Maxime Morge
SMAC team, Laboratoire d'Informatique Fondamentale de Lille
e-mail: `Maxime.Morge@lifl.fr`

Y. Demazeau et al. (Eds.): Adv. on Prac. Appl. of Agents and Mult. Sys., AISC 88, pp. 97–102.
springerlink.com   © Springer-Verlag Berlin Heidelberg 2011

of *social welfare*. Within this paper, we propose *Casanova*, a distributed method to solve SMP. We seek to provide agent behaviors leading negotiation processes to socially optimal assignments. We propose a realisation of the minimal concession strategy applied to SMP. Our strategy has useful properties: it preserves the privacy, it improves the optimality of the solution and the equity amongst partners.

2 Stable Marriage Problem

SMP were first studied by [6] in order to find optimal assignments. In a SMP there are two finite sets of participants: the set of men and the set of women.

Definition 1 (SM). A **stable marriage problem** of size n (with $n \geq 1$) is a couple $SM = \langle X, Y \rangle$ where:

- $X = \{x_1, \ldots, x_n\}$ is a set n men ranking women in a strict and complete order forming his preference list. $\forall 1 \leq i \leq n$, $x_i = (y_i^0, \ldots, y_i^{n-1})$
- $Y = \{y_1, \ldots, y_n\}$ is a set n women ranking men in a strict and complete order forming her preference list. $\forall 1 \leq i \leq n$, $y_i = (x_i^0, \ldots, x_i^{n-1})$

A person z_1 prefers a partner t_2 to another partner t_3 if and only if t_2 precedes t_3 on z_1's preference list (denoted $t_2 \succ_{z_1} t_3$).

A matching is just a complete one-to-one mapping between the two sexes such that a man x is mapped to a woman y if and only if y is mapped to x.

Definition 2 (Matching). Let $SM = \langle X, Y \rangle$ be a stable marriage problem of size n (with $n \geq 1$). A **matching** for SM is a n-uplet $M = \langle m_1, \ldots, m_n \rangle$ of n marriages where each m_i (with $1 \leq i \leq n$) is a couple $(x_i, y_i) \in X \times Y$ such that the matching is complete, i.e. each individual is married. Formally, $\forall x \in X \ \exists! y \in Y \ (x, y) \in M$. The partner of the agent z in accordance with the matching M is denoted $p_M(z)$.

We want to marry men and women together such that there are no two people of opposite sex who would both rather have each other than their current partners, i.e. finding a stable matching.

Definition 3 (Stable matching). Let $SM = \langle X, Y \rangle$ be a stable marriage problem of size n (with $1 \geq n$). and M a matching for SM. M is **stable** iff: $\forall (x_i, y_i) \in M \ \nexists (x_j, y_j) \in M \ x_j \succ_{y_i} x_i$ and $y_j \succ_{x_i} y_i$.

A typical objective in SM is to find an assignment that is optimal with respect to a metric that depends on the preferences of the agents. For this purpose, we assume that individual agents evaluate their satisfaction using utility functions mapping assignments to numerical values.

Definition 4 (Utility function). Let $SM = \langle X, Y \rangle$ be a stable marriage problem of size n (with $n \geq 1$), $z = (t_i^0, \ldots, t_i^k, \ldots, t_i^{n-1})$ an individual agent and T be the potential partners of z. The **utility function** of the agent z is a function $u_z : T \to \mathbb{R}$. If the matching assigns z with t_i^k, then $u_z(t_i^k) = \frac{(n-1)-k}{n-1}$.

The social welfare theory is used to evaluate the matching, considering the welfare of each person [1]. In this study, we derive from this theory four notions adapted for the stable marriage problem.

Definition 5 (Social welfare). Let $SM = \langle X, Y \rangle$ be a stable marriage problem of size n (with $n \geq 1$) and M a matching for SM.

- The **utilitarian welfare** considers the welfare of the whole society: $sw_u(X \cup Y) = \Sigma_{z \in X \cup Y} u_z(p_M(z))$.
- The **male welfare** considers the welfare of the men: $sw_u(X) = \Sigma_{x \in X} u_x(p_M(x))$.
- The **female welfare** considers the welfare of the women: $sw_u(Y) = \Sigma_{y \in Y} u_y(p_M(y))$.
- The **equity welfare** considers the fairness among partners' welfare in every marriage: $sw_e(X \cup Y) = 1 - \frac{|sw_u(X) - sw_u(Y)|}{n}$.

Utilitarian welfare can be used to measure the quality of a matching from the viewpoint of the system as a whole. The equity welfare may be a suitable indicator when we have to satisfy both the men and the women.

Gale and Shapley described in [6] a centralized algorithm (GS) that always finds a stable matching for any instance of the SMP. They also noted that this algorithm produces a matching in which each man has the best partner he can have in any stable matching. GS involves a sequence of proposals from men to women. It starts by setting all persons free. GS iterates until all the men are engaged. Each man x always proposes marriage to his most-preferred woman, y. When y is already married (e.g. with x_2) she discards the previous proposal with x_2 and x_2 is set free. Afterwards, x and y are engaged to each other. Woman y deletes from her preference list each man x_3 that is less preferred than x. Conversely, man x_3 deletes y from his preference list. Finally, if there is still a free man a new proposal is started. Otherwise, the algorithm terminates. This algorithm is commonly known as the men-propose algorithm because it can be expressed as a sequence of "proposals" from the men to the women. [6] established the existence of a stable marriage thanks to GS that constructs a men-optimal (resp. women-optimal) stable matching, i.e. it optimizes the *male welfare* (resp. *women welfare*).

Example 1. Let us consider the SM $\langle X, Y \rangle$ of size 3:

$$x_1 = (y_2, y_1, y_3) \qquad\qquad y_1 = (x_2, x_1, x_3)$$
$$x_2 = (y_3, y_2, y_1) \qquad\qquad y_2 = (x_3, x_2, x_1)$$
$$x_3 = (y_1, y_3, y_2) \qquad\qquad y_3 = (x_1, x_3, x_2)$$

The output of the men-propose GS algorithm is $M_1 = \langle (x_3, y_1), (x_1, y_2), (x_2, y_3) \rangle$. In accordance with M_1, $sw_u(X \cup Y) = 3$, $sw_u(X) = 3$, $sw_u(Y) = 0$ and $sw_e(X \cup Y) = 1$. We can notice that a stable matching exists even if it is not found by the GS algorithms: $M_3 = \langle (x_1, y_1), (x_2, y_2), (x_3, y_3) \rangle$. In accordance M_3, $sw_u(X \cup Y) = 3$, $sw_u(X) = 1.5$, $sw_u(Y) = 1.5$ and $sw_e(X \cup Y) = 0$.

A distributed version of the GS algorithm (DisEGS) has been proposed by [3]. Each man (and woman) is represented by an agent which exchange messages (*propose*, *accept* and *delete*) as to reproduce the GS algorithm and find the same stable assignment. Contrary to classical GS, each agent keeps its own preferences, which represents a interesting step towards privacy.

3   Casanova Algorithm

In this study, we consider matchings as emergent phenomena resulting from local agent negotiations. The Casanova algorithm is a negotiation strategy to reach a matching in a SMP. Contrary to DisEGS, we do not distinguish men and women. Both men and women send concurrently proposals and reply with acceptance or rejections, which represents the main difficulty of this study.

According to Casanova, agents start the negotiation with the best potential partners. During the negotiation, an agent concedes minimally as soon as its optimal partners has refused. A concession is minimal for an agent since there is no other preferred partner which has not yet refused.

The strategy starts by setting the agent free and the concession level equals to 0. At each run, the agent starts by sending proposals to the sub-list of agents corresponding its concession level. During the first step, the agent sends a *proposal* to the optimal partner. During the second run, the agent addresses proposals to the two preferred agents, and so on. When the agent receives a *proposal*, it only accepts the ones corresponding to its concession level, called acceptable proposals. In this case, the agent gets divorced with its current partner if it is required and it gets engaged with its new partner. It is worth noticing that the agents are allowed to divorce for a preferred partner if and only if the agent is not engaged but married (in order to avoid deadlock). When the agent receives an *acceptance*, the agent *confirms* or withdraws depending whetever or not its current partner is the sender of the acceptance. As previously, the agent is allowed to divorce if he has some regrets, i.e. the potential partner is preferred to the current partner. When the agent receives a *withdrawal*, the agent get divorced. We can notice that the agents count the response to its proposals. If all of them are received and the agent is still free, it must concede, i.e. go further in its preference list to add acceptable partners. When the agent receives a *divorce* notification, the agent takes it into account.

Casanova outputs a stable matching. Suppose it is not the case, i.e. there is an agent A that prefers an agent B (that it's not matched to) and at the same time B also prefers A over the one B is matched with. According to the concession level of A (resp. B), A would propose to (resp. B would accept) B (resp. A) for a partnership.

Example 2. Let us consider the Casanova strategy implemented by the multi-agent system set up as in Ex. 1. As a reminder, y_3's preferences are (x_1, x_3, x_2). Initially, y_3 is free and her concession level is equal to 0. So, the only acceptable partner is x_1. In our example, the local negotiations lead to the stable matching $M_3 = \langle (x_1, y_1), (x_2, y_2), (x_3, y_3) \rangle$ such that $sw_e(X \cup Y) = 0$.

4   Evaluation

Casanova has been implemented with Jason [2]. In order to evaluate Casanova, we run it for some random SMP instances [7] where n, the number of potential partners, is between 2 and 100 (see. Fig. 1). For each instance of SMP, we run 10 times Casanova. Firstly, we compare the value of the equity welfare, the male welfare and

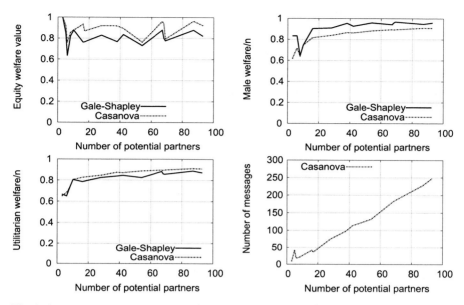

Fig. 1 Comparison of Casanova to GS (or DisEGS) results.

the utilitarian welfare with the one obtained with the help of the GS (or DisEGS) algorithms. Secondly, we counts the number of messages received by each agent. First, we observe that the output of Casanova is a stable marriage which is more equitable and more optimal (from the viewpoint of the system as a whole) that the one returned by the GS or DisEGS algorithms but it less optimal from the from the viewpoint of the men. Additionally, our preliminary results show that the number of messages received by each agent is linear with respect to the size of the problem.

5 Related Works

The principle of Casanova is based on the minimal concession strategy [9, 8, 4]. Each agent starts from the partner that is best for it and, if this latter refuses, the agents concedes by considering less preferred potential partners. Differently from the game-theoretical approach [9], our approach does not assume that the agent knows the preferences of the latter [8]. We say that a proposal is a minimal concession since there is no other proposals which are preferred. Contrary to [8, 4], the deployment of the minimal concession in this paper is not limited to a bilateral negotiation. Finally, we apply the Occam's razor since we do not employ argumentation-based reasoning but a simpler reasoning.

6 Conclusion

In this paper we have presented a realisation of the minimal concession strategy applied to the SMP. According to this strategy, agents start the negotiation with their preferred partners. During the negotiation, an agent concede minimally as soon as its optimal partners has refused. A concession is minimal for an agent since there is no preferred partner which has not yet refused. Our realisation of the minimal concession strategy has useful properties. Firstly, it preserves the privacy since the agents do not reveal explicitly their preferences. Secondly, the approach improves the optimality of the matching and its equity.

We need to realize more experiments for evaluating other metrics of social welfare and for comparing with other MAS approaches.

Acknowledgements. We would like to thank Bruno Beaufils, Jomi Fred Hubner, Antoine Nongaillard, and Santiago Villareal for their help on this work.

References

1. Arrow, K., Sen, A., Suzumura, K.: Handbook of Social Choice and Welfare. Elsevier, Amsterdam (2002)
2. Bordini, R.H., Hubner, J.F., Wooldridge, M.: Programming Multi-Agent Systems in AgentSpeak Using Jason. John Wiley & Sons Ltd., Chichester (2007)
3. Brito, I., Meseguer, P.: Distributed stable marriage problem. In: 6th Workshop on Distributed Constraint Reasoning at IJCAI 2005, pp. 135–147 (2005)
4. Bromuri, S., Urovi, V., Morge, M., Toni, F., Stathis, K.: A multi-agent system for service discovery, selection and negotiation (demonstration). In: 8th International Joint Conference on Autonomous Agents and Multiagent Systems (AAMAS 2009), pp. 1395–1396. IFAAMAS (2009)
5. Chevaleyre, Y., Dunne, P.E., Endriss, U., Lang, J., Lematre, M., Maudet, N., Padget, J., Phelps, S., Rodrguez-Aguilar, J.A., Sousa, P.: Issues in multiagent resource allocation. Informatica 30, 3–31 (2006)
6. Gale, D., Shapley, L.S.: College admissions and the stability of marriage. American Mathematical Monthly 69, 9–14 (1962)
7. Gent, I.P., Prosser, P.: An empirical study of the stable marriage problem with ties and incomplete lists. In: ECAI 2002, pp. 141–145 (2002)
8. Morge, M., Mancarella, P.: Assumption-based argumentation for the minimal concession strategy. In: McBurney, P., Rahwan, I., Parsons, S., Maudet, N. (eds.) ArgMAS 2009. LNCS, vol. 6057, pp. 114–133. Springer, Heidelberg (2010)
9. Rosenschein, J.S., Zlotkin, G.: Rules of encounter: designing conventions for automated negotiation among Computers. The MIT press series of artificial intelligence. MIT Press, Cambridge (1994)
10. Shoham, Y., Leyton-Brown, K.: Multiagent Systems: Algorithmic, Game-Theoretic, and Logical Foundations. Cambridge University Press, Cambridge (2009)

Searching Flocks in Peer-to-Peer Networks

Hugo Pommier, Benoît Romito, and François Bourdon

Abstract. Storage in peer-to-peer networks must be reliable and dependable. This reliability is partially dependent of data placement. In some approaches, information pieces are moving in the network to find a placement that optimizes different criteria. But this mobility may be a drawback for the localization of those pieces. In this paper, we propose to measure the impact that mobility of network objects has on their localization.

1 Introduction

Efficient and robust data storage into peer-to-peer networks requires the set-up of several mechanisms to ensure that data should always be retrieved. Indeed, peer-to-peer networks are dynamic systems where peer failures and disconnections are common events which can make the service unavailable or provoke unrecoverable data losses. Fault tolerance can be obtained by using erasure coding [6] where a data is split into m blocs. After an encoding step, $m + n$ fragments are generated. With this kind of code, the initial data can be recovered by combining any subset of m fragments among the $m + n$. Thus, the system can tolerate n faults. To ensure system durability, lost fragments are regenerated when a threshold is reached. This maintenance process requires a monitoring that can be centralized [3] or decentralized [10]. The decentralized approach is best suited for open environments but requires that fragments stay close in the network [4]. In [7], the authors introduce the concept of fragments mobility to provide a data placement that can react and adapt itself to changes in the logical network. In this approach, each fragment of a document is embedded into a mobile agent that evolves in

Hugo Pommier · Benoît Romito · François Bourdon
Université de Caen Basse-Normandie / GREYC - CNRS UMR 6072,
Bd. du Maréchal Juin, 14032 Caen CEDEX France
e-mail: {hpommier,bromito,fbourdon}@info.unicaen.fr

Y. Demazeau et al. (Eds.): Adv. on Prac. Appl. of Agents and Mult. Sys., AISC 88, pp. 103–108.

a fully-decentralized unstructured peer-to-peer network. The monitoring of fragments is decentralized and as a consequence, fragments of a same document are moving together in group to keep a high level of locality between them. This group motion is obtained by locally applying simple rules and the emergent behavior is comparable to bird flocks. Finally, each document is turned into a flock of fragments that can move into the network trying to satisfy a certain objective function. In this paper, we are only interested in this approach and particularly on the flocks localization process which is also vital for information availability. A lot of information retrieval algorithms have been proposed for unstructured peer-to-peer networks. Flooding [12] and expanding ring [5, 11] algorithms are very efficient in terms of search success rate but have an important operating cost due to a high number of generated messages. To face this messages explosion problem, another family of search algorithms relies on random walks for their lightness and relative efficiency. In [5] authors modulate the number of random walkers and show a drastic reduction of messages in specific cases. On the contrary, [1] show better results when the number of walkers is variable and related to a document popularity value. In [9] the search algorithm is inspired by ants stigmergy properties and traces of hosted documents are disseminated into the network and decreasing with distance. Consequently, a random walk is oriented toward the peer with the higher level of traces for a given file. To our knowledge the efficiency of random walks has not been evaluated when data is moving. Mobile objects localization raises some questions. It is legitimate to ask if mobility of objects lowers the localization performances or if mobility speed has an impact on the search results. That is why we propose in this paper to measure the performances of two flock searching algorithms based on random walks. In section 2 we briefly describe the flocking architecture introduced in [7]. We then propose in section 3 two flock searching algorithms that we evaluate in section 4.

2   The Bio-inspired Information Model

The information model of [7] depicts an architecture built on two layers. The first one, called network management layer, is in charge of the overlay construction and management. The second one provides behaviors for the mobile agents. In this approach, each fragment of a file is put into a mobile agent making its own decisions.

Peer-to-Peer Network. The network layer is based on the Scalable Membership Protocol (SCAMP) [2]. SCAMP constructs a random directed graph having a mean degree converging to $(c+1)\log(n)$ with c a design parameter. Consequently, the resulting graph remains connected if the link failure probability is smaller than $\frac{c}{c+1}$. On top of that, the degree grows slowly with system size so the network's neighborhood size scales well.

Mobile Agents. This layer hosts a multi-agents system in which each fragment of information (obtained by an erasure code) is kept in an associated cognitive mobile agent making its own decisions. This cognition allows the agents to autonomously move from peers to peers following flocking rules similar to those proposed by Craig Reynolds [8]. Reynolds identified three rules for each agent to follow: cohesion, separation and alignment. In this model, flocking is not a quality of any individual bird but an emergent behavior from interactions between all agents. By combining Reynolds rules with a maximal distance between two elements in a network estimated by the Round Trip Time (RTT), it is possible to reproduce flocking behavior between fragments. The details of the algorithm can be seen in [7].

3 Searching Algorithms

Random Walk Based Agents. We build a distributed Markov chain on the overlay. SCAMP is connected and thus a searching agent can reach all nodes in the network. Let $dep_x = V_x \cup \{x\}$ be the set of possible moves for an agent on the peer x with V_x the set of x's neighbors. We define P_{xy} the transition probability, from a peer x to a peer $y \in dep_x$:

$$\sum_{y \in dep_x} P_{xy} = 1 \qquad \text{with} \qquad P_{xy} = \frac{1}{\text{Card}(dep_x)} = \frac{1}{(c+1)\log(n)+1}$$

$(c+1)\log(n)+1$ comes from the mean degree of the SCAMP network. The distribution is uniform and each state can be reached with the same probability. We assume that peers degree in the network is constant in time. As a consequence, the Markov chain's transitions are symmetric and the distribution is stationary.

Pheromone Based Searching Agents. A random walk is a memoryless process, then it is possible for an agent to visit more than one time the same peer. This phenomenon slows down the network coverage speed of the search agent. That's why, we propose a second algorithm using the agents ability to drop pheromones on peers to indicate its moves. We want to keep the random aspect of the search but we want to mark areas that have already been visited. When an agent arrives on a peer, it drops a certain amount of pheromones. To direct its search, an agent selects a node with the smaller amount of pheromones. A flock is a mobile system. Consequently, it is possible for a search agent to visit the whole network and fail to locate the targeted flock. That's why, the vertices marking should be temporary. To this end, we perform a pheromones evaporation rate at each timestep that depends of the network size.

$$evapo_x = \frac{\rho_x}{n}$$

with x a peer, ρ_x the amount of pheromones on x, and n the number of peers in the network. It ensures that no traces of pheromones are left on a peer after a time equivalent to n moves in the network.

4   Experimental Results

We have implemented the flocking and the search algorithms on a discrete-events and multi-agents simulator. For each measure, we launched 100 consecutive agents in a 400 peers network.

Number Of Hops. In Fig. 1 each curve describes the mean number and the standard deviation of hops required by an agent to locate flocks of different size. The search agents always move at each simulation cycle and a flock with a $\frac{1}{x}$ speed moves every x simulation cycles. The Fig. 1(a) and Fig. 1(b) shows the results when the flock speed is respectively $\frac{1}{1}$ and $\frac{1}{10}$. We can see that the pheromone-based search performs significantly better when the flock size is between 15 and 35 fragments. We can also see that both algorithms perform better as the flock size increases. Finding the same behavior on all variations of speed suggests that the speed parameter plays no role on the file search process and whatever the method used. These remarks are confirmed by the curves of Fig. 2. Indeed, the Fig. 2 shows, for each algorithm, that the number of hops required to locate a flock is not influenced by flock speed.

Network Coverage and Search Success Rate (SSR). As seen in Fig. 3(a) using pheromones increases the speed of the exploration. To measure the SSR, we have limited the number of hops of a search agent and we measured the percentage of success searches. We made those experiment on 15 and 30 fragments flocks. We can see on Fig. 3(b) that the pheromone-based algorithm performs better for small flocks. We can also observe on Fig. 3(c) that the flock of size 30 is found faster.

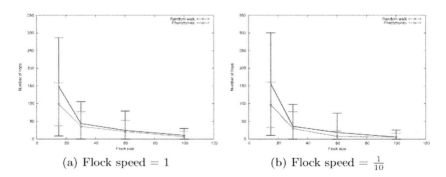

(a) Flock speed $= 1$ (b) Flock speed $= \frac{1}{10}$

Fig. 1 Number of required hops to locate flocks of different size.

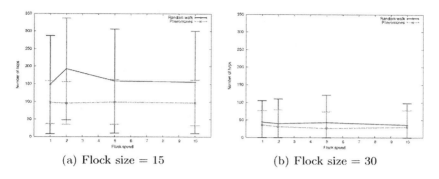

(a) Flock size = 15   (b) Flock size = 30

Fig. 2 Number of required hops to locate flocks of different speed.

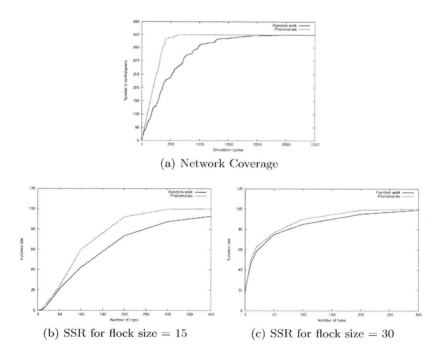

(a) Network Coverage

(b) SSR for flock size = 15   (c) SSR for flock size = 30

Fig. 3 Performances of search algorithms

5   Conclusion and Future Work

We proposed in this paper two search algorithms to locate objects that are moving into a peer-to-peer network following flocking rules. The first algorithm is a random walk and the second drops pheromones into the network to increase its searching success. We initially wanted to observe their behaviors in response to the mobility of a flock. The experiments revealed a weak influence of motion's speed on the results of a search. The second aspect was to measure their performances. We saw that the probability to find a flock when the number of hops is limited is better for the pheromones algorithm. We also saw that the mean number of hops to find a flock is changing with its size. When the flock is small, it is better to use a pheromones-based agent. In future work, we will study the minimal threshold required to use pheromones-based agents. Our search algorithms contact only one agent in a flock. But to reconstruct a data in a (m, n)-erasure coding scheme it is necessary to contact m agents. Future work will study the required mechanisms to retrieve those m fragments once a fragment has been found. One solution would be to disseminate the request to a selected subset of agents in this flock. Future work also includes the deployment of our search algorithms in a physical network and considering other peer-to-peer topologies.

References

1. Bisnik, N., Abouzeid, A.: Modeling and analysis of random walk search algorithms in p2p networks. In: HOT-P2P 2005, pp. 95–103 (2005)
2. Ganesh, A.J., Kermarrec, A., Massoulié, L.: SCAMP: Peer-to-Peer Lightweight Membership Service for Large-Scale Group Communication. In: Crowcroft, J., Hofmann, M. (eds.) NGC 2001. LNCS, vol. 2233, pp. 44–55. Springer, Heidelberg (2001)
3. Ghemawat, S., Gobioff, H., Leung, S.: The Google file system. In: SOSP 2003, pp. 29–43 (2003)
4. Giroire, F., Monteiro, J., Pérennes, S.: P2P storage systems: How much locality can they tolerate? In: LCN 2009, pp. 320–323 (2009)
5. Lv, Q., Cao, P., Cohen, E., Li, K., Shenker, S.: Search and replication in unstructured peer-to-peer networks. In: ICS 2002, pp. 84–95 (2002)
6. Plank, J.: A Tutorial on Reed-Solomon Coding for Fault-Tolerance in RAID-Like Systems. Softw., Pract. Exper. 27(9), 995–1012 (1997)
7. Pommier, H., Romito, B., Bourdon, F.: Bio-Inspired data placement in peer-to-peer networks. Benefits of using multi-agents systems. In: WEBIST 2010, pp. 319–324 (2010)
8. Reynolds, C.W.: Flocks, herds and schools: A distributed behavioral model. In: SIGGRAPH 1987, pp. 25–34 (1987)
9. Ronasi, K., Firooz, M.H., Pakravan, M.R., Avanaki, A.N.: An enhanced random-walk method for content locating in P2P networks. In: ICDCSW 2007, pp. 21–24 (2007)
10. Rowstron, A., Druschel, P.: Storage Management and Caching in PAST, A Large-scale, Persistent Peer-to-peer Storage Utility. In: SOSP 2001, pp. 188–201 (2001)
11. Yang, B., Garcia-Molina, H.: Improving Search in Peer-to-Peer Networks. In: ICDCS 2002, pp. 5–14 (2002)
12. Zeinalipour-Yazti, D., Kalogeraki, V., Gunopulos, D.: Information retrieval techniques for peer-to-peer networks. In: CSE 2004, vol. 6, pp. 20–26 (2004)

Securing Mobile Agents via Combining Encrypted Functions

Hamed Aouadi

Abstract. The use of mobile agents has resulted in a number of new security problems. In literature, many solutions are presented. They split into weak and low cost software solutions and strong and high cost hardware solutions. In our previous researches we proposed a hybrid solution but the presence of the hardware is usually not appreciated. In this paper, we survey the state of the art in mobile agent security, our previous approach and our new solution. In the new proposition, we combine encrypted functions in order to produce a new mechanism of mobile agent security.

1 Introduction

Computer networks have evolved with time. Originally, computer network applications were based upon a client/server model. Network applications have progressed to support distributed computing through the use of remote procedure calls and distributed objects. A further extension to the distributed model comes from the use of mobile agents. A mobile agent is an autonomous software entity that can execute on a host in a network and then suspend itself and transfer its execution to another host. In this paper we provide a survey of the most relevant solutions proposed so far and then we present our contributions. The first solution that we suggested consisted to the use of securing software secured it-self by hardware. Our new solution consist to combine encrypted functions to provide a fully software solution.

2 Security Mechanisms

A number of approaches have been developed to protect mobile code. In the literature there are several classifications of these approaches. In the next sections we will present some of them classified into hardware solutions and software solutions:

2.1 Hardware Approaches

In this case agent execution is only possible if a special peripheral exists. The agent tests the existence of the peripheral on starts or while executing the agent

Hamed Aouadi
Computer Science Department, Faculty of Sciences of Tunis, Tunisia
e-mail: hamed_aouadi@yahoo.fr

Y. Demazeau et al. (Eds.): Adv. on Prac. Appl. of Agents and Mult. Sys., AISC 88, pp. 109–112.
springerlink.com   © Springer-Verlag Berlin Heidelberg 2011

[1]. In this scheme the agent can be tampered if we delete (jump) the test of peripheral from the agent code.

- *Trusted Processing environment*: This approach is proposed by Wilhelm [2] and consists to the entire execution of the agent inside a special peripheral called TPE (Trusted Processing Environment). The agent communicates with the visited site through a secure logical interface and migrates from one TPE to another in asymmetric encrypted form.

- *Smart cards*: The use of smart cards to protect software is proposed by Mana [3]. The idea consists in the segmentation of the agent; some segments will be encrypted with the public key of the card. While executing the agent the site transmits the encrypted segment to the card where it will be decrypted and executed inside it.

2.2 Software Approaches

The purpose of this class of approaches is to provide security of agents only with software mechanisms, so we reduce costs and we provide maintainable products.

- *Obfuscation:* Hohl [4] gives a detailed overview of the threats stemming from an agent encountering a malicious host as motivation for Blackbox security. The strategy behind this technique is simple; scramble the code in such a way that no one is able to gain a complete understanding of its function (specification and data), or to modify the resulting code without detection. The approach also establishes a time interval during which the agent and its sensitive data are valid. After this time elapses, any attempt to attack the agent becomes worthless.

- *Computing with encrypted functions*: The goal of Computing with Encrypting Functions [5] is to determine a method whereby mobile code can safely compute cryptographic primitives, such as a digital signature, even though the code is executed in not trusted computing environments and operates autonomously without interactions with the home platform. The approach is to have the agent platform execute a program embodying an enciphered function without being able to discern the original function; the approach requires differentiation between a function and a program that implements the function. Supposing that a mobile agent has to execute a certain function f, then f is encrypted to obtain $E(f)$, and a program is created that implements $E(f)$. Platforms execute $E(f)$ on a clear text input value x, without knowing what function they actually computed. The execution yields $E(f(x))$, and this value can only be decrypted by the agent owner to obtain the desired result $f(x)$.

3 Our Approaches

In this section, we present our previous researches result which provides a hybrid solution and the new result.

3.1 Hybrid Solution

The purpose of this approach is to provide a maintainable software (the main part) solution that is as robust as hardware solutions [6].

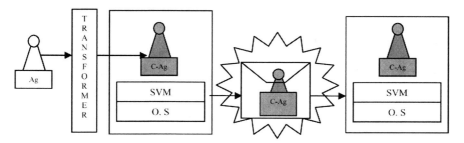

Fig. 1 Hybrid Solution

In this way, we try to replace the Trusted Processing Environment (TPE) of Wilhelm [2] by a software machine that we call SVM (Secure Virtual Machine) [7] that will be it-self protected by a smart card. An agent will be written in an ordinary language then transformed by software called agent transformer on equivalent agent but not understandable and only executable by SVM.

3.2 A New Software Solution

The goal of the new approach is to provide a fully software solution. The main idea consists to use a combination of encrypted functions in order to make them useable in multiple entries. In fact, encrypted functions provide a strong software solution but they are limited to a single entry. In our solution, suppose that we have n entries x_1, x_2, x_3... for every entry x_i we apply an encrypted function f_i; so we obtain n results y_1, y_2, y_3...the obtained results will be combined by a combiner to obtain one result Y. To this result Y we apply another encrypted function F.

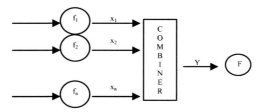

Fig. 2 Combining encrypted functions

The combination of partial results y_i can be obtained by concatenating them with regular number length. For example, if y_i are binary number the will be one bit but if they are unsigned integer the length will be tow bytes and so on. If the y_i have different types, all partial results will be aligned. In this way it will be possible to extract every partial result yi from Y in order to use it in computation states of the endless function F.

4 Conclusion

In the past, having security features that protect mobile agents was considered as impossible. As a matter of fact, much of the attention focused on creating security solutions to protect a host in a mobile agent system. Without security mechanisms to protect a mobile agent, many of the proposed uses for agents would not be possible. In this paper, many possible solutions have been explored, but none of them has been able to solve all of the problems in order to keep a mobile agent secure. At the end of the paper, we suggested tow approaches; the first provide a securing software which is it-self protected by a hardware and the second approach provide a point of start of a full software solution based on encrypted functions.

References

1. Hohl, F.: An approach to solve the problem of malicious hosts, retrieved from http://www.informatik.uni-stuttgart.de
2. Wilhelm, U.G.: A Technical Approach To Privacy Based On Mobile Agents Protected By Tamper-Resistant Hardware. PhD thesis, Ecole Polytechnique Fédérale de Lausanne (1999)
3. Maña, A., Pimentel, E.: An efficient software protection scheme. University of Málaga, Spain (2002)
4. Hohl, F.: Time Limited Blackbox Security: Protecting Mobile Agents from Malicious Hosts. In: Vigna, G. (ed.) Mobile Agents and Security. LNCS, vol. 1419, pp. 92–113. Springer, Heidelberg (1998)
5. Sander, T., Tshudin, C.: Towards mobile cryptography. In: Proceeding 1998 IEEE Symposium on Security and Privacy, Oakland, California, pp. 215–224 (May 1998)
6. Aouadi, H.: Proposition d'une Approche Hybride Sécurisant les Agents Mobiles, Thèse de doctorat en informatique, université de la Manouba, Tunisie (October 2009)
7. Aouadi, H., Ahmed, M.B.: Security enhancements for mobile agents platforms. International Journal of Computer Science and Network Security (IJCSNS) 6(7), 216–221 (2006)

A Multi-Agent System with Distributed Bayesian Reasoning for Network Fault Diagnosis

Álvaro Carrera, Javier Gonzalez-Ordás, Javier García-Algarra, Pablo Arozarena, and Mercedes Garijo

Abstract. In this paper, an innovative approach to perform distributed Bayesian inference using a multi-agent architecture is presented. The final goal is dealing with uncertainty in network diagnosis, but the solution can be of applied in other fields. The validation testbed has been a P2P streaming video service. An assessment of the work is presented, in order to show its advantages when it is compared with traditional manual processes and other previous systems.

1 Introduction

Network and service management is a part of the business core for a telecommunication operator. It involves different processes; troubleshooting is a critical one. Network operation is rather expensive and so, a main concern for all the players of this market. On the other hand, operation excellence strongly affects the customer experience, and thus, may have impact in its retention.

Diagnosis automation supported by artificial intelligent techniques can help to reduce operation expenditure, avoiding unnecessary human intervention. In the case of fault management, automation means to be able to detect, diagnose and repair possible problems in the system. The main objectives of these fault management automation are scalability and ability to infer with incomplete and inaccurate information. The approach proposed in this paper faces these problems with a multi-agent architecture which uses distributed Bayesian inference as reasoning mechanism. Multi-agent paradigm is a very common approach in diagnosis systems [5, 10, 1, 4] and cognitive networks [2, 8].

Álvaro Carrera · Mercedes Garijo
Universidad Politécnica de Madrid, Madrid, Spain
e-mail: a.carrera@dit.upm.es, mga@dit.upm.es

Javier García-Algarra · Javier González-Ordás · Pablo Arozarena
Telefónica I+D, Madrid, Spain
e-mail: algarra@tid.es, javiord@tid.es, pabloa@tid.es

Y. Demazeau et al. (Eds.): Adv. on Prac. Appl. of Agents and Mult. Sys., AISC 88, pp. 113–118.
springerlink.com © Springer-Verlag Berlin Heidelberg 2011

This work extends previous results [3] where a multi-agent system with Bayesian reasoning was proposed for fault diagnosis in Virtual Private Networks. This article presents how the system has evolved for a home scenario and which aspects have been improved.

To properly frame this study, a P2P streaming scenario was chosen. In this scenario, there are a multimedia provider user and a multimedia client user. Many faults may occur both in connection and in services. The system is designed to provide to an end-user or an operator the result of the diagnosis made upon receipt of a notification of a symptom of failure. The result is expressed in percentages representing the certainty of the occurrence of a given hypothesis.

2 Multi-Agent Architecture

This section describes the architecture which has been designed for dealing with fault management in a scenario as presented above. One of the main design goals of this architecture has been the integration of capabilities in order to manage uncertainty. To handle it, diagnostic agents take their decisions based on a Bayesian network which models uncertainty about network faults, as described in sect. 3.1.

The following agent types have been identified:

Bayesian Agents offer reasoning capability to the system. There are two types: Diagnosis Agents and Belief Agents. *Diagnosis Agents* are the main responsibles of the diagnosis. They request and recollect evidences or beliefs to infer the root failure cause. Their behavior is controlled by a Bayesian network that models the information about the scenario. *Belief Agents* offer their expert knowledge about different domains (like one Home Area Network). Each belief agent can be expert in a section of the network, in a concrete service, etc.

Scenario Agents offer real-time information required to know the status of the scenario. There are two types: Monitor agents and Observation agents. *Monitor agents* detect symptoms in the scenario and notify to the appropriate diagnosis agent. *Observation agents* offer services to know the status of scenario components, devices, services, etc.

Support Agents offer a support layer to all other agents. Support agents are classified as Yellow Pages Agent, Knowledge Agents and Storage Agent. *Yellow Pages Agents* offer to all agents a directory to search services or agents and their capabilities. *Knowledge Agents* offer a way to generate and deploy new knowledge in the system on the fly. They execute the Expectation Maximization (EM) learning algorithm [7] to improve the performance of the system and deploy appropriate knowledge in Bayesian agents. *Storage Agents* offer a service to store diagnosis in database to perform self-learning process.

In sake of reusability, most of the agents listed in this section are completely generic. Only scenario agents should be implemented to be adapted.

Figure 1 shows the interactions between agents in the scenario presented in sect. 1. To clarify the diagram, the interaction with Yellow Pages Agent has been omitted.

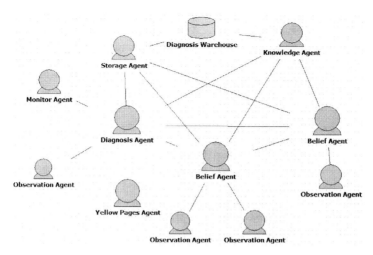

Fig. 1 Agents Hierarchy

3   Bayesian Inference and Learning

Bayesian agents of the system perform two types of inferences:

Centralized inference: The system performs centralized Bayesian inference when request or receive observations or evidences. Every time an agent receives new environment information, it performs local inference. After this, when an agent has resolved all their possible local actions; it could request a set of beliefs to other Bayesian agent, then this pair of agents performs Distributed Bayesian inference.

Distributed inference: When an agent receives a request of a set of beliefs, it should update its own Bayesian network (more exactly, its Bayesian inference engine) to share beliefs using the method described in section 3.1.

3.1   Distributed Bayesian Reasoning

In this study, a distributed way to perform Bayesian reasoning is presented. This system performs Bayesian inference to deduce the root failure cause using a multi-agent architecture, as it is shown in the previous sections (see sect. 2). As the complexity of the diagnosis scenarios grows, scalability may become a problem. To support this, the system is able to distribute the inference process in several smaller Bayesian networks [9] to delegate parts of diagnosis in agents specialized in different problems, regions, services, etc. As mentioned above, the inference process combines two inference strategies. This section details further each one.

Centralized inference is a method in which some kind of evidences or observations are used to calculate the probability of a hypothesis to be true. Distributed inference adds to this method the chance to use beliefs instead of observations. In other words, centralized inference uses as inputs discrete observation (for example true or false, yes or no, etc.), but distributed inference uses uncertain observations as inputs too (for example, true with confidence equals to 0.65). Thus, beliefs (or uncertain observations) can be shared and propagated remotely to perform distributed inference.

The method used to perform distribute inference is a variation of the Virtual Evidence Method (VEM) algorithm [6]. This method is a simple way to set a desired belief in a target node. It consists in adding a child node to any other node (Target node) in the Bayesian model and using equation 1 (see below), thus getting control of the target node. To reach the desired value of the target node, the conditional probability table of the child node should be modified following the instructions shown below. These changes in the model and in the probability tables are performed in execution time by system agents.

This child node should have a fixed structure, two states: trigger state (TS) and non-trigger state (NTS). Conditional probability table of child node should be filled with the following equation:

$$P(ChildNode = TS|TargetNode = SX) = \frac{P_{desired}(TargetNode = SX)}{P_{current}(TargetNode = SX)} \cdot P(aux)$$

(1)

In this equation, there is one degree of freedom, P(aux), that is used to assure the probability is always less than or equals to 1.0. This equation should be instantiated one time per one state of target node. After this process is finished, trigger state is added as an observation to child node, inference is performed and target node has the desired belief in their states.

In this study, flexibility offered by this method is used to perform distributed diagnosis reasoning. During the diagnosis process, agents perform centralized inference until all their possible actions have been solved. Then, they share their beliefs about several hypotheses with other agents that are able to continue with diagnosis process in other relevant areas, in other words, one agent delegates part of the diagnosis in other agents that are specialized in some relevant hypotheses and areas of knowledge.

4 Evaluation

A testbed has been built to evaluate the performance of the system. This testbed simulates the real scenario thus allowing complete management of all systems involved in the study case (see the last paragraph of sect. 1).

In this testbed, to access device configuration and network or service information, several external tools have been used (monitoring and management tools).

We have tested the diagnosis system causing connectivity problems, jitter, dropping packages, configuring the service with wrong data, etc. Then, we evaluate if the system reach valid conclusions and how many time waste in each diagnosis.

Some relevant results are shown below:

Table 1 shows the diagnosis results of the system. Each row represents the following confidence: reliable (0.9 - 1), likely (0.3 - 0.9) and uncertain (0.01 - 0.3).

Table 1 Diagnosis Results Ranking

Diagnosis Confidence	Diagnosis Result	
	Correct	Incorrect
Reliable	51%	0%
Likely	26%	7%
Uncertain	8%	8%

These results can be improved using more data as input in the self-learning process. Actually, we are recollecting more data from testbed to generate more accurate Bayesian networks.

Table 2 Diagnosis Time Ranking

Diagnosis duration (in seconds)	Percentage
10-30	79%
30-50	15%
other	6%

Table 2 shows the time wasted since a symptom is detected to diagnosis is finished. Comparing this work with previous works [3], now the system improves the diagnosis time comparing with the previous approach and the system delegates portions of diagnosis in several physical places.

The system wastes less time because the agents interchange less messages through the core network using the beliefs request capability offered by the distributed Bayesian inference, and delegating portions of diagnosis offers extra flexibility to access to private regions or domains (like Home Area Network), because these portions of diagnosis can be offered like services and consumed externally from any diagnosis agent.

5   Conclusions and Future Work

In this paper, we have demonstrated how a multi-agent system with distributed Bayesian reasoning for network fault diagnosis can be applied to a relevant P2P network scenario. One of the most relevant results are the flexibility of the architecture, that can be adapted easily to other scenarios. Furthermore, the use of probabilistic approaches provides an answer even when the system has barely environment information. And finally, the scalability is provided by the multi-agent paradigm and by the distributed Bayesian inference.

In the future, we plan to further explore some of the challenges identified. One of them is to improve the parallelism between agent tasks to improve the system performance. Another is to add autonomic recovery capabilities to the system, once a conclusive diagnosis has finished.

References

1. Dheedan, A., Papadopoulos, Y.: Multi-agent safety monitor. hull.ac.uk (2003) (2009)
2. Fortuna, C., Mohorcic, M.: Trends in the development of communication networks: Cognitive networks. Computer Networks 53(9), 1354–1376 (2009)
3. García-Algarra, F.J., Arozarena-Llopis, P., García-Gómez, S., Carrera-Barroso, A.: A lightweight approach to distributed network diagnosis under uncertainty. In: INCOS 2009: Proceedings of the 2009 International Conference on Intelligent Networking and Collaborative Systems, pp. 74–80. IEEE Computer Society, Washington, DC, USA (2009)
4. Hongyan, S., Xuefeng, J.: Study on Large-Scale Rotating Machinery Fault Intelligent Diagnosis Multi-Agent System. IEEE, Los Alamitos (2008)
5. Lee, G.J., Clark, D.D., Smith, A.C.: CAPRI: A Common Architecture for Distributed Probabilistic Internet Fault Diagnosis. Ph.D. thesis. MIT, Cambridge (2007)
6. Pan, R., Peng, Y., Ding, Z.: Belief update in bayesian networks using uncertain evidence. In: 18th IEEE International Conference on Tools with Artificial Intelligence ICTAI 2006, vol. (3), pp. 441–444 (2006)
7. Peña, J.: An improved bayesian structural EM algorithm for learning bayesian networks for clustering. Pattern Recognition Letters 21(8), 779–786 (2000)
8. Smart, P.R., Sieck, W., Braines, D., Huynh, T.D., Sycara, K., Shadbolt, N.R.: Modelling the Dynamics of Collective Cognition: A Network-Based Approach to Socially-Mediated Cognitive Change (2010)
9. Xiang, Y., Poole, D., Beddoes, M.P.: Multiply sectioned bayesian networks and junction forests for large knowledge-based systems. Computational Intelligence 9(2), 171–220 (1993)
10. Zhang, G., Huang, S., Yuan, Y.: The Study of Elevator Fault Diagnosis Based on Multi-Agent System, pp. 1–5. IEEE, Los Alamitos (2009)

Using TraSMAPI for Developing Multi-Agent Intelligent Traffic Management Solutions

Ivo J.P.M. Timóteo, Miguel R. Araújo, Rosaldo J.F. Rossetti,
and Eugénio C. Oliveira

Abstract. Intelligent Traffic Management is undoubtedly a promising solution to tackle modern cities' problems related to the growth of the urban traffic volume as it is a non-invasive approach when compared to interventions to the road network structure. Among possible solutions aiming at Intelligent Traffic Management, we believe that Multi-Agent Systems are the most appropriate metaphor to deal with complex domains such as road networks and traffic management and control systems. However, we feel that traffic management and control, particularly intelligent traffic control, is an issue that has not yet been addressed to its full potential. Therefore, we propose in our approach to use TraSMAPI, a tool that offers the possibility of developing real-time Multi-Agent solutions over microscopic simulators, as a basis for the development of intelligent traffic management systems aiming at the creation of revolutionary solutions in the field of traffic and transport systems.

1 Introduction

Modern cities are experiencing an incredible growth in terms of both population and financial power of each individual. This financial independence allows more and more citizens to have private vehicles and to take the ease of mobility and communication for granted. In fact, the flow of goods and people within cities directly related to the road network is now reaching previously unthinkable levels. This makes road networks one of the most important assets of any city as the quality of that flow directly affects the city's economic dynamics, health levels related to air pollution and the well-being of citizens related to the satisfaction resulting from the liberty of movements. However, urban road networks are not entirely prepared to face the new problems that arise with the incredible volume of urban traffic in large cities.

Ivo J.P.M. Timóteo · Miguel R. Araújo · Rosaldo J.F. Rossetti ·
Eugénio C. Oliveira
Department of Informatics Engineering, Artificial Intelligence and Computer Science Lab,
Faculty of Engineering, University of Porto, Rua Dr Roberto Frias S/N,
4200–465 Porto, Portugal
e-mail: {ivo.timoteo,miguel.araujo,rossetti,eco}@fe.up.pt

Y. Demazeau et al. (Eds.): Adv. on Prac. Appl. of Agents and Mult. Sys., AISC 88, pp. 119–128.
springerlink.com   © Springer-Verlag Berlin Heidelberg 2011

These problems are more profound than they first appear to be, as the bottle-neck for further improvements on the traffic flow and the ability to deal with lar-ger volumes of traffic seems to be in the appropriate management and control of the traffic flow, especially at intersections, rather than on infrastructure. In fact, the typical solutions to traffic congestion problems, such as building new roads, increasing the number of lanes, setting up speed limits as well as dedicated lanes, is not only usually non-significant in urban areas but can even lead to negative feedback worsening congestion in road networks. This phenomenon has been demonstrated by the Braess's paradox [1, 2] and confirmed by several experiments and real-life situations [3, 4 and 5]. Also supporting the effective traffic manage-ment approach is the fact that structural intervention to the road network in urban areas is generally impossible or unviable.

Effective Traffic Management can be approached through various methods from scheduling the utilization of certain lanes, changing the traffic lights program according to the time of the day or day of the week, having police agents control-ling traffic in special occasions, and so on. However, we feel that one of the major flaws in traffic control systems is that they are very "static" in their approach us-ing little to none information from their environment and having previously col-lected data as a foundation for their solutions. In modern cities with great traffic volumes, traffic behaves rather chaotically [6] and past data concerning the typical behavior starts to lose validity. This demands for future traffic management sys-tems to be adaptive according to their surrounding environment and possibly pos-sess learning abilities to better respond to future problems. This adaptive behavior and/or learning ability can be considered intelligent and will be referred in this paper as Intelligent Traffic Management.

Intelligent Traffic Management is the main focus of our work and our ultimate goal. When analyzing our domain of study, road networks, its actuators (i.e. traffic lights, variable speed signs, GPS, and so forth), its sensors (i.e. induction loops, video cameras, radars, and so on), as well as its users, we can perceive it, quite in-tuitively, as a swarm of interacting agents. Users are definitely intelligent and self-ish, looking for the best solution for themselves without any concern of the global repercussions. The actuators, however, tend to be, as pointed out before, static and non-adaptive, resulting in trivial solutions which could be greatly improved.

We might have sought for centralized solutions to this problem but given the dimension and complexity of such a domain, it raised many problems both in terms of processing power and error recovery. Also, simply adding more control-ling agents would not be efficient if it had to be done in the main system every time. Another important aspect is that we believe traffic management can be summarized as finding local solutions, accounting for information from nearby agents through communication and expecting some kind of swarm behavior.

These reasons lead towards a distributed solution where every agent is autono-mous, locally aware of its environment, adaptive to its neighboring environment and able to communicate with other agents in order to exchange important infor-mation about the system. In fact, such a solution has been widely used by the community working on traffic analysis as suggested by Schleiffer [7].

When dealing with real-life systems that have to be fully functional without interruptions, real-life experiments can only be made in the late phases of development. This implies that the development of such solutions needs to be supported by studies in simulators. In this work we present the TraSMAPI MAS platform, which provides real-time interaction with microscopic traffic simulators, collects metrics and statistics and offers an integrated framework to develop Multi-Agent Systems. Using TraSMAPI we can focus solely on the creation of agents for intelligent traffic management without having to implement any interaction with the simulator. TraSMAPI allows the user to devise and implement agents no matter the simulator to be used, allowing the same solution to be tested in different simulators. For the case-studies in this paper, we have used TraSMAPI and the microscopic simulator SUMO [8].

This paper will start by focusing on the use of Multi-Agent Systems (MAS) in Traffic control and management. In section 3, we present a typical architecture of a Multi-Agent Solution using TraSMAPI, as well as a deeper look into TraSMAPI's Multi-Agent System framework architecture. Then we shall present a simple case study in section 4, focusing on the implementation of the Multi-Agent System solutions using TraSMAPI.

2   MAS in Traffic and Transport

Before continuing to the development of Multi-Agent System solutions to traffic management, it is necessary to discuss on key concepts related to autonomous agents, their relationships and how suited they are to the field of traffic and transport. These models are being increasingly used within analysis frameworks as an effective tool to aid the understanding of complex and stochastic phenomena. Traffic and transportation systems have profited from the adoption of the multi-agent metaphor and have also stimulated much research on and development of agent-based technologies as they are, in fact, complex environments filled with intelligent agents. Indeed, as we shall see in the following discussion, most applications of Multi-Agent Systems to traffic and transport focus on traffic control and management, although other forms are also addressed by the scientific community as well as by practitioners.

To the best of our knowledge, some former attempts to apply agent-based techniques to address transportation issues date back to the 90's. For instance, Haugeneder and Steiner [9] proposed a co-operative agent-based architecture as a means to improve traffic management and control. Owing to their characteristics and concepts, multi-agent systems have a natural aptitude to cope with a wide range of issues in contemporary traffic and transportation scenarios [10].

Not amazingly, most works report on applying agent-based techniques to control systems and traffic management to make those systems more autonomous and responsive to recurrent traffic demand (e.g. [11]). The analysis of Intelligent Traffic Systems through this approximation has also been investigated (e.g. [12, 13]), and some other works report on applications to freight transport and optimization of resource use (e.g. [14]). Another work has been reported in the literature, which provides a fairly good survey on the application of agent-based approaches to transport logistic [15]. Nonetheless, the challenging issue of modeling more realistically the decision-making process underlying travelers' behavior has encouraged an increasing use of agents for such a purpose. For example, drivers are endowed with cognitive abilities to plan a trip accounting for a mental model of the world and an expectation of the utility their choices would bring about (e.g. [16, 17, 18 and 19]). In this same direction, agent concepts have also proved to be very useful in fostering the improvement of the activity-based analysis of travel demand and of advanced travelers' information systems [20].

In this work we propose a multi-agent architecture using TraSMAPI to underlie the implementation of a tool to analyze and to test with different traffic control strategies and management policies effectively. The Multi-Agent System architecture herein proposed is conceived in a way it allows agents to be generic for any kind of solution and that can interact both autonomously and cooperatively through a specific interaction protocol that fosters short-term tactic as well as long-term strategic control and management decisions.

3   The Development of MAS Using TraSMAPI – General Approach

The development of agent-based traffic solutions is usually drawn back by the great effort the developer has to spend in designing interactions with simulators and communication protocol between agents. However, TraSMAPI is herein presented as a tool that offers complete integration between the Multi-Agent System framework and the simulator interaction module allowing the developer to focus solely on the creation of the solution agents. Furthermore, TraSMAPI is independent from any simulator, which allows the developers to cross information obtained in different simulators without changing their solution code. This specific characteristic is very useful when analyzing results since it allows the solutions to be evaluated on a more simulator-independent basis.

This section will focus on the architecture of TraSMAPI Multi-Agent System framework, comparing it to other Multi-Agent System frameworks that could be used and finally describing the typical architecture of a Multi-Agent Solution.

3.1   The TraSMAPI MAS Framework Architecture

TraSMAPI offers a Multi-Agent System framework which supports agent-to-agent as well as broadcast communication (see figure 1). It manages the flow of activity of the agents to guarantee synchronism and connects directly to a statistics module which can be very helpful in the creation of heuristics and past knowledge useful for learning algorithms and decision making. Such a MAS framework allows the creation of new agents by following a common interface, which implements the basic MAS-Agent interactions that allow the MAS framework to control the flow of activity and ask for agents' action in each time step. Each agent will run in a different process. The MAS framework behaves as a moderator between the agents and controls the flow of information between the agents and the simulation environment.

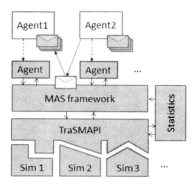

Fig. 1 A MAS solution using the TraSMAPI framework

Given that this is a simulation framework, each time step is usually higher than the average conversation time between agents. Due to performance issues, agents are put "asleep" and "awaken" whenever there is a message addressed to them. This allows the system to detect whenever a communication is complete and to increment the current time step.

The agents themselves can be created with a reference to an object with direct access to the simulation environment so that they can query and change the environment autonomously. However, it is ensured that two agents cannot be accessing or altering information from the simulation environment simultaneously in order to guarantee that there are no concurrency problems.

The communication in the Multi-Agent System Framework is based on an asynchronous message system. The agents have their own message queue in which other agents, and the moderator, might leave messages. These messages are variable and highly dependent on the implementation and goals of the application (e.g. messages can be information requests, answers to requests previously made, advice, general messages sent by the moderator to control the flow of the negotiation, and more). All the messages should be created using a generic message interface so that the MAS framework can manage the communication.

[21] can be consulted for more detail on the implementation of TraSMAPI.

3.2 TraSMAPI MAS Framework vs. Other MAS Frameworks

Questions might be raised on the motives that led the creation of the TraSMAPI MAS framework and why to use it instead of other widely distributed MAS frameworks. The basic motive is that TraSMAPI MAS framework was developed to be completely integrated with TraSMAPI which enables their users to simply call TraSMAPI methods and Statistic module methods when coding their agents in a fully transparent way. This might seem trivial but the gain in productivity is evident. If we were using more generic frameworks we would have to expend a fair amount of effort in the configuration of the interaction with TraSMAPI and on the definition of the communication protocol.

Also, the TraSMAPI MAS framework is intended for simulation and not real world implementation. When simulating, at first, we are looking for abstract solutions and fast prototyping in order to prove our concept. The TraSMAPI MAS framework allows the developer to start coding his solution agents and to test with them immediately. This makes it almost the ideal brainstorming tool. Naturally, if the developer intends to implement the solution in a real world situation, then it should be developed and tested using a more powerful, generic MAS framework.

3.3 A Typical MAS Solution Using TraSMAPI

The development of solutions using TraSMAPI can be simply interpreted as the creation of agents according to our objectives. In fact, the development of MAS solutions using TraSMAPI is fairly simple and follows a well-defined process.

Firstly, it is necessary to initialize TraSMAPI and the MAS framework, and then you add all of your agents to the MAS framework. The agents are created using the *interface agent*. Agents should create TraSMAPI objects to serve as proxies for the objects in the simulation (e.g. create the Traffic Light object to serve as proxy for the traffic light in the simulation). *Proxy objects* are abstractions to make the manipulation and extraction of information transparent to the developer and there is such a generic proxy object for every element in the simulation with which the agent can interact.

In order to allow the MAS framework to control the flow of the simulation and the interaction between agents and between agents and the simulator the agents have to implement two methods. One is the *action* method, called at every simulation step which is simply the MAS framework telling the agent to act. The action method is used to evaluate the environment, choose the behavior and actuate upon the environment (using the proxy object). The other method is *newMessage* which is called every time a new message is sent to the agent. Agents can send messages during the execution of both methods. Since TraSMAPI is openly distributed, it is possible to add new modules and to expand existing ones.

4 Traffic Lights Control in a Grid Road Network

The creation of a simple Multi-Agent solution using TraSMAPI and its main features is illustrated hereafter. We are not interested in obtaining ground-breaking results but rather confirm a classic and expected result: adaptive traffic light agents achieve better results than static, pre-programmed, traffic lights.

4.1 Simulation Environment

The simulation environment is a grid formed by three horizontal roads and three vertical roads with traffic lights in every intersection (see figure 2). The road sections (the length of each section is 250 meters) start with one lane but when approaching the intersection they split into three lanes, each allowing turning right, left or going forward, respectively, where the probability of turning in each direction follows a normal distribution. All traffic lights implement the same cycle, where red and green times are calculated by the agents. As vehicles turning left do not have any priority granted, the cycle also considers four more seconds for the lanes turning in that direction.

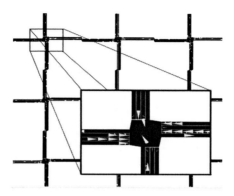

Fig. 2 Simple grid network for the simulation of a Multi-Agent solution

4.1.1 The Proposed Approach

In our solution, agents should have an adaptive behavior considering the stopped and incoming vehicles in each lane in order to minimize the waiting time of incoming vehicles. Furthermore, they will communicate with their neighbors in order to improve the flow of the network as a whole and not only locally. The communication should be informative, as an advice, or asking for help. In the beginning of the simulation they will broadcast messages and receive direct answers to build knowledge on which agents are their neighbors (and on their position). We then compare the results obtained with the results of the pre-scheduled traffic lights (with controllers having the same schedule) for different densities of traffic.

4.1.2 Agent's Decision Module

The decision module decides between both traffic light states based on the result of an evaluation function. The lanes with the higher value need to be opened so that the vehicles can flow.

The evaluation function used, F , is a linear combination of other evaluation functions:

$$F = \sum_{i=0}^{n} ki * Ei$$, where ki is the coefficient by which the evaluation function

Ei is multiplied.

The different evaluation functions, Ei, were directly derived from different metrics such as the sum of the number of vehicles waiting in each lane of each road, the sum of the incoming vehicles in each lane of each road, warning messages from neighbors, and the amount of seconds have passed since the last state change. The coefficients ki were tuned so as to yield the desired agent behavior.

4.1.3 Preliminary Results and Discussion

We considered two scenarios in our study. First, we implemented a Multi-Agent solution under a uniform distribution of vehicles in which every incoming lane to the system contributes with an equal number of vehicles. Finally, an asymmetric distribution of vehicles in which the vertical lanes contributed with three times more vehicles than the horizontal lanes was also considered.

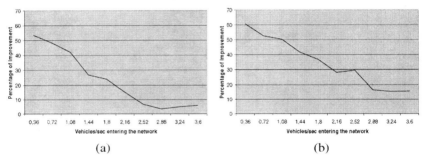

(a)   (b)

Fig. 3 Vehicle distribution: (a) uniform distribution scenario; (b) asymmetric distribution scenario

The results plotted in graphs of figure 3 above represent the improvement in terms of stopped time, given in percentage, relative to the pre-scheduled solution. The density of vehicles is presented by the number of vehicles per each second entering the network.

The results obtained were significant in terms of the improvement achieved relative to the pre-scheduled traffic light. For small traffic densities the traffic light agents, having an adaptive behavior in relation to incoming lanes and communicating to warn others of incoming vehicles, have achieved the expected outstanding improvements.

5 Conclusions

Intelligent traffic management can be considered one of the most promising solutions to contemporary traffic problems. This fact makes it a very interesting topic on which to develop new traffic solutions. It is, however, very hard to test these solutions in the real world given the implications to the road network stability and the importance it has to the citizens' welfare, the city economy and air pollution levels. This supports the need to simulate urban environments in order to test recently developed solutions. We believe that TraSMAPI offers the tools needed to start that development offering an integrated MAS framework, statistics module and an application programming interface with various microscopic simulators. This allows quicker development of solutions and the possibility to explore new possibilities more effortlessly. The successful case study demonstrates the use of the features of TraSMAPI and indicates that it can be used to develop complex traffic management systems. The very next steps in our research will include a deeper investigation on pros and cons of implementing MAS-based and distributed solutions to traffic management and control. Security and safety aspects will also be studied, where other performance measures should be accounted for.

References

[1] Braess, D.: Über ein Paradoxon aus der Verkehrsplanung. Unternehmensforschung 12, 258–268 (March 1968)

[2] Braess, D., Nagurney, A., Wakolbinger, T.: On a paradox of traffic planning. Journal Transportation Science 39, 446–450 (2005)

[3] Yang, H., Bell, M.G.H.: Models and algorithms for road network design: a review and some new developments. Transport Reviews 18(3), 257–278 (1998)

[4] Easley, D., Kleinberg, J.: Networks, p. 71. Cornell Store Press (2008)

[5] Knödel, W.: Graphentheoretische Methoden und ihre Anwendungen, pp. 57–59. Springer, Heidelberg (1969)

[6] Nagel, K., Rasmussen, S.: Traffic at the edge of chaos. arXiv:adap-org/9502005v1 (1995)

[7] Schleiffer, R. (Guest Editor): Transportation Research Part C 10, Pergamon (2002)

[8] Krajzewicz, D., Hertkorn, G., Rössel, C., Wagner, P.: SUMO: Simulation of Urban Mobility. In: Proceedings of the 4th Middle East Symposium on Simulation and Modelling, MESM 2002, pp. 183–187. SCS European Publishing House (2002)

[9] Haugeneder, H., Steiner, D.: A multi-agent approach to cooperation in urban traffic. In: Proc. of the Workshop of the Special Interest Group on Cooperating Knowledge Based Systems, CKBS 1993, pp. 83–99 (1994)

[10] Schleiffer, R.: Intelligent agents in traffic and transportation. Transportation Research 10C, 325–329 (2002)

[11] Hernández, J.Z., Ossowski, S., Serrano, A.G.: Multiagent architectures for intelligent traffic management systems. Transportation Research 10C, 473–503 (2002)

[12] Rickert, M., Nagel, K.: Experiences with a simplified microsimulation for the Dallas/Fort Worth area. International Journal of Modern Physics C 8, 483–504 (1997)

[13] Wahle, J., Bazzan, A.L.C., Klügl, F., Schreckenberg, M.: The impact of real-time information in a two-route scenario using agent-based simulation. Transportation Research 10C, 399–417 (2002)

[14] Adler, J.L., Blue, V.J.: A cooperative multi-agent management and route-guidance system. Transportation Research 10C, 433–454 (2002)

[15] Davidsson, P., Henesey, L., Ramstedt, L., Törnquist, J., Wernstedt, F.: Agent-based approaches to transport logistics. In: Proc. of the 3rd Workshop on Agents in Traffic and Transportation, pp. 14–24. AAMAS, New York (2004)

[16] Dia, H.: An agent-based approach to modelling driver route choice behaviour under the influence of real-time information. Transportation Research 10C, 331–349 (2002)

[17] Nagel, K., Marchal, F.: Computational methods for multi-agent simulations of travel behaviour. In: 10th IATBR Conference (2002)

[18] Rossetti, R.J.F., Liu, R., Cybis, H.B.B., Bampi, S.: A multi-agent demand model. In: Proc. of the 13th Mini-Euro Conference and the 9th Meeting of the Euro Working Group Transportation, pp. 193–198 (2002)

[19] Rossetti, R.J.F., Bordini, R.H., Bazzan, A.L.C., Bampi, S., Liu, R., Van Vliet, D.: Using BDI agents to improve driver modelling in a commuter scenario. Transportation Research 10C, 373–398 (2002)

[20] Dia, H., Purchase, H.: Modelling the impacts of advanced traveler information systems using intelligent agents. Road and Transport Research 8, 68–73 (1999)

[21] Timóteo, I., Araújo, M., Rossetti, R., Oliveira, E.: TraSMAPI: An API Oriented Towards Multi-Agent Systems Real-Time Interaction with Multiple Traffic Simulators. In: 13th International IEEE Annual Conference on Intelligent Transportation Systems, Madeira Island, Portugal, September 19-22 (2010)

EPIS: A Grid Platform to Ease and Optimize Multi-agent Simulators Running

E. Blanchart, C. Cambier, C. Canape, B. Gaudou, T.-N. Ho, T.-V. Ho, C. Lang, F. Michel, N. Marilleau, and L. Philippe

Abstract. This paper presents the work done during the first year of the EPIS project. This project deals with the process of conducting multiple and parallel multi agents-based simulations (MABS) on a cluster or a grid in order to generate sufficient data for scientific use (*e.g.* in the case of a sensibility analysis of a simulation). We provide a new, general and user-friendly approach to marry MABS and High-Performance Computing (HPC). We, thus, propose a workflow and an associated HPC infrastructure. These two permit to easily deploy a lot of simulations on a cluster without any prior parallelizing work. The method wants to be as generic as possible: no particular MABS targeted, no overhead and HPC compliance work has to be done only once. Moreover the user is guided by a web interface that handles the workflow.

Keywords: Multi-agent simulation, distributed simulation, parallelization, High-Performance Computing.

Eric Blanchart
UMR 210 Eco&Sols (INRA, IRD, SupAgro), IRD, Montpellier, France

Christophe Cambier
UMI 209 UMMISCO, IRD, UCAD, Dakar, Sénégal

Clive Canape
Institut de Recherche pour le Développement (IRD), Montpellier, France

Benoit Gaudou
UMR 5505 IRIT, CNRS, Université de Toulouse, Toulouse, France

Ho The Nhan and Ho Tuong Vinh,
UMI 209 UMMISCO, IRD, Institut de la Francophonie pour l'Informatique (IFI), Hanoi, Vietnam

Christophe Lang and Laurent Philippe
LIFC, Univesité de Franche-Comté, 16 route de Gray, 25030 Besançon cedex, France

Fabien Michel
LIRMM, CNRS, Université Montpellier II, 161 rue Ada 34392 Montpellier Cedex 5, France

Nicolas Marilleau
UMI 209 UMMISCO, Institut de Recherche pour le Développement (IRD), Bondy, France
Contact author: e-mail: nicolas.marilleau@ird.fr

Y. Demazeau et al. (Eds.): Adv. on Prac. Appl. of Agents and Mult. Sys., AISC 88, pp. 129–134.
springerlink.com   © Springer-Verlag Berlin Heidelberg 2011

1 MABS and HPC

In [12], Shannon identified two steps which success implicitly relies on being able to perform numerous simulation runs: (1) *Experimentation. Executing the simulation to generate the desired data and to perform sensitivity analysis*; (2) *Analysis and Interpretation. Drawing inferences from the data generated by the simulation runs.*

In this respect, whatever the quality of a simulation team, having (1) enough computing resources and (2) the ability of using them to produce a sufficient number of runs is a critical issue considering the success of a simulation study. Indeed, in most cases, numerous runs should be done to study the different aspects of a simulation model (robustness, sensitivity, impact of initial conditions, output statistical analysis, verification and validation, etc.).

This is especially true about the modeling and simulation of Multi-Agent Systems (MAS). Because they describe the trajectory of each agent, MAS models embed dynamics that could lead to gigantic solution spaces, even when only few parameters are used to describe the system [9]. Despite the interest of exploring MABS models through multiple simulation runs, this work is seldom done.

This exploration is even more problematic in the case of heavy simulations. In the SWORM project [2], a simulator aiming at reproducing the earthworms influence on the soil structure and the nutrient availability by simulation has been developped. Even with an optimized source code, 1 week is needed for only one simulation.

Meanwhile, High Performance Computing (HPC) is today a hot topic and many computing resources are actually available through grids or clusters of computers. So, there is obviously a major issue considering the use of HPC by MABS. Until now, MABS using HPC mainly focus on speeding or scaling up MABS relying on implementations designed so that they are already compliant with HPC (*e.g.* [1]). Still, almost all the MABS platforms are not HPC compliant. In such cases, one has to first work on how to deploy his regular MABS on a HPC architecture. This requires HPC programming skills and could thus be a hard task if planned on the fly. So, practitioners often do not even consider this opportunity and translating regular MABS into HPC compliant ones is still a major issue.

Addressing this issue, efforts have been done in the scope of the RePast platform [8]. [7] proposes a middleware that allows the distribution of RePast sequential models. [4] also uses a middleware approach together with an Aspect Oriented Programming (AOP), again in order to minimize *code intrusions* in the original model.

In this paper, we address these drawbacks by using a more general approach. Linking MABS and HPC, the idea is to work at the platform level rather than the model level. So, our proposal relies on two main features: A workflow and a HPC infrastructure supporting it. The workflow is intended to guide MABS platform developers so that they easily make their platform compliant with the proposed HPC infrastructure. The main idea is that the infrastructure will only be a means to easily deploy a lot of simulation runs over a cluster of nodes, without any prior parallelizing work. So, (1) our approach does not target a particular MABS platform, (2) there is no overhead as it is only about deploying multiple model instances and (3) the HPC compliance work has to be done only once as it works at the platform level.

The outlines of this paper are as follows. Section 2 presents the context of our work. Then an overview of the framework is proposed in Sections 3 and 4. We illustrate our framework with the example of the SWORM simulator [2] coming from soil sciences.

2 Overview of Grid Computing Framework Dedicated to Simulation Domains

Two main approaches can be used to reduce experiment computation time to get results by taking advantage of the power of a grid or a cluster. The first one intends to *parallelize experiment plans* on a grid. The second one intends to *distribute a unique simulation* on a grid. We develop below the first approach which is up to now the only one used in the EPIS project. For a description of the second one, readers can refer to [11].

An experiment plan is composed of a set of simulations qualified by a set of valuated parameters. Parallelizing a plan aims at deploying and running its simulations on several nodes (processors). For example, if a Sworm experiment plan defines thousand simulations to be run on a grid (composed of 5 nodes), two hundred simulations could be affected to each node. Each simulation is independent of each other (*i.e.* simulations does not exchange any data). Note that, this approach is rather simple to apply. A real gain can be observed if many simulations are needed: this approach does not reduce the compute duration of one simulation, it permits only to take advantage of a grid architecture to run simultaneously several simulations.

Several works try to provide frameworks that allow to parallelize one simulation. In this context OpenMole [10] decomposes a simulator into few independent modules. The schedule of these modules is organized according to a workflow defined by a script. This script written in Groovy language is interpreted by the OpenMole framework which starts and manages the simulation. This work focuses on simulator optimization and forget user aspect. On contrary, projects such as SimExplorer [3] (an extension of OpenMole) or GPGCloud [6] take a particular attention to the user interface. These works try to develop a user-friendly graphical user interface permitting to define experiment plans and to execute them on a Grid. Nevertheless, these tools suffer of either a lack of genericity or a lack of simplicity. For example, the use of SimExplorer needs to setup the software and to have skills in Groovy. Moreover GPGCloud does not allowed to run simulators coming from another platform.

The aim of our work is to provide a platform associating genericity with simplicity. It is a user friendly portal (dedicated to non-computer scientists) allowing grid computing, especially parallel simulation running without the complexity of the grid using. The grid is also hidden behind a web access.

3 Epis Workflow

Let consider a modeler faced to a complex problem, such as the one tackled by the Sworm model. He will thus implement his own simulator or reuse the existing Sworm, depending on his needs, on his own computer. To have significant results, he wants to play many and many experiments of the model. Due to the complexity of the studied system, the simulator needs a huge computer power to give intended results (*e.g.* 1 week is needed to play the simple Sworm experiment).

The modeler thus wants to get benefits from the computational power of the cluster to explore the influence of the simulation parameters. He connects to the EPIS portal through its preferred web browser. Then, he selects a simulator (*e.g.* Sworm) in a list or upload a new one, and creates a new experiment.

After creating the experiment, the modeler has to specify his experiment plan by defining the ranges of parameters, constraints, outputs and so on. In order to make the configuration easier for any researcher, we have developed a user-friendly web interface dedicated to drive the modeler through the definition of an experiment plan. Figure 1 summarizes the data exchanges. The interface allows the modeler to determine the variation domain of each parameter of the simulator and which simulation outputs he wants to observe. An illustration of that is, for the Sworm simulator, to identify a fixed population of agents representing earthworms and the range of soil parameter values that determines different kinds of soil. This kind of experiment plans can exhibit the impact of soil structure on earthworm dynamics and behavior.

Once the user has clicked on the "send" button, the interface produces, from the modeler's experiment plan, an XML file describing all the simulations that must be launched. Then the web server calls a script (in an EJB component on the application server) to transform this XML file into an SGE file (file that can be executed on the cluster). These two files are then sent via an SSH tunnel to the cluster and the jobs are launched. Once the jobs are performed, the output XML files are stored on the application server and the web server allows the user to download the results.

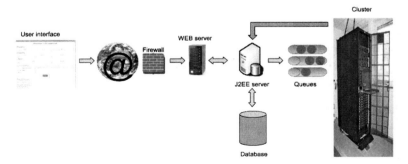

Fig. 1 Workflow of the EPIS project

4   Framework Architecture

As shown on Figure 1, the proposed framework is based on standard components: web server, application server, database, and queue manager for the cluster.

As usual, the web server (a Tomcat server[1]) is in charge of the presentation part of the framework. The presentation part covers all the web pages needed to download the model, to give the set of studied parameters, to start the simulation and to return the results to the user. The download pages are generic to all simulators.

The application server (a Jonas server[2]) is the core of the framework. It is in charge of uploading the data from the web server, of recording them in the database, of starting the simulations on the cluster, of gathering the results from the simulator runs and of presenting them to the user. Aside from the web server, the application also provides an access for heavy clients, *i.e.* with a web service interface.

The database contains all the data needed for the simulation runs: the model (or the simulator if the platform is not supported by the portal), the description files, the parameter files and the resulting data. Conceptually a model is stored with its name, description, identifying number, a number identifying its simulation platform, the set of its parameters and the set of its outputs (practically these two sets are stored in their own tables).

The cluster part is in charge of submitting the runs to the queue manager. Most of the exchanges between the application server and the cluster part are done remotely, from the application server, by using remote commands as *scp* or *ssh*. As several runs will be issued from one parameter set, a global management of the job submissions is done by the SGE *Sun Grid Engine* management system. SGE prevents the cluster overhead by limiting ressource using

5   Conclusion

The EPIS framework is an interesting solution giving a simple access to high performance computing. It allow scientists to get earlier their results or to play bigger experiments.

Without prior knowledge in computer sciences, users are able to determine experiment plans, and to play it on a grid or a cluster. They are guided by a web interface that handles a simple workflow (from the simulator upload to the parallel running of simulations).

The presented platform is based on technologies coming from Web and distributed systems. The framework is based on a J2EE application server and a SGE queue manager for the cluster. The J2EE application server aims at managing web portal and controlling the cluster and jobs (simulation) pushed in cluster queue.

We propose a modular and extensible architecture permitting, up to now, to: (i) select a simulator; (ii) create an experiment plans; (iii) start a huge number of simulations on a cluster, (iv) upload and set up new simulators. Tomorrow, this architecture will be improved by adding new modules dedicated to, for example,

---

[1]`http://tomcat.apache.org/`
[2]`http://wiki.jonas.ow2.org`

experiments result analysis. In addition, users are able to deploy and try themselves simulators they have developed. So, more simulators will be plugged in the portal and offered to scientific communities.

But the major contribution of the EPIS project (maybe done during its second year) will be in the domain of agent-based distributed simulation. Up to now, we focus on the experiment plan parallelizing. Another interesting way intends to distribute a single simulation over several nodes of a cluster. A such simulation needs to propose specific simulator architecture and algorithms ensuring the clock synchronizing, the MAS environment coherency, and so on. Some results have been proposed in for example [5] and [11]. But, this way must be investigated further.

References

1. Aaby, B.G., Perumalla, K.S., Seal, S.K.: Efficient simulation of agent-based models on multi-gpu and multi-core clusters. In: Simutools 2010: Proceedings of the 3nd International Conference on Simulation Tools and Techniques/ OMNeT++ 2010 Workshop (2010)
2. Blanchart, E., Marilleau, N., Chotte, J., Drogoul, A., Perrier, E., Cambier, C.: SWORM: an agent-based model to simulate the effect of earthworms on soil structure. European Journal of Soil Science 60(1), 13–21 (2009)
3. Chuffart, F., Dumoulin, N., Faure, T., Deffuant, G.: Simexplorer: Programming experimental designs on models and managing quality of modelling process. International Journal of Agricultural and Environmental Information Systems (IJAEIS) 1, 55–68 (2010)
4. Cicirelli, F., Furfaro, A., Giordano, A., Nigro, L.: Distributed simulation of repast models over hla/actors. In: Turner, S.J., Roberts, D., Cai, W., El-Saddik, A. (eds.) 13th IEEE/ACM International Symposium on Distributed Simulation and Real Time Applications, Singapore, October 25-28, pp. 184–191. IEEE Computer Society Press, Los Alamitos (2009)
5. Hassoumi, I., Marilleau, N., Lang, C.: Mise en place et évaluation d'un algorithme de répartition de charge pour les plateformes de simulations distribuées basées sur les systèmes multi-agents. In: JFSMA: Défis Sociétaux, Madhia, Tunisie, pp. 85–94 (2010)
6. Kato, Y., Yamaki, H., Asai, Y.: GPGCloud: Model sharing and execution environment service for simulation of international politics and economics. In: Yang, J.-J., Yokoo, M., Ito, T., Jin, Z., Scerri, P. (eds.) PRIMA 2009. LNCS, vol. 5925, pp. 616–623. Springer, Heidelberg (2009)
7. Minson, R., Theodoropoulos, G.K.: Distributing repast agent-based simulations with hla. Concurrency and Computation: Practice and Experience 20(10), 1225–1256 (2008)
8. North, M.J., Tatara, E., Collier, N., Ozik, J.: Visual agent-based model development with Repast Simphony. In: Agent 2007 Conference on Complex Interaction and Social Emergence, pp. 173–192. Argonne National Laboratory, Argonne, IL, USA (2007)
9. Parunak, H.V.D.: Pheromones, probabilities, and multiple futures. In: Bosse, T., Jonker, C., Geller, A. (eds.) MABS 2010. LNCS, vol. 6532, pp. 44–60. Springer, Heidelberg (2011)
10. Reuillon, R., Chuffart, F., Leclaire, M., Faure, T., Dumoulin, N., Hill, D.: Declarative task delegation in OpenMOLE. In: HPCS, Caen, France, pp. 55–62 (2010)
11. Sébastien, N.: Distribution et parallelisation de simulations orientées agents. Ph.D. thesis, University of La Réunion (2009)
12. Shannon, R.E.: Introduction to the art and science of simulation. In: WSC 1998: Proc. of the 30th conference on Winter simulation, pp. 7–14. IEEE Computer Society Press, USA (1998)

Formal Specification of Holonic Multi-Agent Systems: Application to Distributed Maintenance Company

Belhassen Mazigh, Vincent Hilaire, and Abderrafiaa Koukam

Abstract. In complex systems, multiple aspects interact and influence each other. A vast number of entities are present in the system. Traditional modeling and simulation techniques fail to capture interactions between loosely coupled aspects of a complex distributed system. The objective of this work is to extend a Holonic methodology by using a formal specification language based on two formalisms: Generalized Stochastic Petri Net (GSPN) and Z language. Such a specification style facilitates the modeling of organizations and the interactions between them with both reactive and functional aspects. We illustrate the suitability of our generic approach by applying it to a Distributed Industrial Maintenance Company.

1 Introduction

In Industrial Maintenance Company (IMC) a vast number of entities interact and the global behaviour of this system is made of several emergent phenomena resulting from these interactions. The characteristics of this system have increased both in size and complexity and are expected to be distributed, open and highly dynamic. Multi-Agent Systems (MAS) are well adapted to handle this type of systems. Indeed, the agent abstraction facilitates the conception and analysis of distributed microscopic models [1].

Using any holonic perspective, the designer can model a system with entities of different granularities. It is then possible to recursively model subcomponents of a complex system until the requested tasks are manageable by atomic easy-to-implement entities. Holonic MultiAgent Systems (HMAS) provides terminology and theory for the realisation of such dynamically organising agents. They transfer modularity and recursion to the agent paradigm. The different organisations which make up an IMC must collaborate in order to find and put in place various strategies

Belhassen Mazigh
Faculty of sciences, Department of Computer Science, 5000, Monastir, Tunisia
e-mail: belhassen.mazigh@gmail.com

Vincent Hilaire · Abderrafiaa Koukam
Université de Technologie de Belfort Montbéliard, Belfort, France
e-mail: vincent.hilaire@utbm.fr, abder.koukam@utbm.fr

Y. Demazeau et al. (Eds.): Adv. on Prac. Appl. of Agents and Mult. Sys., AISC 88, pp. 135–140.
springerlink.com © Springer-Verlag Berlin Heidelberg 2011

to maintain different production sites. Several constraints should be integrated in the process of strategy search and decision taking before mobilizing intervention teams. To satisfy some of these constraints, we propose a formal holonic approach for modelling and analysis all the entities that constitutes an IMC.

The objective of this work consists in consolidating an Agent-oriented Software Process for Engineering Complex Systems called ASPECS [2] by using a formal specification and analysis of the various organizations and the interactions between them. This type of analysis, will allow checking certain qualitative properties, as well as a quantitative analysis to measure the indicators of performance.

After a brief presentation of the framework, the maintenance activities in a distributed context are presented. A quick overview of the ASPECS process and modelling approach will be presented in section 3. The analysis and conception phase of the ASPECS process and their associated activities are then described in Section 4, while they are applied to the IMC case study. Section 5 present formal specifications of the various organizations and the interactions between them based in composition of GSPN tool and Z language. Finally, Section 6 summarises the results of the paper and describes some future work directions.

2   Industrial Maintenance Company Distributed Context

In distributed context, the maintenance activities are divided on these two following structures:

  ▫ Central Maintenance Team (CMT) which realizes the process of reparation;
  ▫ Mobile Maintenance Team (MMT) which carries out inspections, replacement and several other actions on the various production sites.

To ensure the maintenance of several production sites, many teams specialized in various competence fields should be mobilized. Those in charge of handling these resources should overcome complex logistical problems thereby the need to develop aiding methods and tools for decision making to efficiently manage this type of organisations.

3   A Quick Overview of the Used ASPECS Process

Such as it was proposed by Gaud [2] and Cossintino [3], ASPECS is a step-by-step requirement to code software engineering process based on a metamodel, which defines the main concepts for the proposed HMAS analysis, design and development. The target scope for the proposed approach can be found in complex systems and especially hierarchical complex systems. The main vocation of ASPECS is towards the development of societies of holonic (as well as not-holonic) multiagent systems. ASPECS has been built by adopting the Model Driven Architecture (MDA) [4]. In Cossentino and al. [5] they label the three metamodels "domains" thus maintaining the link with the PASSI metamodel. The three definite fields are:

▫ The *Problem Domain*. It provides the organisational description of the problem independently of a specific solution. The concepts introduced in this domain are mainly used during the analysis phase and at the beginning of the design phase.

▫ The *Agency Domain*. It introduces agent-related concepts and provides a description of the holonic, multiagent solution resulting from a refinement of the Problem Domain elements.

▫ The *Solution Domain* is related to the implementation of the solution on a specific platform. This domain is thus dependent on a particular implementation and deployment platform.

Our contribution will relate to the consolidation of the *Problem Domain* and the *Agency Domain*. We propose a formal specification approach for analysis the various organizations and the interactions between them facilitating therefore the *Solution Domain*.

4 Holarchy Design of Distributed IMC

In this section, we use the ASPECS methodology to describe partially the analysis phase, the design of the agent society and propose a holonic structure of the IMC case study. This approach has enabled us to establish the Holonic structure of the IMC (Fig. 1).

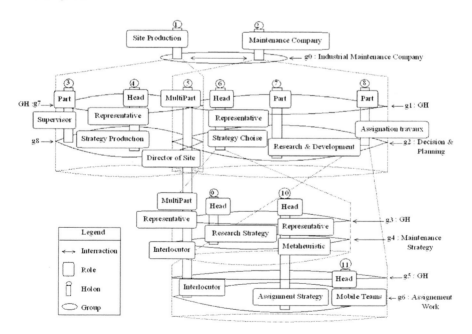

Fig. 1 Holonic Structure of the IMC

Groups (*g1, g3, g5* and *g7*) are holonic ones (HG). At the 0 level of the holarchy, two super-holons *H1* and *H2* play the role of the "*Site Production*" and "*Maintenance Company*" in *g0* group: *IMC*. Thus, *g0* is an instance of a "*Maintenance Company Organisation*". Each of these two super-holons contains an instance of the IMC organization (group *g8* and *g2*). Inspired by monarchic government type, holon members playing respectively the roles of "*Strategy Production*" and "*Strategy Choice*" (*H4* and *H6*) are automatically named *Head* and *Representative* of the other members. Holon Part *H8* playing the role of "*Assignment Work*" is decomposed and contains an instance of the "*Assignment Work Organization*". Its government is inspired by the Apanarchy where all the members are implied in the process of decision-making (all holons are *Head*). Holon Head *H6* playing the role of "*Strategy Choice*" is decomposed and contains an instance of the "*Maintenance Strategy Organization*" with the same government as the *H8* holon. The atomic holon *H5* play the role of *Multipart* as it is shared by two couples of super holons (*H1, H2*) and (*H6, H8*). This holon represents the environnemental part of the application.

5   Formal Specification and Verification of the IMC Organisation

We use our specification formalism ZGSPN introduced in [6] for efficiency modelling and analysis of IMC organisations. This specification formalism combines two formal languages: Z [7] and Generalized Stochastic Petri Nets (GSPN) [8]. Our approach consists in giving a syntactic and semantic integration of both languages. Syntactic integration is done by introducing a Behaviour schema into Z schema. The semantic integration is made by translating both languages towards the same semantic domain. To validate our approach, we have limited our work to the specification of the *Assignment Work Organization* which is a part of the holonic structure of the system studied with two MMT. The choice of the intervening teams depends on the following information: the availability of the MMT, the distance at which the MMT is from the production site and spare parts stock level of MMT. We suppose that our system can be in three different states: Mobile Team(i) Available (MTA(i)), Mobile Team(i) on Production Site (MTPS(i)) and Mobile Team(i) with Critical Level of Stock (MTCLS(i)). For this reason, we use a free [9] or built type to describe the system state:

STATE_SYS ::= MTA(i) | MTPS(i) | MTCLS(i) such as i = (1, 2)

State System and invariants: The system to be specified is described by its state and following average times, estimated by the *Planning organization*, such as: t_{DMMTi} the average time Displacement of Mobile Maintenance Team(i) to reach Production site, associated to transition $T(i)$; $t_{RepMMTi}$ the average time for intervention of Mobile Maintenance Team(i), associated to transition $T'(i)$; t_{SD} the time limit to which Maintenance Team must arrive on a production site; t_{RepCMT} the average time for Repairing the defective parts by Central Maintenance Team, associated to transition $T''(i)$. Other parameters are introduced to supplement the specification such as: Ci the level stock of Mobile Maintenance Team(i)); Cmin(i) the minimum level stock of Mobile Maintenance Team(i) (below this value, MMT(i) must re-enters to the IMC)); m and n the initial state of stocks. The state system is presented with **Sys** schema. In the initial state the MMT are available and the spare parts stock level is at its maximum (m and n). The initial state is presented with **InitSys** schema.

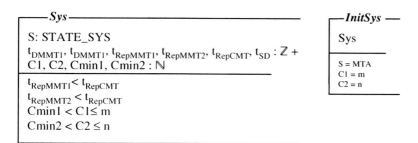

$$\boxed{\begin{array}{l} \text{\underline{\textit{Sys}}} \\ \text{S: STATE\_SYS} \\ t_{DMMT1}, t_{DMMT1}, t_{RepMMT1}, t_{RepMMT2}, t_{RepCMT}, t_{SD} : \mathbb{Z}\,+ \\ C1, C2, Cmin1, Cmin2 : \mathbb{N} \\ \hline t_{RepMMT1} < t_{RepCMT} \\ t_{RepMMT2} < t_{RepCMT} \\ Cmin1 < C1 \le m \\ Cmin2 < C2 \le n \end{array}}$$

$$\boxed{\begin{array}{l} \text{\underline{\textit{InitSys}}} \\ \text{Sys} \\ \hline S = MTA \\ C1 = m \\ C2 = n \end{array}}$$

InterventionsMobilesTeams

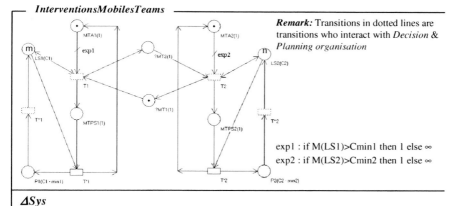

Remark: Transitions in dotted lines are transitions who interact with *Decision* & *Planning organisation*

exp1 : if M(LS1)>Cmin1 then 1 else ∞
exp2 : if M(LS2)>Cmin2 then 1 else ∞

ΔSys

$M(MTA1) = 1 \wedge M(LS1) > Cmin1 \wedge t_{DMMT1} \le t_{SD} \Rightarrow S' = MTA1 \wedge (\lambda_1 = 1/ t_{DMMT1})$
$M(MTPS1) = 1 \Rightarrow S' = MTPS1 \wedge (\lambda'_1 = 1/ t_{RepMMT1})$
$M(MTA1) = 1 \wedge M(LS1) > Cmin1 \wedge t_{DMMT1} > t_{SD} \Rightarrow S' = MTA1 \wedge (\lambda_1, \lambda'_1, \lambda''_1) = 0$
$M(MTA2) = 1 \wedge M(LS2) > Cmin2 \wedge t_{DMMT2} \le t_{SD} \Rightarrow S' = MTA2 \wedge (\lambda_2 = 1/ t_{DMMT2})$
$M(MTPS2) = 1 \Rightarrow S' = MTPS2 \wedge (\lambda'_2 = 1/ t_{RepMMT2})$
$M(MTA2) = 1 \wedge M(LS2) > Cmin2 \wedge t_{DMMT2} > t_{SD} \Rightarrow S' = MTA2 \wedge (\lambda_2, \lambda'_2, \lambda''_2) = 0$

*Operations and interrogations: "**Interventions Mobiles Teams**"* is a behavioural schema since it allows the description of the reactive aspect of the system which is no more than its status change in response to demand requested production site. The predicates part allows also modifying the observation *S* to make it compatible with the system status at a given time. Each place represents a system state. For example, when *MTA1* place (respectively *MTA2*) contains token, and that *LS1* place contains tokens strictly higher than a critical level *Cmin1* (respectively *Cmin2* for *SL2*) and that the average time so that *MMT1* arrives at the destination site production remains lower than equal to one pre-estimated limiting time T_{SD}, consequently the marking of *MTA1* place remains unchanged (respectively for *MTA2*) and a rate $\lambda_1=1/ t_{DMMT1}$ will be associated to *T1* transition from the GSPN (respectively $\lambda_2=1/ t_{DMMT2}$ for *T2*). Finally, *exp1* and *exp2* expressions translate the fact that if the level of the inventories of *MMTi* teams with reached critical level, it will not have the possibility of intervening on any site of production (probably it will turn over to IMC).

6 Conclusion

In this article we showed that HMAS is well adapted to analyse and design an IMC holarchy. The meta-model utilized can be exploited in the implantation stage with the advantage of having formally validated its structure and its behaviour by using composition formalisms Approach. Our future works will focus on a finer analysis of this system type and on a formal modelling of the various scenarios associated with the analysis stage. The notion of multi-views should be integrated. Indeed, the search for and the choice of strategy depends on the point of view of the person or the team required to take decisions according not only the constraints linked to the system but also to their environments. At the same time, it will be interesting to use HMAS to propose a multi-view holarchy introduced in [10] and consequently integrate it in the different existing meta-models.

References

1. Gruer, J.P., Hilaire, V., Koukam, A.: Multi-agent approach to modeling and simulation of urban transportation systems. In: 2001 IEEE International Conference on Systems, Man, and Cybernetics, vol. 4, pp. 2499–2504. IEEE, Los Alamitos (2001)
2. Gaud, N.: Systèmes Multi-Agents Holoniques: de l'analyse à l'implantation. PhD thesis, Université de Technologie de Belfort-Montbéliard (2007)
3. Cossentino, M., Gaud, N., Hilaire, V., Galland, S., Koukam, A.: A holonic metamodel for agent-oriented analysis and design. In: Mařík, V., Vyatkin, V., Colombo, A.W. (eds.) HoloMAS 2007. LNCS (LNAI), vol. 4659, pp. 237–246. Springer, Heidelberg (2007)
4. Object Management Group. MDA guide, v1.0.1, OMG/2003-06-01 (2003)
5. Cossentino, M., Gaud, N., Hilaire, V., Galland, S., Koukam, A.: ASPECS: an agent-oriented software process for engineering complex systems How to design agent societies under a holonic perspective. Auton. Agent Multi-Agent Syst. 20, 260–304 (2010)
6. Mazigh, B.: Formal specification using Z and GSPN. Technical Report, the Department of Computer Science, Monastir University (2006)
7. Lightfoot, D.: Formal specification using Z. The Macmillan Press, Basingstoke (1991)
8. Mazigh, B.: Modélisation et évaluation des systèmes de productions par les réseaux de Petri stochastiques généralisés. PhD thesis, Université de Haute Alsace, Mulhouse. France (1994)
9. Arthan, R.D.: On Free Type Definitions in Z. In: Published in the Proceedings of the 1991 Z User Meeting. Springer, Heidelberg (1992)
10. Rodriguez, S., Hilaire, V., Koukam, A.: Towards a holonic multiple aspect analysis and modeling approach for complex systems: Application to the simulation of industrial plants. Journal Simulation Modelling Practice and Theory 15, 521–543 (2007)

Controlling Bioprocesses Using Cooperative Self-organizing Agents

Sylvain Videau, Carole Bernon, Pierre Glize, and Jean-Louis Uribelarrea

Abstract. This paper presents an Adaptive Multi-Agent System (AMAS) to deal with the control of complex systems, such as bioprocesses, toward user-defined objectives. This control is made under a double constraint: no model of the controlled system can be used and the information available is limited to the values of the observable variables. Thanks to their observations, agents of the AMAS self-organize and create an adequate control policy to lead the system toward its objectives. The developed system is described and then tested on examples extracted from a prey-predator problem. Finally, the results are detailed and discussed.

1 Introduction

The automatic regulation of a bioprocess is a complex task. Such a system features real-time constraints and a very limited amount of available measures, while the quantity of interactions and reactions occurring remains very high, and mostly unknown [1]. In this paper, the meaning of regulation and control of a bioprocess is the action of keeping a quasi-optimal environment in order to enable the expected growth of the microorganisms, while limiting or suppressing any product with toxic characteristics. This point underlines the needs to offer an adaptive approach able to deal with this regulation, without being outdated when the system dynamics are changing [2].

Sylvain Videau · Carole Bernon · Pierre Glize
IRIT, University of Toulouse III, 118 route de Narbonne, 31062 Toulouse cedex 9, France
e-mail: `videau@irit.fr, bernon@irit.fr, glize@irit.fr`

Jean-Louis Uribelarrea
LISBP, INSA, 135 avenue de Rangueil, 31077 Toulouse cedex 4, France
e-mail: `jean-louis.uribelarrea@insa-toulouse.fr`

Y. Demazeau et al. (Eds.): Adv. on Prac. Appl. of Agents and Mult. Sys., AISC 88, pp. 141–150.
springerlink.com   © Springer-Verlag Berlin Heidelberg 2011

A widespread approach to regulate highly dynamic systems involves a model of the system to control [2, 6]. However, in the case of biological processes, these models are rarely available and both very difficult and time consuming for the experts to create. At this point, the system responsible of the control seems to have two choices: controlling without any knowledge of the system to control, or trying to generate a model of the system to control while controlling it. The first alternative can be useful and is actually widely applied in the real world, with for example, proportional-integral-derivative (PID) controllers and their improvements [9, 11]. PID controllers are often in charge of regulating a specific variable, e.g., temperature or pressure, and use mathematical functions to decide on which modification to apply. However, such an approach is mainly available at the variable scale, meaning that a full control of a process implies using several controllers which act on their own on different variables without any communication between them.

On the other hand, trying to model a bioprocess while controlling it, without relying on any detailed biological knowledge of what is happening in the bioprocess, is extremely difficult. If the model aims to be biologically relevant, this difficulty arises from the management of the time lags between actions and observation of their results, as well as the noise appearing when observing them. Furthermore, systems trying to match the observation of the current system with an existing model, extracted from a database, lack genericity to be applied on a wide range of bioprocesses. Moreover, techniques coming from artificial intelligence are often deprecated because of their black box aspect, entailing uncertainty about their predictions abilities, and often restricting them to the regulation of specific variables. Globally, the main problem appears to be the lack of adaptability of the controller, which has to follow the dynamics of the bioprocess while remaining generic enough to be used on very distinct bioprocesses without an extensive instantiation phase.

MAS models to control and monitor some aspect of industrial processes already exist [10] but need to be instantiated to the process they control and represent. More specifically, some works have already underlined the usefulness of MAS for modeling bioprocesses and overcoming classical drawbacks of state-of-the-art control approaches [4]. However a few actually exist. Even if some works are interested in assisting when improving efficiency of bioprocesses [5] or discovering faults [8], as far as we know, the aim of none is to control these bioprocesses as understood here. The use of MAS to control complex systems, especially biological ones, represents more a set of general design solutions to control processes rather than an implementation of a control approach adequate to self-adapt to different kinds of problems without an important phase of instantiation.

To summarize, an efficient and generic way to control complex systems will not involve the use of models of the system to control, and implies learning and self-adaptation in order to follow the dynamics of the system.

This work details the use of an Adaptive Multi-Agent System (AMAS) which relies on the cooperation of its agents to self-adapt to any modification encountered [3], making this approach especially suited to deal with highly dynamic systems such as bioprocesses. Adaptation in an AMAS results from the ability of its agents to self-organize by continuously trying to reduce the criticality of the most annoyed agent in this AMAS. This behavior ensures that an AMAS will converge toward the production of an adequate global function. In this work, an AMAS is built to control complex systems, such as bioprocesses, by modeling a control policy instead of the controlled system itself.

This paper is organized as follows. First, section 2 introduces the Control Adaptive Multi-Agent System (CAMAS) and details the behavior of the involved agents as well as their interactions. Then, section 3 focuses on a prey-predator problem and explains the results obtained for different experiments. Finally, section 4 discusses the conclusions and prospects that this work offers.

2 Controlling with an Adaptive Multi-Agent System

The general structure of the Control AMAS (CAMAS) proposed is described in this section, before detailing the behaviors and the interactions of the agents composing it.

2.1 General Structure of the CAMAS

We consider a bioprocess as a black box on which only a few sensors are available to give partial information on what is happening in the bioprocess. Controlling a bioprocess consists in leading some of the measurable values toward user-defined objectives. So, the main goal of the Control AMAS is to learn the contexts of control to apply on the bioprocess in order to sustain this evolution. This CAMAS has to act without relying on a model of the bioprocess, as mentioned in section 1, meaning that it is only able to observe the evolution of some specific variables, on which sensors are available, in order to decide on the actions to take.

The CAMAS is composed of three distinct kinds of agents following a perception-decision-action lifecycle which cooperate according to the AMAS theory described in [3]. The basic idea underlying this cooperation consists, for every agent in an AMAS, in always trying to help the agent which encounters the most critical situation from its own point of view. Figure 1 gives the structure of a CAMAS controlling a system in which three variables are observable, two of which can be modified. The three different types of agents involved are shown, as well as the links modeling the existing interactions between them. The next sections will provide a more in-depth description of these agents and interactions.

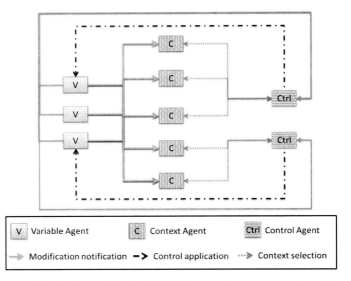

Fig. 1 Agents and their relationships in a CAMAS controlling a system where three variables can be observed, of which two can be modified.

2.2 Behavior of Agents

2.2.1 Variable Agents

The *Variable Agents* represent the link between the system to control and the CA-MAS. Actually their goal is to notify the other agents of any modifications in the values of the real system's variables that the *Variable Agents* consider relevant. Each of the observable variables of the process is represented by a unique *Variable Agent*. For example, to apply a CAMAS on a bioprocess, each observable variable of the bioprocess such as the temperature or the quantity of substrate, has to be associated with a *Variable Agent*.

Furthermore, these agents are able to evaluate their own criticality. This criticality estimates the difference between the current and expected values of a variable and represents the degree of satisfaction of an agent. Its value ranges from 0 to 100 with 0 expressing the highest possible satisfaction. This value can be set from physical considerations (e.g., the quantity of some element that cannot become negative) or extracted from the objective defined by a user (if (s)he wants a specific variable to reach a given value, this value can then be used as reference for the highest criticality the variable can reach).

2.2.2 Context Agents

The *Context Agents* have the most complex behavior in the CAMAS. Their goal is to represent a situation leading to a specific control. They do not aim to model what is biologically happening inside the system to control, but rather aim at modeling the part of the control to apply in the current situation to reach the objectives. When such an agent finds itself in its triggering situations, it notifies the *Control Agents*, by submitting an action to apply in order to reduce the criticality of the system according to its own knowledge.

This action is composed of linear functions, representing speed of variation of a value during a specific time. Depending on the problem, an action can be as simple as a direct modification of value, or a complex combination of several functions during several time steps.

In order to know when its action is relevant, a *Context Agent* relies on two different sets of information. First, a collection of input values, representing the range of variables values on which this *Context Agent* considers its action is worth to be applied. Each one of these ranges is paired with a speed representing the evolution of the quantity of the variable on this specific range, allowing distinguishing several distinct input values while considering the same bounds. This element enables the *Context Agent* to know if it has to be triggered or not. Then, a *Context Agent* possesses a set of forecasts, which describes the impact of the action proposed on the criticality of the different variables of the system. Therefore, observing the set of forecasts ensures that the action of a *Context Agent* will not increase the criticality of the variables while selected. Those three characteristics, input values, action and forecasts are modified during the life of a *Context Agent*. Thanks to the feedback it receives, such an agent adjusts these values according to its behavior.

Finally, each *Context Agent* is defined by an inner state related to its current role in the MAS organization. A total of three different inner states exist: disabled, enabled and selected. The agents can switch from a state to another thanks to the messages they receive from other agents in the system, such as a notification of a criticality increase, or the modification of the speed of a variable matching the inputs of the *Context Agent*.

- A *disabled agent* considers itself non-relevant in this specific situation. Basically, it is not in a triggering state.
- An *enabled agent* thinks that it is relevant and potentially deserves to be selected to apply its action. It then computes its forecasts and sends them, as well as its action, to the corresponding *Control Agent*.
- A *selected agent* is validated by a *Control Agent* and its action is applied to control the system. This selected *Context Agent* has then to observe the consequences of its action in order to reinforce or update its forecasts and its action.

Over time, the number of *Context Agents* is prone to several changes depending on the states of the system and on the relevance of the existing *Context Agents* to represent the current situation. If several agents are close enough, considering their triggering states and their action, they are merged together into a single *Context Agent*. On the other hand, if there is no *Context Agent* describing the current state of the system, the third kind of agents, named *Control Agent*, can ask for its creation.

2.2.3 Control Agents

Control Agents form the last category of agents. One unique *Control Agent* is linked to every *Variable Agent* existing in the system on which an action is physically possible to control this system. This suggests that it may exist some *Variable Agents* which do not receive any messages from *Control Agents*. However, all the *Control Agents* receive the notifications of criticality changes coming from all of the *Variable Agents*.

Control Agents are designed to select the most relevant action among the ones submitted by *Context Agents*. This comparison is done thanks to the forecasts of the *Context Agents*, and can lead to the lack of any selection if all the potential agents are predicted to bring a situation worse than the current one. In this case, or if there is no *Context Agent* in a valid or in a selected state, a *Control Agent* can create another *Context Agent*. The input values of this newly created *Context Agent* are computed from the current set of values of the variables, and its action is extracted from the ones of existing agents and tries to mimic those offering the biggest reduction of the criticality of the most critical *Variable Agent*.

3 Controling a Prey-Predator System

For showing the viability of the approach, this CAMAS is then instantiated to a problem exhibiting a dynamics appearing often in bioprocesses: a prey-predator problem.

3.1 Why Such a System?

Instantiating the CAMAS on a prey-predator problem makes it possible to apply the developed approach to a system showing a dynamics close to those encountered in real bioprocesses involving several populations of cells. Controlling a mathematical model represented by the Lotka-Volterra equations [7] enabled us to extract several distinct sub-problems of increasing difficulty. This control is done by allowing a dynamic modification of two parameters during the simulation. These two parameters represent the access points for the CAMAS on the system to control. The goal of this experiment is to show how the CAMAS can control these two values to lead both the quantity of the prey and predator populations toward two different objectives which cannot be reached without applying modifications. The CAMAS has to control the following mathematical system:

$$dX/dt = aX - bXY$$
$$dY/dt = cXY - dY$$

This system involves the following variables:

- X, prey quantity;
- Y, predator quantity;
- a, growth rate of prey;

- b, impact of the predation on prey;
- c, rate of growth of the predator population according to the size of the prey population;
- d, death rate of predator.

The objectives are values that X and Y must attain, while the parameters a and c are the access points for the CAMAS. Objectives and initial conditions are described in the table 1.

Table 1 Initial conditions of the prey-predator problem.

Name	Type	Initial Value	Objectives
X	Variable	2	5
Y	Variable	2	15
a	Access Point	1	-
b	Parameter	1	-
c	Access Point	1	-
d	Parameter	1	-

3.2   *Experimental Results*

3.2.1   With Hand-Made Contexts

Preliminary results were obtained by using hand-made *Context Agents*. In this example, the input values and action of each context are defined before the simulation, and remain unmodified during its run. Thirty two *Context Agents* were empirically designed, sixteen for each of the two parameters used as access points, designed to match the different combinations of inputs involving values below or above the objectives, with different speeds. These *Context Agents* are sufficient enough to cover the possible values that the variables can take in this specific example, making it impossible to be in a case in which there is no *Context Agent* available.

The main goal of such an example is to underline the usability of the concept of contexts to control the prey-predator system.

Figure 2 highlights this point, as the CAMAS manages to drive the two variables toward their objectives.

3.2.2   With Automatic Contexts

This example extends drastically the first one, by implementing the adaptive behavior of the agents. Indeed, even if the CAMAS relies on a reduced set of *Context Agents*, it is able to create new *Context Agents* during the simulation, and each of these *Context Agents* can modify its inputs values, as well as its action. As a consequence, this example implements the full extent of the agent's design described in section 2, and allows to observe the impact of its self-organization on the results. In

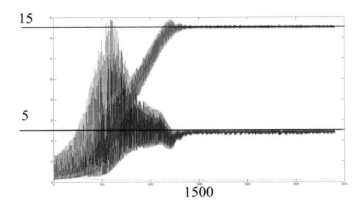

Fig. 2 Results obtained with hand-made contexts.

this example, the CAMAS starts with an empty set of *Context Agents*. The results are detailed in figure 3.

A significant improvement of the amplitude of the result can be observed, with the stabilization of the values on their objectives (with an error of 1.5 %) and a reduction of the noise occurring when the variables are driven from their initial values to the expected ones. However, the variables seem to reach the value of their objectives a few steps of simulation later, this delay can be explained by the learning phase taking place during the control.

Considering the amount of *Context Agents* created, the first example described in section 3.2.1 involved thirty two *Context Agents*, while this example led to the creation of twenty two *Context Agents* from an empty set. Globally, the implemented behaviors led to better results, while involving less *Context Agents*.

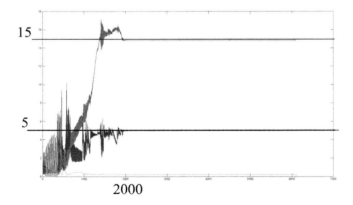

Fig. 3 Results obtained with automatically generated contexts.

3.2.3 Evolution of the CAMAS

Several problems are currently tested, with a focus on the introduction of a more efficient initialization of the action of newly created *Context Agents*, and the addition of modifications of objectives in the middle of a simulation. Their aim is to gradually increase the complexity of the situations that the CAMAS has to deal with, in order to tune, if needed, the behavior of the agents before applying the approach on a real bioprocess.

4 Conclusion and Prospects

This paper presented a Multi-Agent System to control complex systems by creating a model of an adequate control policy. Using an Adaptive Multi-Agent System brings genericity and limits the impact of the drawbacks coming from the current state-of-the-art approaches such as the need to rely on a model of the system to control. Each agent composing the AMAS follows a local and cooperative behavior, driven by the use of the criticality, a value representing the degree of satisfaction of this agent. Its main goal is to decrease as much as possible the maximum criticality it can observe. Thanks to the AMAS theory and to the implementation of the agent's behavior, this local objective leads (if it is possible) to a reduction of the maximum criticality of the system.

Three different kinds of agents, composing the Control Adaptive Multi-Agent System, were designed in order to control complex systems such as bioprocesses, without relying on their model. Their interactions involve the definition of concepts like forecasts, actions, and input values, which are needed in order to adapt to the dynamics of the system to control, and learned from environmental feedbacks or thanks to the communications between agents. The resulting global behavior of the CAMAS is the ability to reach user-defined objectives by observing the available measurable variables, without any extensive knowledge of the system to control.

The CAMAS was then applied on a prey-predator problem where the two populations have to reach a specific value, and the details of the results were presented. This experiment allows studying the reaction of the CAMAS to control a system exhibiting a behavior widely encountered in bioprocesses. These results strengthen the usefulness of the adaptive behavior of the agents, improving the hand-made results while enabling the CAMAS to dynamically control systems despite modifications encountered during the run. This first step toward the application on a real-world bioprocess validates the approach sustaining the CAMAS.

However, there is still room from improvements in some aspects of this approach. For example, future works will confront a CAMAS to a mathematical model of a bioprocess dynamics, in order to evaluate the scaling of our approach on a more complex problem, while moving its application closer to a real world bioprocess.

Finally, this CAMAS will be confronted to benchmark problems, in order to allow its comparison with existing control systems, and to measure its genericity with non biologically related problems.

References

1. Alford, J.S.: Bioprocess Control: Advances and Challenges. Computers & Chemical Engineering 30, 1464–1475 (2006)
2. Astrom, K.J., Wittenmark, B.: Adaptive Control. Addison-Wesley Longman Publishing, Boston (1994)
3. Capera, D., Georgé, J.P., Gleizes, M.P., Glize, P.: The AMAS Theory for Complex Problem Solving Based on Self-organizing Cooperative Agents. In: 12th IEEE Int. Workshops on Enabling Technologies, Infrastructure for Collaborative Enterprises, Linz, Austria, pp. 383–388. IEEE Comp. Society, Los Alamitos (2003)
4. Davidsson, P., Wernstedt, F.: Software Agents for Bioprocess Monitoring and Control. Journal of Chemical Technology and Biotechnology 77, 761–766 (2002)
5. Gao, Y., Kipling, K., Glassey, J., Willis, M., Montague, G., Zhou, Y., Titchener-Hooker, N.: Application of Agent-Based System for Bioprocess Description and Process Improvement. Biotechnology Progress 26(3), 706–716 (2010)
6. Whitaker, H.P., Yamron, J., Kezer, A.: Design of Model Reference Adaptive Control Systems for Aircraft. Tech. Rep. R-164. MIT, Cambridge (1958)
7. Lotka, A.J.: Contribution to the Theory of Periodic Reactions. The Journal of Physical Chemistry 14(3), 271–274 (1910),
   http://pubs.acs.org/doi/abs/10.1021/j150111a004,
   doi:10.1021/j150111a004
8. Ng, Y.S., Srinivasan, R.: Multi-agent Based Collaborative Fault Detection and Identification in Chemical Processes. Engineering Applications of Artificial Intelligence 23(6), 934–949 (2010)
9. Scott, G.M., Shavlik, J.W., Ray, W.H.: Refining PID Controllers Using Neural Networks. Neural Computation 4(5), 746–757 (1992), http://www.mitpressjournals.org/doi/abs/10.1162/neco.1992.4.5.746, doi:10.1162/neco.1992.4.5.746
10. Van Tan, V., Yoo, D.S., Shin, J.C., Yi, M.J.: A Multiagent System for Hierarchical Control and Monitoring. Journal of Universal Computer Science 15(13), 2485–2505 (2009)
11. Visioli, A.: Tuning of PID Controllers with Fuzzy Logic. IEE Proceedings - Control Theory and Applications 148(1), 1–8 (2001),
    http://link.aip.org/link/?ICT/148/1/1, doi:10.1049/ip-cta:20010232

Towards Context-Based Inquiry Dialogues for Personalized Interaction

Helena Lindgren

Abstract. ACKTUS is a semantic web application for modeling knowledge to be integrated in support systems for health care, and for designing the interaction with the end user applications. This paper presents the ongoing work on integrating argument-based inquiry dialogues between agents that include contextual factors in the reasoning. Practical applications of agent-based and interactive dialogue systems are rare in the medical and health domain, partly due to its safety-critical nature. The purpose of the work presented in this paper is to demonstrate the added value a personalized dialogue system can provide a clinician in clinical practice in terms of learning and decision making as supplement to a regular decision-support system in the dementia domain.

Keywords: argumentation, agents, inquiry dialogue, personalisation, learning.

1 Introduction

Clinical decision support systems (CDS) are developed based on clinical practice guidelines (CPG) for the medical and health domains typically with the purpose to increase the quality of care, disseminate CPGs, decrease costs and support work processes. Problems arise when the domain knowledge is incomplete and there is lack of consensus on work procedures and on how to translate research results into clinical practice. The general professional often has to make the best out of limited knowledge, experiences, time resources, etc. A semantic web application (ACKTUS) is being developed that provides a modeling environment for integrating different types of knowledge and for modeling the interaction with a knowledge system [1]. The aim is to provide personalized support to users such as non-expert clinicians at the point of care and to construction and mining workers in monitoring and

---

Helena Lindgren
Department of Computing Science, Umeå University, SE-90187 Umeå Sweden
e-mail: helena@cs.umu.se

Y. Demazeau et al. (Eds.): Adv. on Prac. Appl. of Agents and Mult. Sys., AISC 88, pp. 151–161.

preventing work related injuries. Apart from the clinical knowledge expressed in CPGs and other evidence-based medical knowledge sources, we also allow knowledge such as rules-of-thumb and best practice knowledge to be integrated. Preference orders among knowledge sources, level of expertise in knowledge domains and other contextual factors are taken into account in the reasoning about e.g., diagnosis in a patient case. The interaction design of the knowledge systems is an important aspect in the process of formalizing the knowledge, since the main purpose of the ACKTUS applications is to promote learning and skill development in the end user. Part of this interaction can be designed using a set of agents that participate in a dialogue with the user. In this paper we describe the argumentation-based framework in which we are integrating dialogues as supplement to traditional form-based designs for data collection and for presentation of results of inferences made by the system. Just as the domain expert through ACKTUS is provided an environment for modeling the domain knowledge and how data collection forms are composed, the expert may also be able to model the conditions for agent-based dialogues with the end user. Therefore, we explore the argument-based inquiry dialogue system described by Black and Hunter [2] and exemplify the integration of preference orders and values for the purpose to personalising a physician's interaction with a CDS in a medical scenario from the dementia domain.

The paper is organized as follows. The ACKTUS model is described and extended to agent design in Section 2, paving way to illustrating our application of inquiry dialogues for learning and decision making in Section 3, and an example is given from the dementia domain in Section 3.1. Finally, Section 4 concludes the paper, providing implications for future work.

2 ACKTUS and Agent Design

The service-oriented system architecture of ACKTUS includes an ontology that captures 1) components used for tailoring interaction with the resulting knowledge applications, 2) components of argument-based reasoning and 3) components for modeling the user agents as actors in a reasoning process. Services are provided that handle e.g., storage, visualization, reasoning and terminology issues. In this section ACKTUS is described in the perspective of agent design, and extended with notions useful for agent-based dialogues.

Incorporating the knowledge in an argumentation framework allows the modeling of different aspects of knowledge and defeasibility in a reasoning process where knowledge is subject to changes over time. Argumentative reasoning is typically viewed as a process of valuing different standpoints in a subject, where some arguments fail in the process, while other become supported and justified. Depending on available information and knowledge structures, agents can put forward claims and conflicting arguments drawn from different knowledge repositories in a pattern of defeasible reasoning. *Argumentation schemes* constitute an important structure in argumentation theory, which enable the application of general patterns of reasoning to arguments expressed in a local context of argumentation [3, 4]. Argumentation schemes are described as reasoning patterns that provide a structure of inference in the valuation of arguments. There are commonly *critical questions* (CQ) associated to a scheme that function as activators of arguments for different purposes (e.g., [5]). CQs are typically regarded as defeaters of the argument instantiating a scheme, but can also identify valid lines of reasoning that further support the argument [5]. For modeling the different aspects of knowledge and interaction flows, the ACKTUS ontology integrates *argumentation schemes* as a protocol for reasoning and *interaction protocol* as a type of protocol for interaction, of which CQ is a sub-type (Table 1, see [1] for a description of the interaction model). The ACKTUS protocols function as tools, which the user interacts with in order to execute activity. The argumentation schemes implement knowledge sources such as clinical practice guidelines and can be explicitly selected by a domain expert to be included in a *reasoning context* [1]. As a result, the constraints on the reasoning as performed in interaction with the system can be explained by the underlying knowledge sources. This facilitates transparency and learning in the interaction with an ACKTUS application. In agent-based dialogues with the end user, the reasoning contexts can be used for identifying the topic and associated rules as subsets of knowledge bases.

The ontology also organizes the nodes used for representing and visualizing knowledge, which is partially based on the argument interchange format (AIF) [6]. AIF is a draft for the ongoing development of a formalism to be used for sharing, editing and visualizing arguments over the world wide web. AIF contains nodes representing information and scheme applications (i.e., rules) where the edges represent data and information supply in an argumentation graph. The basic components of AIF are scheme application nodes (S-node), which represent rules and instantiate schemes, and information application nodes (I-node), representing the premise and conclusion components of an S-node in a reasoning process [6] (Table 1).

The ACKTUS ontology is also used for organizing the characteristics of user agents such as knowledge domain, preferences, motives, etc., and can be used for adapting reasoning and interaction. Consequently, the contextual information can also be used in argument-based dialogues involving actors as agents. In order to structure the reasoning as a dialogue between agents, ACKTUS can be further described using notions commonly used in the agent research community (e.g., [7]). A summative description of the central ACKTUS ontology components is provided in Table 1. For our purposes we add to the table the following notions useful for agent-based dialogues, including our interpretation of these within the ACKTUS framework: *belief*, *goal*, *plan*, *argument* and *message*.

Deliberation and argument aggregation are two tasks which are intertwined in ACKTUS since all facts and rules are equipped with strengths. However, in the case when two decision options are in conflict due to their strengths, additional strategies are needed to resolve the situation. In the decision-support system for dementia current strategy is to provide an overview of the conflicting arguments including their supporting and contradicting arguments with their strength so that the physician is supported in making an informed decision about diagnosis. What is currently underway is to integrate rules that determine which argument is acceptable (or arguments in the case of multi-diagnosis). These rules are based on expert physicians'

Table 1 Description of concepts used in ACKTUS, extended with additional concepts related to agent design, which are marked with *.

Concept	Description
Interaction Protocol	represents a pattern of interaction with a knowledge system
Assessment Protocol	is a set of Interaction Protocols
Argumentation Scheme	represents a pattern of reasoning, consisting of a conclusion descriptor and a set of premise descriptors. It commonly implements the content of a knowledge source
Reasoning Context	is a set of Argumentation Schemes
Critical Question	(CQ) is a type of Interaction Protocol, which activates an Argumentation Scheme or a Reasoning Context
Interaction Object	is a type of Interaction Protocol, which functions as a structured placeholder for information concerning a phenomenon, possibly the focus for argumentation or data collection activity
S-node	is a scheme-application node in AIF that implements an Argumentation Scheme. Corresponds to an argument (or a defeasible rule) and has a set of premises and a conclusion (I-nodes)
I-node	is a (defeasible) piece of information or knowledge about an item (fact), structured by an Interaction Object
Scale	is a dictionary of values and can be of different types, depending on what is measured or valued
Inference	is the application of a (defeasible) rule (S-node), which generates new (defeasible) beliefs (I-nodes)
Actor	is an agent enrolled by an organization, acting as a professional in a domain, using knowledge. A human actor has body function and body structures, besides values, preferences and motives
Belief *	is a defeasible fact or a defeasible rule, represented by I-nodes or S-nodes in AIF
Goal *	is the answering of a critical question, which is activated by an agent in a dialogue
Plan *	can be derived from the activated reasoning context, consisting of sub-plans following the relations between reasoning contexts and schemes
Argument *	consists of a claim (I-node) and a set of grounds (I-nodes and S-nodes), supporting the claim. An argument has also a strength drawn from a Scale of values. An argument is a part of a dialogue and is constructed from beliefs and rules (as part of a theory about a knowledge domain)
Message *	is one of the types Tell or Ask passed from one agent to another. Ask represents a critical question and Tell represents a belief or a rule

valuation of knowledge sources and the reliability of the sources providing the information about the patient. However, this method is not tailored to individual users' level of expertise, preferences or need for personalized support. Since personalization has been shown to be highly desirable in evaluation studies (e.g., [8]), the work presented in this paper aims at integrating an alternative and supplementary method to resolve conflicts in the form of agent-based dialogues that follows the user's reasoning process while promoting learning.

3   Dialogue for Learning and Decision Making

Dialogue games are commonly used to describe and characterize argument-based dialogues involving one or more agents [2]. Dialogue games are typically organized by a limited set of allowed acts, or moves, with rules (representing a protocol) directing how the moves can be done at each point in the dialogue, the outcome of a move, and when a game is terminated. The purpose of a dialogue game can be different, corresponding to the motives the agent or agents have with their participation. Walton and Krabbe defined the following five types [9]: 1) *Persuasion* - participants aim to resolve conflicts of opinion, 2) *Negotiation* - participants aim to agree on a method for collaboration that resolves their conflicting interests, 3) *Deliberation* - participants aim to jointly decide on a plan of action, 4) *Information seeking* - participants aim to share knowledge, 5) *Inquiry* - participants aim to jointly discover new knowledge. Systems have been developed for the different types, where the system for deciding upon organ viability for transplantation uses a deliberative dialogue [5] and examples are provided from the cancer domain using inquiry dialogues in [2].

Using deliberative dialogues for deciding upon plans of actions is a natural approach when reasoning about interventions in a teamwork setting. However, since our focus currently is on integrating support for individuals' learning and diagnostic decisions as a part of a CDS the inquiry type will be explored in this paper. The motivation for using a multi-agent dialogue perspective in our work is that this approach allows for modeling the conditions for knowledge development in individuals as well as the system, in spite of ambiguous and incomplete domain knowledge. We assume that the participants are cooperative and reliable, aiming at disseminating and increasing knowledge and finding optimal decisions and actions in patient cases. We assume also that one single medical actor does not possess all knowledge required for providing a patient optimal care, but that the knowledge is distributed over a team of professionals with different viewpoints of a clinical situation. For space reasons in this paper, we assume that there are two agents participating in a dialogue, where the user of a CDS and the agent, implemented as part of the CDS and which is monitoring the CDS knowledge, represent the two actors. Indirectly, the implemented knowledge in the CDS represents the knowledge of an expert in the domain, without assuring that this expert possesses all knowledge relevant for a clinical situation (which is also the case in real clinical situations).

We describe in this section the application of Black's and Hunter's argument-based inquiry dialogue system, which is adapted to ACKTUS mainly in that we consider individuals' preferences as central for personalization and tailored feedback in a learning process. We also make use of the ACKTUS *reasoning contexts* to identify the topic of a dialogue (a concept identifiable in the ACKTUS ontology) and to activate a corresponding subset of the knowledge base to form the belief base of the CDS agent. This way the dialogues can follow the user's trail of thinking in that the user choses topics, e.g., in a differential diagnostic reasoning process. Our adaptation will be described from the perspective of a practical diagnostic example, and the interested reader finds the formal definitions in [2].

Walton and Krabbe describe the inquiry dialogue type as initiated from an initial situation of general ignorance and striving for growth of knowledge and agreement [9]. Each agent has the goal to 'find a proof or destroy one' [9]. Black and Hunter describes an inquiry dialogue system where they define the following two subtypes of inquiry dialogues [2]; *argument inquiry* and *warrant inquiry* dialogue. In the argument inquiry dialogue the participating agents jointly search for a 'proof', which takes the form of an argument for the topic of the dialogue. In the warrant inquiry dialogue the 'proof' takes the form of a dialectical tree and can act as a warrant for the argument at its root. In none of the cases an agent is able on its own to construct the argument or the dialectical tree, based on its limited set of beliefs.

A major difference between the two types of inquiry dialogues is that in an argument inquiry dialogue the agents are not allowed to determine the acceptability of the arguments constructed, while in the warrant inquiry dialogue determining acceptability is the purpose. Therefore, argument inquiry dialogues are often embedded within warrant inquiry dialogues. Another difference is that the topic of an argument inquiry dialogue is a defeasible rule, while the topic of a warrant inquiry dialogue is a defeasible fact. There are three legal moves defined for the two types of dialogue: *open* ($\langle x, open, dialogue(\theta, \gamma)\rangle$), *assert* ($\langle x, assert, \langle \Phi, \phi \rangle \rangle$) and *close* ($\langle x, close, dialogue(\theta, \gamma)\rangle$). The format used for moves in the example dialogues follows the format described in [2], where x represents the agent, $\langle \Phi, \phi \rangle$ is an argument, $\theta = wi$ and γ represents a defeasible fact for a warrant inquiry, and in the case of an argument inquiry $\theta = ai$ and γ represents a defeasible rule. Each agent has a possibly inconsistent belief base and it is assumed that all agents have the same role [2]. By making a query store (which is loaded with for the topic relevant sub-topics) and each agent's commitment store (loaded with asserted knowledge during the dialogue) public, the agents can make use of common knowledge in the dialogue. A dialogue is terminated when all participants have made a close move, which guarantees that all relevant information has been taken into consideration.

Black and Hunter provide a protocol for modeling inquiry dialogues and a strategy for generating dialogues (choosing among candidate moves in a dialogue), which uses an adapted version of Garcia and Simari's Defeasible Logic Programming (DeLP) for representing agents' beliefs [10]. DeLP is adapted by making the sets of strict rules and facts empty and define a defeasible fact. This way all knowledge becomes defeasible, which is suitable for their as well as our purposes. A defeasible rule is denoted $\alpha_1 \wedge \ldots \wedge \alpha_n \to \alpha_0$ where α_i is a literal for $0 \leq i \leq n$. A defeasible fact is denoted α where α is a literal.

Black and Hunter also associate a preference level with a defeasible rule or fact in the formation of a belief, although they do not account for the source of this preference level. The preference ordering is used in the comparison of two arguments. We replace the numbers used in [2] with an explicit ranking among knowledge sources to make the comparison in our example transparent and associate this ranking to defeasible rules. Furthermore, we use an ordered set of values to associate strength to defeasible facts. A belief is a pair (ϕ, S) where ϕ is either a defeasible rule or a defeasible fact. If ϕ is a defeasible rule then $S \in S_0 = \{ cpg, cons, rot \}$ and if ϕ is a defeasible fact then in our example $S \in S_1 = \{true, false\}$ or $S \in S_2 = \{poss, excl\}$. S_2 is a subset of values expressed in a consensus guideline that we will use in the example for expressing the confidence in a hypothesis [11] (*Scale* in Table 1). Clinical practice guidelines (*cpg*) are considered more reliable than consensus guidelines (*cons*), while both are considered more reliable than a 'rule-of-thumb' (*rot*), which is often based on fragmented experiences of an individual professional. Therefore, the following additional beliefs about the strength of knowledge sources are integrated in the CDS agent's (A2) belief base: $(cpg > cons, cpg > rot, cons > rot)$, where $>$ is a binary relation meaning 'strictly preferred to'. It is also known that the value *excluded* (*excl*) is strictly preferred to (is stronger than) *possible* (*poss*). For space reasons, a subset of the agents' belief bases will be used in the following example from the dementia domain, and consequently, it is a simplified procedure for establishing the presence of the diagnosis in question.

3.1 Dialogue Example from the Dementia Domain

Consider the case when a physician investigates a possible case of dementia and has the initial hypothetical diagnosis a possible Lewy Body dementia (DLB). The physician (A1) has an incomplete belief base containing the following: (p, x, a, b, y, (DLB, *poss*), (a $\rightarrow$ y, *rot*), (x $\rightarrow$ (DLB, *poss*), *rot*)), where the literals represent the findings vascular symptoms (p), extrapyramidal symptoms (x), memory deficit (a), aphasia (b) and dementia (y). A1 also has two rules of thumb, which are in this example fragmented knowledge that is not sufficiently based on medical literature.

The physician (A1) makes the initial move to open a warrant inquiry dialogue (wi_0) at timepoint 1 (move m_1 in Table 2) with the topic of the argumentation $\langle$DLB, $poss\rangle$. The CDS agent (A2) has a belief base drawn from the CDS, which recognizes the topic and retrieves the reasoning context dealing with DLB and the following associated rules: (x $\wedge$ y $\rightarrow$ $\langle$DLB, $poss\rangle$, *cons*), (p $\rightarrow$ $\langle$DLB, $excl\rangle$, *cons*). As a result a query store is loaded with the following: (p, x, y, $\langle$DLB, $poss\rangle$, $\langle$DLB, $excl\rangle$). The query store can be used by both agents to select sub-topics for moves in the dialogue.

A2 then initiates an argument inquiry dialogue (ai_1) as a nested dialogue within the initial dialogue in order to sort out if the physician has grounds for the claim with an *open* move. By this A2 puts focus on creating the argument for the claim. A1 contributes to ai_1 with a belief about x and makes an *assert* move, which is not sufficient for creating the argument. A2 tries to close the dialogue but A1 can select as the next move to assert also y. If A2 would like to investigate what this belief

Table 2 Example of a warrant inquiry dialogue with nested argument inquiry dialogues, where t represents timepoints, CS is commitment store and m_t is a move at timepoint t.

t	A_1 CS	m_t	A_2 CS
1		$\langle 1, open, dialogue(wi_0, \langle DLB, poss\rangle)\rangle$	
2		$\langle 2, open, dialogue(ai_1, x \wedge y \rightarrow \langle DLB, poss\rangle)\rangle$	
3	(x, *true*)	$\langle 1, assert, \langle\{(x, true)\}, x\rangle\rangle$	
4		$\langle 2, close, dialogue(ai_1, x \wedge y \rightarrow \langle DLB, poss\rangle)\rangle$	(x, *true*)
5	(y, *true*)	$\langle 1, assert, \langle\{(y, true)\}, y\rangle\rangle$	
6		$\langle 2, open, dialogue(ai_{11}, a \wedge b \rightarrow y)\rangle$	(y, *true*)
7	(a, *true*)	$\langle 1, assert, \langle\{(a, true)\}, a\rangle\rangle$	
8		$\langle 2, close, dialogue(ai_{11}, a \wedge b \rightarrow y)\rangle$	(a, *true*)
9	(b, *true*)	$\langle 1, assert, \langle\{(b, true)\}, b\rangle\rangle$	
10		$\langle 2, assert, \langle\{((a \wedge b \rightarrow y), cpg), (a, true), (b, true)\}, y\rangle$	(b, *true*)
11	(a $\wedge$ b $\rightarrow$ y, *cpg*)	$\langle 1, close, dialogue(ai_{11}, a \wedge b \rightarrow y)\rangle$	
12		$\langle 2, close, dialogue(ai_{11}, a \wedge b \rightarrow y)\rangle$	
13		$\langle 1, close, dialogue(ai_1, x \wedge y \rightarrow \langle DLB, poss\rangle)\rangle$	
14		$\langle 2, assert, \langle\{(a \wedge b \rightarrow y), (a, true), (b, true), (x, true), (x \wedge y \rightarrow$ $\langle DLB, poss\rangle, cons)\}, (DLB, poss)\rangle$	
15	(x $\wedge$ y $\rightarrow$ $\langle$DLB, *poss*$\rangle$, *cons*)	$\langle 1, close, dialogue(ai_1, x \wedge y \rightarrow \langle DLB, poss\rangle)\rangle$	
16		$\langle 2, close, dialogue(ai_1, x \wedge y \rightarrow \langle DLB, poss\rangle)\rangle$	
17		$\langle 1, close, dialogue(wi_0, \langle DLB, poss\rangle)\rangle$	
18		$\langle 2, open, dialogue(ai_2, p \rightarrow \langle DLB, excl\rangle)\rangle$	
19		$\langle 1, assert, \langle\{(p, true)\}, p\rangle\rangle$	
20		$\langle 2, assert, \langle\{(p \rightarrow \langle DLB, excl\rangle, cons), (p, true)\}, (DLB, excl)\rangle$	(p, *true*)
21	(p $\rightarrow$ $\langle$DLB, *excl*$\rangle$, *cons*)	$\langle 1, close, dialogue(ai_2, p \rightarrow \langle DLB, excl\rangle)\rangle$	
22		$\langle 2, close, dialogue(ai_2, p \rightarrow \langle DLB, excl\rangle)\rangle$	
23		$\langle 1, close, dialogue(wi_0, \langle DLB, poss\rangle)\rangle$	
24		$\langle 2, close, dialogue(wi_0, \langle DLB, poss\rangle)\rangle$	

is based on, A2 can open a new nested dialogue (ai_{11}) with the move *open* (m_6 in Table 2). The query store is then loaded with a and b. A1 can chose to assert a, then A2 tries to close and A1 makes another assert move, contributing with b. Then A2 can create the argument about y (m_{10} in Table 2). This CPG-based rule is included in the commitment store and becomes integrated in the belief base of A1.

Back to the argument inquiry dialogue (ai_1) after two close moves, A1 tries to close the dialogue, and A2 can assert the grounds for the topic of the warrant dialogue and create the argument for a possible DLB (m_{14} in Table 2). This alternative and stricter, consensus-based rule to assert a possible DLB is added to the commitment store and belief base of A1. The outcome of this dialogue (ai_1) is the constructed argument $Arg_1 = \langle\{(a \wedge b \rightarrow y), (a, true), (b, true), (x, true), (x \wedge y \rightarrow \langle DLB, poss\rangle, cons)\}, (DLB, poss)\rangle$, which then can be used to determine the support for the topic of the warrant inquiry dialogue. At this point, there is no other move for A1 to select except to try to close the top-level warrant dialogue.

A2 can now open a new argument inquiry dialogue(ai_2) concerning p, which contradicts the topic of the warrant dialogue (m_{18} in Table 2). A1 asserts p, and then A2 can make an assert move with a counter-argument to the topic of the dialogue (m_{20} in Table 2): $Arg_2 = \langle\{(p \rightarrow \langle DLB, excl\rangle, cons), (p, true)\}, (DLB, excl)\rangle$. This argument is also included in the commitment store, and consequently also in A1:s belief base. Based on the knowledge that the value *excluded* is strictly preferred to the value *possible* and that the knowledge sources are equally reliable, Arg_2 defeats the root argument. The dialectical tree that is constructed has two arguments; the root node with the sub-argument that supports the root node (Arg_1: DLB is possibly present) and the argument node that defeats the root node (Arg_2: DLB is excluded).

3.2   Outcome and Potential Consequences of the Dialogue

To summarize the outcome of the dialogue, the physician A1 has gained knowledge about how assessing a possible DLB (which is represented by the inclusion of three rules of higher reliability than the existing rules to the agent's belief base as part of a user model of the physician A1) and it is known that there must be another cause of symptoms than DLB in the patient case. At this point the CDS agent A2 can chose to open a new warrant inquiry dialogue to investigate the support for an alternative diagnosis which is supported by both p, a, b, x, and y. In this new dialogue, the constructed arguments about DLB can be reused (e.g., if the presence of Alzheimer's disease is the new topic, a key task is to exclude DLB). Consequently, in a complex situation of differential diagnosis, several warrant inquiry dialogues will be needed in order to reach a final decision about diagnosis, which ultimately, is the physician's responsibility. In the interaction with the CDS, the agent who may contribute with knowledge to the analysis initiates dialogues, and the activated subsets of rules are restricted by the topics of the dialogues, structured as reasoning contexts in ACK-TUS. Consequently, the domain expert who models the content of the reasoning contexts also models what to be reasoned about in the dialogue-based interaction with the end user system, and based on which knowledge sources as supplement to the end user's belief base.

The CDS agent could have taken the faster way to defeating A1:s claim, by attacking the claim in its first step with the counter argument. However, by initiating argument inquiry dialogues with the physician A1 instead, it can be sorted out on what grounds the physician holds this hypothesis and an opportunity for learning is created. In order to further enlightening the physician, the CDS agent may after the dialogue is completed, propose to the user to follow the CDS:s preference order among sources (*cons* > *rot*) and consequently a preference order among rules (x $\wedge$ y $\rightarrow$ $\langle DLB, poss\rangle$, *cons*) > (x $\rightarrow$ $\langle DLB, poss\rangle$, *rot*). If the physician approves of this priority, a practical consequence could be that in a later situation when the physician the next time proposes a suspected possible DLB, the physician can use the preference order when selecting the grounds for the argument, or be reminded to use the guideline-based rule on this topic if the physician falls back into old routines and forgets the new knowledge, e.g., in a stressful situation or when a long time has passed since the previous occasion when DLB was a potential diagnosis.

4 Conclusions

The service-oriented web application ACKTUS is used for modeling knowledge and tailored interaction with support systems in the health domain. The possibility to extend ACKTUS applications with agent dialogues was explored to provide end users with personalized reasoning support that also tailors support for learning as supplement to the general summative form-based analyses of patient cases. The adaptation and integration of the argument-based inquiry dialogues described by Black and Hunter into the ACKTUS framework was proposed for modeling dialogues to be used in a cooperative knowledge building reasoning process between agents in a patient case. We have described how existing knowledge structures in ACKTUS such as reasoning contexts and argumentation schemes can be used to guide the agents, and how the outcome of different warrant inquiry dialogues can be reused in a differential diagnostic reasoning process. Arguments can be compared and valued based on individuals' preference orders among knowledge sources, or based on reliability levels expressed by knowledge sources that are associated to beliefs. In this process agents representing users increase their belief bases, which can be extended further with preferences about gained knowledge in dialogue with the user and reused to promote learning in a longer perspective.

Ongoing work concerns how the dialogues described in this paper can be created and visualized using critical questions and argumentation schemes, which are represented in natural language, as an interaction layer between the formal dialogue execution and the user for achieving as natural and intuitive interaction as possible.

Acknowledgements. The project is partly funded by VINNOVA (The Swedish Governmental Agency for Innovation Systems) and Emil and Wera Cornell foundation.

References

1. Lindgren, H., Winnberg, P.: A Model for Interaction Design of Personalised Knowledge Systems in the Health Domain. To appear in Proc. 5th International workshop on Personalisation for e-Health (2010)
2. Black, E., Hunter, A.: An inquiry dialogue system. Autonomous Agents and Multi-Agent Systems 19(2), 173–209 (2009)
3. Walton, D.: Argumentation Schemes for Presumptive Reasoning. In: Mahwah, N.J. (ed.) Erlbaum (1996)
4. Bex, F., Prakken, H., Reed, C., Walton, D.: Towards a formal account of reasoning about evidence: argumentation schemes and generalisations. Artif. Intell. Law 11(2-3), 125–165 (2003)
5. Tolchinsky, P., Modgil, S., Cortés, U.: Argument schemes and critical questions for heterogeneous agents to argue over the viability of a human organ. In: AAAI Spring Symposium Series; Argumentation for Consumers of Healthcare, pp. 377–384 (2006)

6. Chesñevar, C., McGinnis, J., Modgil, S., Rahwan, I., Reed, C., Simari, G., South, M., Vreeswijk, G., Willmott, S.: Towards an Argument Interchange Format. The Knowledge Engineering Review 21(4), 293–316 (2006)
7. Fox, J., Glasspool, D., Modgil, S.: A Canonical Agent Model for Healthcare Applications. IEEE Intelligent Systems 21(6), 21–28 (2006)
8. Lindgren, H.: Towards personalized decision support in the dementia domain based on clinical practice guidelines. To appear in User Modeling and User-Adapted Interaction (2011)
9. Walton, D.N., Krabbe, E.C.W.: Commitment in dialogue: Basic concepts of interpersonal reasoning. SUNY Press (1995)
10. García, A.J., Simari, G.R.: Defeasible logic programming an argumentative approach. Theory and Practice of Logic Programming 4(1-2), 95–138 (2004)
11. Lindgren, H., Eklund, P.: Differential Diagnosis of Dementia in an Argumentation Framework. Journal of Intelligent & Fuzzy Systems 17(4), 387–394 (2006)

Modelling Driver Interdependent Behaviour in Agent-Based Traffic Simulations for Disaster Management

David Handford and Alex Rogers

Abstract. Accurate modelling of driver behaviour in evacuations is vitally important in creating realistic training environments for disaster management. However, few current models have satisfactorily incorporated the variety of factors that affect driver behaviour. In particular, the interdependence of driver behaviours is often seen in real-world evacuations, but is not represented in current state-of-the art traffic simulators. To address this shortcoming, we present an agent-based behaviour model based on the social forces model of crowds. Our model uses utility-based path trees to represent the forces which affect a driver's decisions. We demonstrate, by using a metric of route similarity, that our model is able to reproduce the real-life evacuation behaviour whereby drivers follow the routes taken by others. The model is compared to the two most commonly used route choice algorithms, that of quickest route and real-time re-routing, on three road networks: an artificial "ladder" network, and those of Lousiana, USA and Southampton, UK. When our route choice forces model is used our measure of route similarity increases by 21%-93%. Furthermore, a qualitative comparison demonstrates that the model can reproduce patterns of behaviour observed in the 2005 evacuation of the New Orleans area during Hurricane Katrina.

1 Introduction

Evacuation of large areas due to disasters requires effective management and realistic training environments are increasingly being used in order to teach operators how to manage traffic in the safety of a simulated environment. However for training to be effective, the simulated environment must have the flexibility to respond to the variety of actions operators can make, such as

David Handford · Alex Rogers
School of Electronics and Computer Science, University of Southampton, UK
e-mail: {djh07r,acr}@ecs.soton.ac.uk

Y. Demazeau et al. (Eds.): Adv. on Prac. Appl. of Agents and Mult. Sys., AISC 88, pp. 163–172.
springerlink.com   © Springer-Verlag Berlin Heidelberg 2011

setting up road blocks or diversions, in addition to simulating situations with limited real-life data. To this end, the use of agent-based models of individual drivers has proven to be an effective way of modelling traffic systems [3], in which the aggregation of driver behaviours reproduces real-life patterns of overall traffic behaviour. During real-life evacuations, studies have shown patterns of traffic behaviour in which a perceived degree of physical danger causes drivers to choose routes similar those of others, so as to avoid being isolated [5, 12]. In the evacuation of the New Orleans area during Hurricane Katrina in 2005 this interdependence in driver behaviour led to situations in which, despite there being two possible escape routes, a disproportionate number of drivers used just one over the other.

However, route choice behaviour in current state-of-the-art evacuation simulations (such as MATSIM [9] or PARAMICS [3]) incorporate limited driver behaviours and thus it is difficult to reproduce real-life patterns of traffic behaviour. In PARAMICS, the behaviour of drivers who are "unfamiliar" with an area is to take the route that they believe to be quickest to the exit with a preference for using major roads over minor roads [11]. Both MATSIM and PARAMICS also offer route choice behaviour, known as dynamic route planning (used in PARAMICS for the behaviour of drivers "familiar" with an area). Here the drivers are able to re-plan their escape routes, factoring in real-time knowledge of the congestion on other roads. However, this causes a repulsive behaviour between drivers, the opposite of what is actually observed in real-life evacuations, where drivers desire to use routes which they believe others will be using [5]. Neither of the offered algorithms are able to replicate the observed patterns of behaviour in evacuations and thus their use for disaster management simulation is significantly impaired [4].

To address this shortcoming, in this paper we present a novel agent-based route choice forces model. Our approach is inspired by pedestrian behaviour models, where research into the interdependence between individual's decisions is more mature. As in the crowd social force model, a variety of factors or "forces" act on and influence an agent's decisions [6]. Within this model, forces become virtual signposts at junctions, the directions of which are determined from utility-based path trees. When an agent reaches a junction they probabilistically choose between the routes, based on the utility they believe they will gain from following each particular route, determined from the magnitude of the force. For evacuation simulation two forces are defined: the desire to take the quickest route to safety and a varying desire to be with others depending on the driver's particular level of panic. The utility-based path tree for the force representing the desire to be with others is created using a method inspired by floor field modelling in crowd behaviour [8]. Prior to running the evacuation simulation, drivers are simulated using their non-evacuation routes out of town, along which they leave a trail. These trails represent knowledge of routes used by others in the past. Within the path tree, the stronger the trail along a route, the more utility will be received for using it. The level of utility a driver believes they will gain varies with their particular desire to be with others.

Thus in more detail, this paper extends the current state-of-the-art in driver route choice models in the following ways:

- We develop a probabilistic agent route choice mechanism known here as the route choice forces model, in which decisions are influenced by a set of forces representing the factors which influence a driver's behaviour. We incorporate real-life observed evacuation behaviours as a force representing a driver's desire to be with others.
- We define a metric to evaluate how effectively an algorithm can replicate a driver's desire to use the route of others during evacuations. We use this metric to benchmark our model against two existing route choice algorithms (shortest time and real-time re-routing) using three road networks: a simple ladder network and the cities of Lousiana, USA and Southampton, UK. We show that our model increases the metric by 21%-93% over using other algorithms, in addition to conforming to qualitative observations from evacuation during Hurricane Katrina in 2005.

The rest of the paper is arranged as follows: Section 2 describes the context within which the model is developed. Sections 3 presents the model itself. Section 4 describes the metric and empirically evaluates the model. Finally, Section 5 discusses the model's further development.

2   The Evacuation Setting

Our route choice force model forms a behavioural component within a disaster management and traffic operator training simulator in development at BAE SYSTEMS. This simulator is being developed to allow operators to manage traffic flows in the event of emergencies. Our model is being developed alongside the 3D visualiser component, which allows controllers to view the state of the roads through virtual CCTV cameras, and a driver behaviour component which models the tactical level behaviours such as car following and lane changing. An event-driven queue-based mesoscopic traffic simulation model, based on the agent-based MATSIM traffic simulator [2] and implemented in C++, is used to model the individual movement of the cars as they evacuate from a start zone to a predetermined safe zone. A region is represented by a road network defined by a set of roads with lengths and speed limit connecting a set of junctions. Within the queue model each road section has a corresponding queue, implemented as a FIFO queue with restrictions on entering and exiting. Drivers are represented by agents with the goal of reaching a safe destination by planning a route and then travelling through the road network. The only choice an agent must decide upon is which way to go once it reaches a junction, which is achieved using our route choice forces model.

3   The Route Choice Forces Model

Within our route choice forces model a variety of forces are defined which represents factors that act on and influence an agent's decisions. The model takes inspiration from the social forces model developed in crowd modelling, in that each force has a certain magnitude of appeal to a particular agent [6, 10]. Where agents need to make choices about their route at junctions, the forces become virtual signposts where they inform agents of a route to follow and the utility they gain from following that route. The agent then uses a mechanism to probabilistically decides which signpost to follow. The route represents a force's directions and the utility value represents its magnitude. These are determined using a utility-based path tree particular to a force. For evacuation simulation two forces are defined: a desire for the quickest exit route and a desire to be with others varying on their level of panic.

In order to create a force which represents a driver's desire to use routes they believe others will, a utility-based path tree is created in which the driver gains utility by using the route of others. To determine these utility values our model uses a concept similar to chemotaxis/stigmergy used in floor field modelling of pedestrian behaviour, in which agents lay down abstract trails as they move through the environment to which others are attracted [7, 8]. Here these trails represent the knowledge a driver will have prior to the evacuation of the usual routes used by others in the past. Prior to running the actual evacuation simulation, agents are simulated using their usual, non-evacuation, routes out of the evacuation area (using the quickest route algorithm). When each agent reaches their destination, each road they have used in their route is taken and the utility value for using their road in future is increased, making it more desirable. Dijkstra's algorithm is then used to create non-cycle highest utility routes through the network, creating the utility-based path tree.

The utility values are normalised by dividing by the total number of cars that have reached the destination. Thus utility value u_{ld} for route section l given the destination d, is given by this formula:

$$u_{ld} = \frac{|\{V : V \in R_d, l \in V\}|}{|R_d|} \tag{1}$$

where R_d is all routes that cars have taken to destination d.

Now, we define the mechanism used by the agent to probabilistically decide which road to follow when at a junction. To this end, the two forces are weighted with coefficients representing how strongly they are acting upon an agent, k_a for the driver attraction force and k_s for the quickest route force, with k_a being used to represent an agent's level of panic. The direction of force $\mathbf{f}$ at a particular junction is given by its path tree and the magnitude, given that the path tree defined route is given by the set of roads R_f, is:

$$|\mathbf{f}| = k_f \sum_{l \in R_f} u_{ld} \tag{2}$$

where k_f is the agent's personal coefficient for force $\mathbf{f}$, for the two forces defined here these are k_a and k_s, and u_{ld} is the utility of using link l given destination d.

The score for road i at junction j given destination d is calculated by combining all forces in the direction of that road such that:

$$Score(i_{jd}) = \sum_{\mathbf{f} \in F} |\mathbf{f}| \tag{3}$$

where F is a set of all the forces which act upon the driver in the direction of road i.

Using the score, the drivers then use a probabilistic model to decide which road to take. In our model a normalisation function is used, however a range of possible functions could be used such as the softmax activation function [1]. The probability of a particular route i being selected by a driver with destination d at junction j is given by:

$$p_{ijd} = \frac{Score(i_{jd})}{\sum_{a_d \in N_j} Score(a_d)} \tag{4}$$

where N_j is all the roads leading off junction j that it is possible for a driver to take.

4 Empirical Evaluation

In order to investigate the occurrence of the route attraction pattern a metric is presented enabling the comparison of different types of route choice algorithms via a measurement of similarity between evacuating drivers' routes. For each evacuated driver a count is increased on each road which they have used, such that the count n_l represents the number of evacuation routes which have used road l and N is the set of all counts. If the routes are equally distributed across the road network then each road will have an equivalent level of usage. However, if the routes are concentrated on a few particular roads, then the road usage count will be more varied. The metric is therefore defined as $stdev(N)$. With road usage count n_l for road section l being given by:

$$n_l = \frac{|\{V : V \in R, l \in V\}|}{|R|} \tag{5}$$

where R is all set of routes which have been taken to an evacuation safe point. Three road networks are now defined,

- A theoretical construct of a "ladder", as shown in Figure 1(a). In which drivers start at the bottom and escape at the top through one of the two exit points. The drivers have the choice of using one or other of the main

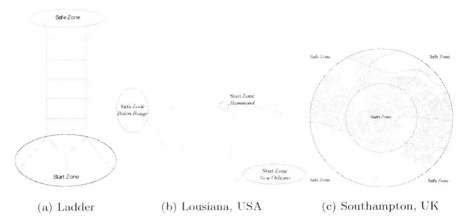

(a) Ladder   (b) Lousiana, USA   (c) Southampton, UK

Fig. 1 The three road networks on which the algorithms are evaluated.

roads to reach a safe point. The ladder is used to demonstrate occurrence of drivers desiring to be with others.
- A road network crafted from the roads in Lousiana, USA, shown in Figure 1(b). The map is generated from OpenStreetMap data for Lousiana, but only including the major evacuation routes. Drivers evacuate from New Orleans and Hammond, through the city of Baton Rouge, similar to the actual routes taken by residents during the 2005 Hurricane Katrina.
- A network that represents the full road network of the Southampton, UK area, shown in Figure 1(c), generated from OpenStreetMap data. Using this road network the performance of the algorithm over a large-scale area can be observed.

Using these road networks, the evaluation compares three different algorithms,

- **Route choice forces model.** A number of runs of our route choice forces model with different coefficient weightings for the forces. Two forces act upon the agent: shortest route time and driver attraction.
- **Shortest time.** A behaviour algorithm which has the simplistic behaviour of finding a quickest path tree to the exit points, with a preference of using major roads. This algorithm is the same as the one used in PARAMICS for "unfamiliar" drivers [11].
- **Real-time rerouting.** A dynamically re-routing algorithm which at regular intervals re-calculates the quickest path trees to take into account the delays caused by congestion. This algorithm is the same as the one used in PARAMICS for "familiar" drivers [11].

We first consider the results on the ladder road network. Figure 2 shows the road network usages and Figure 3 plots the metric of similarity for evacuating routes, both after 200 seconds of simulation time for the three different

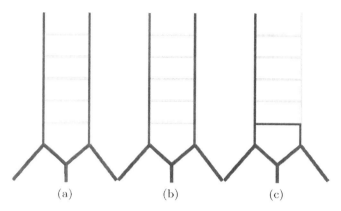

Fig. 2 Road network usage for the ladder road network after 200 seconds when using (a) shortest time, (b) real-time rerouting and (c) route choice forces model. Road usage is represented by the thickness of the line.

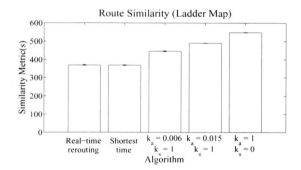

Fig. 3 Similarity in routes when using different route choice algorithms.

algorithms. The route choice forces model is used with coefficient values of $k_s = 0$ for the shortest time force and $k_a = 1$ for the driver attraction force. As show in Figure 2, when using the shortest time or real-time re-routing algorithms drivers symmetrically use both legs of the ladder. However, when using our model traffic asymmetrically uses only one of the legs of the road network to escape, demonstrating the occurrence of driver desires to be with others. Figure 3 shows our algorithm with three different coefficients being used, as well as the two algorithms we are benchmarking against. When comparing our model to the shortest time algorithm, the metric shows that driver routes have become more similar by up to 49%. Varying the coefficients is shown to give a degree of control over the driver's level of panic causing the attraction behaviour, from usual to evacuation situations.

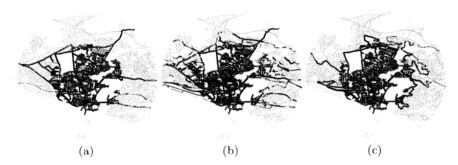

(a)   (b)   (c)

Fig. 4 Road network usage for the Southampton, UK network after 300 seconds when using (a) shortest time, (b) real-time rerouting and (c) route choice forces model. Road usage is represented by the thickness of the line.

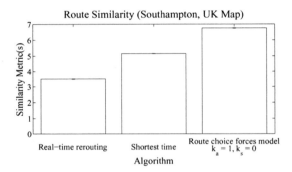

Fig. 5 Similarity in routes when using different route choice algorithms.

The road network usage maps for Southampton and Lousiana are shown in Figure 4 and Figure 6 respectively. These graphs show the road usage after 300 seconds and 1000 seconds respectively, the same point at which the metric is calculated using the routes of the escaped drivers. Figure 4 shows the differences in behaviour when using the different algorithms can be seen directly. When agents are using the shortest time algorithm traffic takes 6 routes out of the town. When using a real-time re-routing algorithm this number increases to around 13 and routes are more distributed around the road network. When our model is used, the number of routes to the exit drops to 3. As Figure 5 shows, the use of our model over the shortest time model gives an increase in the metric of 31% and over the real-time re-routing model of 93%. Considering the Lousiana road network, Figure 7 shows that, using our model, the metric of route similarity is increased by 21% over the shortest time algorithm and 28% over using real-time rerouting. From Figure 6(c) it can be observed that using our model usage of the roads into the north-west

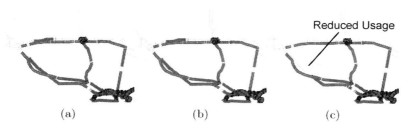

(a)   (b)   (c)

Fig. 6 Road network usage for the Louisiana, USA road network after 1000 seconds when using, (a) shortest time, (b) real-time rerouting and (c) route choice forces model. Road usage is represented by the thickness of the line.

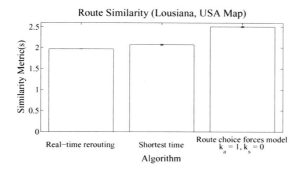

Fig. 7 Similarity in routes when using different route choice algorithms.

corner has decreased from two to one, so that only the southern road is used. This pattern of road usage is the same as was observed during the evacuation of Lousiana, USA during Hurricane Katrina [12].

5 Conclusion

We have presented a route choice model which represents influences on a driver's behaviour as a variety of forces or factors. Within the evacuation context, two factors have been defined: desire to take the quickest route and desire to be with others. We have shown that this model can be used to replicate independent driver behaviours seen in evacuation situations, including those seen in the 2005 evacuation of Lousiana, USA during Hurricane Katrina. Empirical evaluations using a metric of route similarity and the road networks of an abstract "ladder", Lousiana, USA and Southampton, UK, showed that the route choice forces model gives a 21%-49% increase in route similarity over the shortest path algorithm and 28%-93% increase over the real-time rerouting algorithm.

Within the context of the BAE SYSTEMS simulator, future work includes the expansion of the model to include forces relevant to other scenarios, including driver responses to directly observing the routes of others, route planning within unfamiliar environments, variable driver knowledge of routes and driver compliance behaviour to real-time information.

Acknowledgements

This work is jointly funded by BAE SYSTEMS and the EPSRC.

References

1. Bishop, C.: Neural networks for pattern recognition. Oxford University Press, USA (1995)
2. Cetin, N., Nagel, K.: A large-scale agent-based traffic microsimulation based on queue model. In: Proceedings of Swiss Transport Research Conference, Monte Verita, CH (2003)
3. Chen, X., Zhan, F.: Agent-based modelling and simulation of urban evacuation: Relative effectiveness of simultaneous and staged evacuation strategies. Journal of the Operational Research Society 59(1), 25–33 (2006)
4. Church, R., Sexton, R.: Modeling small area evacuation: Can existing transportation infrastructure impede public safety? Technical report, California Department of Transportation (2002)
5. Dow, K., Cutter, S.: Emerging hurricane evacuation issues: Hurricane Floyd and South Carolina. Natural Hazards Review 3(1), 12–18 (2002)
6. Helbing, D., Farkas, I., Vicsek, T.: Simulating dynamical features of escape panic. Nature 407(6803), 487–490 (2000)
7. Henein, C.M., White, T.: Agent-based modelling of forces in crowds. In: Davidsson, P., Logan, B., Takadama, K. (eds.) MABS 2004. LNCS (LNAI), vol. 3415, pp. 173–184. Springer, Heidelberg (2005)
8. Kirchner, A., Schadschneider, A.: Simulation of evacuation processes using a bionics-inspired cellular automaton model for pedestrian dynamics. Physica A: Statistical Mechanics and its Applications 312(1-2), 260–276 (2002)
9. Lämmel, G., Rieser, M., Nagel, K.: Large scale microscopic evacuation simulation. In: Proceedings of the 4th International Conference Pedestrian and Evacuation Dynamics. Springer, Wuppertal (2008)
10. Lin, Q., Ji, Q., Gong, S.: A crowd evacuation system in emergency situation based on dynamics model. In: Zha, H., Pan, Z., Thwaites, H., Addison, A., Maurizio, F. (eds.) VSMM 2006. LNCS, vol. 4270, pp. 269–280. Springer, Heidelberg (2006)
11. Quadstone Limited. Quadstone Paramics V5.0 Technical Notes (October 2004)
12. Wolshon, B.: Empirical characterization of mass evacuation traffic flow. Transportation Research Record: Journal of the Transportation Research Board 2041, 38–48 (2008)

Agent-Based Integrated Decision Making for Autonomous Vehicles in Urban Traffic

Maksims Fiosins*, Jelena Fiosina, Jörg P. Müller, and Jana Görmer*

Abstract. We present an approach for integrated decision making of vehicle agents in urban traffic systems. The planning process for a vehicle agent is broken down into two stages: strategic planning for selection of the optimal route and tactical planning for passing the current street in the most optimal manner. Vehicle routing is considered as a stochastic shortest path problem with imperfect knowledge about network conditions. Tactical planning is considered as a problem of collaborative learning with neighbor vehicles. We present planning algorithms for both stages and demonstrate interconnections between them; as well, an example illustrates how the proposed approach may reduce travel time of vehicle agents in urban traffic.

1 Introduction

The application of multi-agent modeling and simulation to traffic management and control problems becomes more relevant as intelligent assistant functions and car-to-X communication pave the way to a new generation of intelligent networked traffic infrastructure. Typically traffic environments are regulated in a centralized manner using traffic lights, traffic signs and other control elements. Multi-agent traffic systems are modeled with autonomous participants (vehicles), which intend to reach their goals (destinations) and act individually according to their own interests.

Previous research in this area has mostly concentrated on traffic lights regulation methods, traffic lights agent architecture, coordination and decision making mechanisms ([5]). Multi-agent reinforcement learning (MARL) for coordination of traffic lights was applied by Bazzan, Lauer and others ([3], [4]).

Maksims Fiosins · Jelena Fiosina · Jörg Müller · Jana Görmer
Clausthal University of Technology, Julius-Albert Str. 4,
D-38678 Clausthal-Zellerfeld, Germany
e-mail: {maksims.fiosins,joerg.mueller,
       jana.goermer}@tu-clausthal.de

* Supported by the Lower Saxony University of Technology (NTH) project "Planning and Decision Making for Autonomous Actors in Traffic" (PLANETS).

Y. Demazeau et al. (Eds.): Adv. on Prac. Appl. of Agents and Mult. Sys., AISC 88, pp. 173–178.
springerlink.com   © Springer-Verlag Berlin Heidelberg 2011

In contrast, there is less research on individual driver behavior and architectures of "intelligent vehicle" agents: existing research is mostly focused on mesoscopic models for travel demand planning [2] or adaptive cruise control [6].

We consider a structure of decision making of a vehicle agent in an urban traffic environment. A vehicle environment is presented as a directed graph $G = (V, E)$, where nodes and edges represent intersections and streets correspondingly. Denote $N(e_i) \subset E$ a set of edges, which start from the node, where the edge e_i ends. We consider the discrete linear model time $t \in 0, 1, 2, \ldots$.

We suppose that each vehicle j at any time t is located on some edge $e^j(t) \in E$. A relative position of the vehicle j on the edge $e^j(t)$ at time t is defined as a distance to the end of the edge $x^j(t) \in 0, \ldots, d(e^j(t)) - 1$. Let $l^j(t) \in 1, \ldots, l(e^j(t))$ be a lane, $v^j(t)$ be a speed of the vehicle j at time t, $tl^j(t)$ be a time inside a traffic light cycle at the end of the edge $e^j(t)$. The state $s^j(t)$ of the vehicle j is defined by a tuple

$$s^j(t) = < e^j(t), x^j(t), l^j(t), v^j(t), tl^j(t) > . \tag{1}$$

The goal of a vehicle is to reach its destination as quickly as possible.

The planning process for a vehicle agent is broken down into two stages: strategic planning (SP) for selection of the optimal route and tactical planning (TP) for passing the current street in the most optimal manner.

During SP, vehicle agents determine the optimal strategic policy $\pi_{str}^{*j}(e^j(t), l^j(t)) \in N(e^j(t))$, which gives the next edge in the fastest path after the edge $e^j(t)$. Vehicles plan their routes individually, based on historical and actual information about edge travel times, applying a modification of the Stochastic Shortest Path Problem.

During TP, vehicle agents plan their operative decisions together with other agents. Vehicles on one edge plan their actions $a = < \Delta v^j, \Delta l^j >$, where Δv^j is a speed change, $\Delta l^j \in \{-1, 0, 1\}$ is a lane change, in order to minimize travel time of the whole group by applying DEC-MARL to learn the optimal tactical policy π_{tact}^{*j}.

The integrated policy of a vehicle j consists of the strategic and tactical policy $\pi^{*j} = < \pi_{str}^{*j}, \pi_{tact}^{*j} >$.

The paper is organized as follows. In Section 2 we consider underlying planning algorithms: Section 2.1 describes SP, Section 2.2 TP. In Section 3 we provide first experimental results. Section 4 concludes the paper and suggests future work.

2   Planning for the Vehicle Agent

2.1   Strategic Planning

In this section we present the method for SP of a vehicle agent. We modify the algorithm R-SSPPR [7] for calculation of the Stochastic Shortest Path with imperfect information. Agents make their SP individually, without cooperation.

Let T_i^j be dependent random travel times of the agent j through the edges $e_i \in E$, $T^j = \{T_1^j, T_2^j, \ldots, T_{n_e}^j\}$. We assume that the distribution of T^j is unknown, only a sample of travel time realizations $X = \{X_1, X_2, \ldots, X_k\}$ is available, where $X_i = \{X_{i,j}\}, j = 1, \ldots, n_e$ is a set of travel time values for all edges for the i-th historical realization, p_i^j is a probability that i-th realization takes place for the agent j.

Let $I^j(t)$ be the information, available to the vehicle j at time t, which consists of known travel times; it is a set of events $I^j(t) = \bigcup_{e_i \in E_{kn}^j(t)} \{T_i^j = t_i^j\}$, where $E_{kn}^j(t)$ is a set of edges, which travel times are known.

The information $I^j(t)$ is regularly supplemented. Suppose that the travel time at the edge $e_u \in E$ becomes known to the vehicle j at time τ. It can calculate the posterior conditional probabilities $P\{T^j = X_v\}, v = 1, \ldots, k$ by using Bayes' formula:

$$P\{T^j = X_v | I^j(\tau), I^j(\tau')\} = \frac{1}{Z} P\{T^j = X_v | I^j(\tau')\} P\{I^j(\tau) | T^j = X_v, I^j(\tau')\}. \quad (2)$$

where Z is a normalizing constant, ensuring that the sum of all posterior probabilities is equal to 1, $I^j(\tau) = I^j(\tau') \cup \{T_u^j = t_u^j\}$.

Let us denote $\pi_{str}^j(e^j(t), I^j(t)) \in N(e^j(t))$ a decision rule about the edge after $e^j(t)$ for the agent j in its path. For its calculation, we use dynamic programming in this stochastic case with imperfect information. Denote $V_\pi^j(e_i, I^j(t))$ an expected travel time of the vehicle j from the beginning of the edge e_i to the destination edge e_i^d under the decision rule π. The following recurrent equation is true:

$$V_\pi^j(e_i, I^j(t)) = \begin{cases} t_i^j, & \text{if } e_i = e_j^d, \\ t_i^j + E_{\tilde{I}^j}[V(\pi^j(e_i, I^j(t)), \tilde{I}^j)] & \text{otherwise.} \end{cases} \quad (3)$$

where the expectation is taken over all possible future information sets $\tilde{I}^j$.

Then we need to minimize (3) over all possible next edges for calculation of the optimal $V^{*j}(e_i, I^j(t))$ and the optimal policy $\pi_{str}^{*j}(e_i, I^j(t))$.

However, there is some difficulty in calculation of the expectation $E_{\tilde{I}^j}$ over all possible future information sets $\tilde{I}^j$. For this purpose, we need to consider all possible travel times of the edges $e_i \notin E_{kn}^j(t)$. In order to avoid this difficulty, we use the resampling of future values of travel times ([1]). We go with the probability $2^{-\zeta}$ to ζ steps forward and extract according to the probabilities $P\{T^j = X_i | I^j(t)\}$ one value $X_{w,u}$ for the travel time T_u^j, which is added to the set $\tilde{I}^j$. Then the probabilities $P\{T^j = X_i | I^j(t), \tilde{I}^j\}$ are updated according to (2). This procedure is repeated r times, and an average is accepted as the expectation $E_{\tilde{I}^j} V^{*j}(e_k, \tilde{I}^j)$.

SP consists of two stages: pre-planning and routing. During pre-planning, values $V^{*j}(e_i, I^j(t))$ and $\pi_{str}^{*j}(e_i, I^j(t))$ are calculated for all edges and all possible information sets. During routing, the policy $\pi_{str}^{*j}(e_i, I^j(t))$ is used for the optimal routing. We summarize all above mentioned in algorithm 1.

Algorithm 1. Pre-planning stage of the strategic planning

```
 1: e_i ← e_d^j
 2: while prev(e_i) ≠ ∅ do
 3:     I' ← ∪_{u∈N(e_i)} I(e_u)
 4:     for all ζ ∈ 1,…,k do
 5:         I'' ← I' ∪ {T_i^j = X_{ζ,i}}
 6:         for all v ∈ 1,…,k do
 7:             P{T^j = X_v|I''} ← P{T^j = X_v|I'}P{I''|T^j = X_v,I'}
 8:         end for
 9:         for all e_u ∈ N(e_i) do
10:             for all η ∈ 1,…,r do
11:                 V^{(η)}(e_u,I'') ← RESAMPLE(I'')
12:             end for
13:         end for
14:         V*(e_i,I'') ← X_{ζ,i} + min_{e_u∈N(e_i)} avg_{η∈1,…,r} V^{(η)}(e_u,I'')
15:         π*(e_i,I'') ← argmin_{e_u∈N(e_i)} avg_{η∈1,…,r} V^{(η)}(e_u,I'')
16:     end for
17:     e_i ← prev(e_i)
18: end while
```

2.2 Tactical Planning

According to SP, a vehicle enters some edge together with other vehicles. Its TP allows sharing an edge with other vehicles by selecting appropriate speed and lane changes to pass through the edge as quickly as possible.

A state of the vehicle $s^j(t) \in S$ is described by a tuple (1). Vehicle actions consist of pairs $a = <\Delta v, \Delta l> \in A$, which correspond to speed and lane change. So

$$
s^j(t+1) = \begin{cases} < e^j(t), x^j(t) - v^j(t), l^j(t) + \Delta l, v^j(t) + \Delta v, tl^j(t+1) >, \\ \qquad\qquad\qquad\qquad \text{if } x^j(t) - v^j(t) > 0, \\ < \pi_{str}^{*j}(e^j(t)), x^j(t) - v^j(t) + d(e^j(t)), l^j(t) + \Delta l, v^j(t) + \Delta v, \\ \qquad\qquad\qquad\qquad tl^j(t+1) > \text{otherwise.} \end{cases}
\tag{4}
$$

Note that the second situation is only possible if the traffic light signal for the desired direction is green, so $tl^j(t) \in G^j(\pi_{str}^{*j}(e^j(t)))$, where $G^j(e_i) \subset \{0,…,c^j\}$ is a subset of traffic light cycle times, which gives a green signal from $e^j(t)$ to $e_i \in N(e^j(t))$.

We assume that for each state $s^j(t) \in S$ a corresponding reward $r(s^j)$ is available. We further assume that the reward structure is fully additive:

$$
r(s^j(t)) = r^x(x^j(t)) + r^l(l^j(t)) + r^v(v^j(t)) + r^{tl}(tl(t)).
\tag{5}
$$

The position part $r^x(\cdot)$ has bigger values at the end of the edge; the lane part $r^l(\cdot)$ has bigger values for the lane, which has a turn to the next edge in the vehicle route; the speed part $r^v(\cdot)$ has bigger values for bigger speeds; the traffic light part $r^{tl}(\cdot)$ has a big negative value for $tl^j(t) \notin G^j(\pi_{str}^{*j}(e^j(t)))$ and small $x^j(t)$.

Let us apply DEC-MARL algorithm [4] for solving the cooperative task of multiple agents. Let $gr_i(t)$ be a set of agents indices, which are located at edge $e_i \in E$ at time t. Let $s_{gr} \in S^{|gr_i(t)|}$ be a joint state, which includes states of all such agents. A local state-action value function $\tilde{Q}^j(s_{gr}, a^j)$ depends on the action of the j-th agent and joint state s_{gr} of the group. $\tilde{Q}^j(s_{gr}, a^j)$ is updated to ensure a maximum of joint-action Q-functions [4]. A learning procedure is given in algorithm 2.

Algorithm 2. Multi-agent tactical learning algorithm at the edge $e_i \in E$

1: **while** not end of the simulation **do**
2:    **for all** $j \in gr_i(t)$ **do**
3:       observe reward $r^j(s'_{gr})$
4:       $a'^j \leftarrow \tilde{\pi}^{*j}_{tact}(s'_{gr})$, take action a'^j, observe next state s''^j
5:    **end for**
6:    $s''_{gr} \leftarrow \{s''^j\}, \ j \in gr_i(t)$
7:    **for all** $j \in gr_i(t)$ **do**
8:       $\tilde{Q}'^j(s_{gr}, a^j) = \max\{\tilde{Q}^j(s_{gr}, a^j), r(s'_{gr}) + \gamma \max_{a'^j} \tilde{Q}^j(s'_{gr}, a'^j)\}$
9:       $\tilde{\pi}^{*j}_{tact}(s_{gr}) = \begin{cases} a'^j, & \text{if } \max_{a^j} \tilde{Q}'^j(s_{gr}, a^j) > \max_{a^j} Q^j(s_{gr}, a^j) \\ \tilde{\pi}^{*j}_{tact}(s_{gr}) & \text{otherwise} \end{cases}$
10:      $Q(s, a_i) \leftarrow Q'(s, a_i), \ a^j \leftarrow a'^j$
11:    **end for**
12:    $s_{gr} \leftarrow s'_{gr}, \ s'_{gr} \leftarrow s''_{gr}$
13: **end while**

---

3 Experiments and Results

We simulate a traffic network in Hannover, Germany in AimSun, a specialized simulation software for traffic applications. The road network is shown in the Fig. 1.

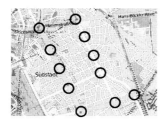

Fig. 1 Street network

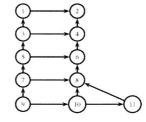

Fig. 2 Considered graph

All intersections are regulated by traffic lights with fixed control plans, known to vehicles. We use the realistic traffic flows, collected in given region of Hannover in morning rush hours. There are traffic flows in all directions; we are interested in the flow $9 \rightarrow 2$; these vehicles use the graph, shown on the Fig. 2, for their decisions.

In our model, we divide each street to cells of 4 m length. The possible speeds of vehicles are: $\{0,5,\ldots,50\}$km/h. One simulation step corresponds to 1/2 sec.

Experimental results are summarized in Table 1. We calculate travel times depending on the ratio to the flows in Hannover in morning hours.

Table 1 Travel times on the route $9 \rightarrow 2$ (sec.) depending on the flows ratio

Planning usage \ Flows ratio	0.5	0.7	0.9	1.0	1.2
Without planning	241.1	256.2	401.4	567.4	934.4
With strategic planning	231.2	242.4	378.1	528.2	856.9
With tactical planning	231.9	246.2	385.0	543.7	870.8
With strategic and tactical planning	225.7	236.1	367.7	518.6	818.5

We conclude that the application of integrated planning allows reducing the travel time of vehicles to about 10%; this is more than SP or TP separately.

4 Conclusions

We proposed an integrated planning process for vehicle agents, which includes both SP and TP. For SP, we showed how to apply existing information for effective solving of a routing problem under imperfect information. For TP we used a modification of DEC-MARL, which allows vehicles to collaborate inside a road segment in order to traverse it in the most quick way. First experiments show that the demonstrated approach allows to reduce the travel time of vehicles.

The proposed approach is used to simulate a larger traffic network in Hannover. In future we will work on an integration of the approach with centralized regulations from TMC, as well as more dynamic agents group formation.

References

1. Andronov, A., Fioshina, H., Fioshin, M.: Statistical estimation for a failure model with damage accumulation in a case of small samples. J. of Stat. Planning and Inf. 139, 1685–1692 (2009)
2. Balmer, M., Cetin, N., Nagel, K., Raney, B.: Towards truly agent-based traffic and mobility simulations. In: Proceedings of AAMAS 2004, pp. 60–67 (2004)
3. Bazzan, A.L., de Oliveira, D., da Silva, B.C.: Learning in groups of traffic signals. Engineering Applications of Artificial Intelligence 23(4), 560–568 (2010)
4. Lauer, M., Riedmiller, M.: An algorithm for distributed reinforcement learning in cooperative multi-agent systems. In: Proc. of 17-th Int. Conf. on Machine Learning, pp. 535–542 (2000)
5. Masterton, T., Topiwala, D.: Multi-agent traffic light optimisation and coordination. In: Thales Research and Technology: White Papers, vol. 2 (2008)
6. Naranjo, J.E., Sotelo, M.A., Gonzalez, C., Garcia, R., de Pedro, T.: Using fuzzy logic in automated vehicle control. IEEE Intelligent Systems 22, 36–45 (2007)
7. Polychronopoulos, G.H., Tsitsiklis, J.N.: Stochastic shortest path problems with recourse. Networks 27, 133–143 (1996)

Towards Urban Traffic Regulation Using a Multi-Agent System

Neïla Bhouri, Flavien Balbo, and Suzanne Pinson

Abstract. This paper proposes a bimodal urban traffic control strategy based on a multi-agent model. We call bimodal traffic a traffic which takes into account private vehicles and public transport vehicles such as buses. The objective of this research is to improve global traffic and reduce the time spent by buses in traffic jams so that buses cope with their schedule. Reducing bus delays is done by studying time length of traffic lights and giving priority to buses, more precisely to buses running late. Regulation is obtained thanks to communication, collaboration and negotiation between the agents of the system. The implementation was done using the JADE platform. We tested our strategy on a small network of six junctions. The first results of the simulation are presented. They show that our MAS control strategy improves both bus traffic and private vehicle traffic, decreases bus delays and improves its regularity compared to a classical strategy called fixed-time control.

1 Introduction

To improve route times of public surface transportation (bus, tramways, shuttles, etc.), cities often use regulation systems at junctions that grant priority to vehicles. These systems are referred to systems equipped with bus priority. The aim of these strategies is to increase the average speed of all vehicles as well as public transport vehicles needed to cross a junction.

The use of these systems is efficient when traffic is light or when they are used to improve a single congested bus route. However, reducing the time of bus journey, although very important for operating a route, is not the primary factor considered by public transport operators whose obligation is to provide passengers services e.g. keeping interval between buses. To take into account the public transport vehicles

Neïla Bhouri · Flavien Balbo
IFSTTAR/GRETTIA, "Le Descartes 2" 2 rue de la Butte Verte,
93166 Noisy Le Grand Cedex
e-mail: neila.bhouri@inrets.fr

Flavien Balbo · Suzanne Pinson
University Paris-Dauphine - LAMSADE, Place du Marechal de Lattre de Tassigny F-75775
Paris 16 Cedex, France
e-mail: balbo@lamsade.dauphine.fr, Suzanne.Pinson@dauphine.fr

Y. Demazeau et al. (Eds.): Adv. on Prac. Appl. of Agents and Mult. Sys., AISC 88, pp. 179–188.
springerlink.com   © Springer-Verlag Berlin Heidelberg 2011

specificity, TRSS (Transportation Regulation Support Systems) were developed. TRSS systems follow a micro-regulation based approach i.e. an approach that models the behavior of each bus [1], [2], [7]. One of the weaknesses of these systems is that the private vehicle traffic flow is hardly taken into account. If it is taken into account, this is only as an external parameter that modify the route times of buses. Another weakness is that traffic light management which is one of the key factors of traffic jams and bus delays, is not included in the TRSS systems.

Our objective is to build a traffic control strategy for bi-modal traffic that is able to regulate both private vehicle traffic and public vehicle traffic. Classical control theory used to regulate bi-mode traffic (public and private vehicles) is confronted with the modeling problem. Traffic flow can be modeled at a macroscopic or at a microscopic level. Microscopic modeling is time-consuming, and it is therefore not well adapted to build real time control strategies for wide urban networks. Macroscopic modeling has been used in [4], [5]. However, macroscopic representation of buses does not allow more than an indirect consideration of the intervals. In these systems the objective was to reduce the time spent in traffic jams so that buses respect their schedule. In [11] a hybrid model was used: macroscopic modeling for private vehicles and microscopic for public transport. The complexity of the strategy shows the limits of these classical modeling approaches to build a bimodal traffic regulation strategy.

Multi-Agent modeling can be a suitable answer to this scaling problem. We note that multi-agent systems are increasingly present in the field of traffic regulation [1] [2] [12] and [13]. The problem of traffic lights coordination on the thoroughfares of the route network has been solved in [8], [9], [10] and [14]. The regulation system represented in [15] is related to traffic assignment using negotiation between vehicles and junctions. We already developed a first prototype that shows promising results [6].

The second section focuses on traffic regulation systems and describes our model: the network model and the identification of the agents with a detailed description of agents, their attributes, their objectives, as well as communication and collaboration protocols. The third section provides the first results of the simulation tests carried out on the Jade platform. Finally, we conclude in the fourth section.

2 Network Modeling

In our model, the urban network is represented by an oriented graph $G = (I, A)$. The nodes $\{I\}$ represent the junctions (or intersections) and the arcs $\{A\}$ represent the lanes that connect the junctions. Two intersections can be connected by one or several arcs depending on the number of lanes on the thoroughfare.

An arc corresponds to a lane. It is characterized by a set of static information (such as its length, its capacity, its saturation output which is the maximum output of exits from the given arc) and dynamic information i.e. the number of vehicles on the arc, the state of the traffic lights at the extremity of the arc, green or red, if the light is green, then the vehicles present on the arc can depart.

A junction is specified by the set E of arcs that enter it and S the set of arcs that leave from it. A junction is managed by a set of stages P. Each of the stages specifies the list of arcs for which the green light is awarded if the stage is active (see figure 1).

The network is used by a number of bus routes. Each route comprises the number of buses from the same origin and in the same direction, and that service a number of predefined commercial bus stops at regular time intervals. The time spent by a bus at a commercial stop is equal to the pre-set time for passengers to mount, plus additional time to regulate the interval, if required.

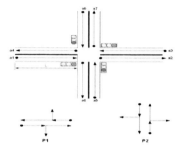

Fig. 1 Example of a junction with 4 arcs and two stages P={P1, P2}. P1 allows for the clearing of the arcs a1 and a3, because the entry flow a1 and a3 can leave the junction at the same green light period. Similarly, P2 clears arcs a5 and a8. The entries and exits of the junction are respectively E={a1, a3, a5, a8} and S={a2, a4, a7, a6}.

2.1 Agent Modeling

In order to identify agents and design the MAS, we present an abstraction of the real system; for every entity of the real world is associated an agent in the virtual world to form a Multi-Agent System (MAS). Homogenous agents are called "agent-type". The developed MAS is made up of the following agent-types:

Junction Agent (JA): is the key agent of our architecture. It is in charge of controlling a junction with traffic lights, and of developing a traffic signal plan. The junction agent modifies the planning of the lights according to data sent by approaching buses.

Stage Agent (SA): the traffic signal plan is elaborated thanks to the collaboration of the junction and stage agents. Each SA is expected to determine the optimal green light split to clear the waiting vehicles on the arcs concerned by the stage. Thus, whatever the complexity of the junction is (and its physical configuration), it is managed by a set of stage agents interacting with the junction agent in order to develop a plan of actions for the traffic lights.

Bus Agent (BA): represents a bus in the real world. It circulates from one arc to another, halts at commercial stops, halts at red lights and obeys the instructions of the bus route agent. The objective of each bus agent is to minimize the time spent at traffic lights (to minimize journey times).

Bus Route Agent (BRA): bus agent only provides a local view of their environment and, in particular, only the journey covered by the BA. Thus, local optimization carried out by bus agents can have a negative impact on the route, notably on its regularity (i.e. the formation of bus queues). To tackle this problem, we propose an agent who has a global view of the route agents, and who can control and modify their behavior in order to guarantee an efficient and regular service.

2.2 Description of Agent Behavior

Bus Agent (BA): In order to minimize the time spent at traffic lights the bus agent interacts with junction agents and its hierarchical superior agent (BRA). All the buses have to provide a regular service and avoid bus queues, in other words, the frequency of buses passing commercial stops must remain stable. To achieve this objective, the BA receives orders from the BRA (for example, stay at the stop for t seconds, if the bus is ahead with respect to the position of the preceding bus).

The BA is composed of a data module, which represents its internal state, and a communication module, that enables exchange with other system agents.

Behavior of a bus agent: Let t_0 be the entering time of the bus agent which behaves in the following way:

- On entering arc i, the BA retrieves information from the arc (the number of vehicles that precede it, the length, capacity, and exit output of the arc). By using these data, the BA calculates a time-space request that is transmitted to the JA in order to prevent an eventual stop at the red light at the following junction. The JA then attempts to satisfy the demand (see junction agent below);
- When approaching a stop, the BA informs the associated BRA. The bus route agent then calculates the duration of the regulation interval and its level of priority and sends it to the bus. The bus must wait during the passenger loading time, as well as the potential regulation time, before leaving the stop.

Calculation of a green light request. This calculation is specified by the interval of time during which the green light is granted to the actual arc so that the bus can pass without stopping at the next junction. Let R be the requested interval: $R = [t_b, t_e]$, with t_b and t_e be the beginning and ending times of the request interval respectively. The calculation of these times is carried out as follows: the bus enters the arc and finds N_v vehicles ahead of it, the vehicles move to the traffic lights lane to wait for the green light thus forming a queue of length F. In order to continue along its route, the queue of vehicles has to be dispersed before it arrives. The green light should thus be granted at the arc at the instant: $t_b = t_o + T - T_F$ with T be the time necessary for the bus to cover the distance *between the beginning of the arc and the end of* the queue, and T_F be the time necessary to disperse the queue F.

This request interval R together with other information (id-number of the bus, its priority, the actual arc of the bus, the next arc to be traveled by the bus) are sent to the JA (at the next junction) who attempts to modify the plan for the lights to satisfy the request.

Junction Agent (JA): The JA is the key agent of our architecture. The JA supervises the group of stage agents (SA) who collaborate together to establish a plan for traffic lights. This plan will, on one hand, maximize the capacity of the junction and, on the other hand, attempt to satisfy, as far as possible, the request interval of the bus. The JA is characterized by static and dynamic data.

The *static data* represent the constraints that characterized the JA. It contains the maximum value of the traffic light cycle (120 seconds). For each cycle, there is an interval of lost time i.e. the period of orange or all red. The all red light is a period during which all the arcs from the same junction have a red light in order to clear the centre of the junction and thus prevent accidents. This fixed period, in conformity with the architecture of the junction, does not depend on the length of the cycle. It is fixed here to a two second period after each stage. It contains also the set of stages of the junction: $P = \{P_1, ..., P_m\}$. The set of stages represents the configuration of the junction (the permitted movements and turns). Determining the stages is a task executed offline by the traffic experts.

They are two types of *dynamic data*, the first is related to the traffic signal plan: it specifies the order of the stages as well as duration of each stage. The second is related to the list of received requests data from the bus agents: each request is specified as follows: $R = (P_i, t_b, t_e, Priority)$, where P_i is the stage that will allow the passage of the given bus, t_b the time when the bus is expected to arrive at the traffic light, t_e, the time when the rear of the bus leaves the arc, and finally 'Priority' is the level of bus priority defined by the bus route agent.

At the end of each cycle, the JA triggers the process of calculating the traffic signal plan for the given cycle. This plan determines the duration of the green light and the ranking of each stage. When the JA receives a request, it records it in the database. The JA then decides to accept or to refuse this request at time t_b. The modification of a traffic signal plan following a priority request by a bus is as follows: 1) Extension of a stage (delay or advance), without exceeding the maximal duration of a stage; 2) Introducing a new stage into the plan.

Calculation of a traffic signal plan. The plan is calculated through the collaboration of the junction agent (JA) and the Stage Agents (SAs). The JA plays the role of a manager in supervising the SAs that act as participants.

The JA begins by forming a group of collaborators called collab_group including the list of stage agents that needs to be managed. JA initializes the variables: $C = CycleMax$, and $t=0$. Variable C controls the size of the calculated cycle. JA sends a message to the stage agents to inform them of the protocol initiated to calculate the traffic light plan. JA sends a message *request* to the agents of the collab_group asking them for the time necessary to clear all the vehicles from their stages, beginning at instant t. Every agent, *i*, of the collab_group calculates its desired green light duration d_i and an index that measures the urgency I_i of the stage, and sends them to the manager. When the manager receives all the responses, the sum *d* of durations is calculated. If $d > C$ then the manager has to solve a conflict since the size of the cycle exceeds the maximum. Conflict is solved when d previously calculated becomes less or equal to C: $d \leq C$. The manager selects the most urgent stage, that is P_j, its duration is d_j. It sends an *accept* message to the stage agent in charge of operating this stage; It withdraws the corresponding stage agent from collab_group; It updates the variables $C=C-d_j$, $t=t+d_j$; finally JA sends request as long as collab_group is not empty.

Conflict resolution. When the sum of green light durations requested by stage agents exceeds the size of the accepted value of the cycle, the manager must restore this sum to the maximal value of the cycle. To achieve a Δt reduction, the manager negotiates with the JAs using a Contract Net Protocol. The cost of the offer is the number of buses penalized if the stage agent reduces its duration of Δt.

Stage Agent (SA): This agent has a collection of both static and dynamic data that represents its internal state. The *Static data* are the list of entry arcs, the list of arcs authorized to clear if the stage is active (or green)

Dynamic data are related to 1) the state of the stage: active or inactive; 2) the duration of green light attributed to the stage; 3) the starting time of stage execution. 'Active' means that the traffic lights controlling the arcs concerned by this stage are green. The vehicles are therefore authorized to depart.

Behavior of the stage agent. The SA participates in the calculation of the traffic signal plan, and is in charge of fixing the optimal duration of green light for the given stage. When the stage agent is asked about the desired duration of green light by the junction agent, this duration d_i and an index I_i that measures the urgency of the stage, are computed and transmitted to the junction agent. If the stage agent receives confirmation from the junction agent, the stage agent stops the process. If the stage agent receives a *cfp (call for propose)* with a cost c, it computes an offer and sends it to the junction agent.

Calculation of the desired duration of the green light
The optimal duration of green light is computed by the following formula:

$$T = \max_{i=1,...,m}\{T_i\} \qquad T_i = w_i \frac{N_i}{D_i} + (1 - w_i).\frac{N_i * L_i}{C_i * V_i}$$

where m is the number of entering arcs at this stage, T_i the time necessary to clear arc i, L_i the length (meter), V_i the average speed (meter/second), N_i the number of vehicles, D_i is the saturation flow and C_i the capacity of the arc i. The number of vehicles N_i and the capacity Ci are expressed in private car unit (pcu) which means that all vehicles on the arc are converted to their equivalent on private vehicles, for example a bus is 2.3 pcu, depending on its length, a truck can be 2, 3 or 4 pcu, etc. D_i is a traffic flow and hence is expressed in pcu/second.

$w_i = N_i/C_i \in [0,1]$ is a parameter that indicates the degree of congestion of the arc. When the arc is congested, $w_i = 1$, which means that only the first part of the equation is used.

Urgency index of a stage. To award priority to a bus, the urgency index of a stage j is defined by the fact that the higher the index, the greater the urgency of the stage:

$$I_j = \sum_{i=0}^{m}(e^{w_i} + e^{b_i})$$

with: w_i the parameter indicating the degree of congestion of arc I, b_i is the number of buses present on arc i, m is the number of arcs entering via stage j and e is the Euler constant in our example. We can note that if there are several buses on arc i (if $b_i > 1$), the term e^{bi} is dominant and therefore gives priority to stages with buses; if $b_i = 0$, the degree of congestion is then taken into account.

Bus Route Agent (BRA). The role of the *route agent* is to supervise bus agents so as to prevent a local level regulation and the creation of bus queues. In other words, this agent can modify the behavior of bus agents in two different ways: 1) *directly*: by keeping those buses, which are ahead in the plan compared to the preceding ones, at the bus stop for a certain period of time; 2) *indirectly*: by modifying bus priorities. This agent has a *global* view of the route it operates on, and can therefore detect bus queues and react to prevent queue formation.

Internal state of the route agent. The route agent encompasses the following data: 1) the set of arcs traveled by the bus on its route; 2) The set of stops on the route: for each stop, its position, and the distance separating it from the next stop; 3) The set of buses on the route; 4) The frequency of buses introduced onto the route. For two consecutive stops A_i and A_j, the route agent maintains the journey time $d_{i,j}$ of the last bus. This helps follow the bus journey and to calculate whether the bus is ahead or late compared to the bus immediately preceding it.

Behavior of the route agent. When a bus agent moves to a stop, the time t taken to cover the distance $L_{i,i-1}$ that separates the two stops A_i and A_{i-1}, is transmitted to the route agent. The route agent then compares t to the time ($d_{i,j}$) taken by the preceding bus and consequently decides whether the bus is ahead or late. The route agent computes the new priority of the bus agent as well as the length of time the bus should wait at the commercial stop if it is ahead [6].

3 Results

To test our bimodal control strategy, we developed a Multi-Agent System prototype on the JADE[1] platform (Java Agent Development Framework). JADE offers Java middleware based on a peer-to-peer architecture with the overall aim to provide a runtime support for agents.

We have tested the strategy on a small network of six intersections (figure 2):

▪ The distance between two adjacent junctions belongs to [200,400] meters.
▪ Each section comprises one or two lanes.
▪ The saturation flow, which is the maximum exit output of the arcs, is identical for each arc and equal 0.5 vehicle/second.
▪ At each entry onto the network, we have installed a source that generates vehicles at a frequency $F \in$ [4 s ... 10 s].
▪ Some of the junctions have two stages while others have three stages.
▪ Two buses enter the network. For Bus 1, the frequency of the generated buses is 80 seconds and 180 seconds for Bus 2.

---

[1] jade.tilab.com/

We have compared the developed MAS strategy to a fixed time strategy with 30 seconds for each stage. We have run the simulation with these two strategies and for half-hour simulation time.

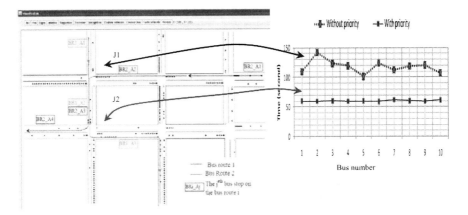

Fig. 2 The simulated network

Fig. 3 Buses travel time with and without bus priority

Figure 3 depicts buses travel time: the higher curve shows buses travel time between the stops BR1_A1 and BR1_A2 when buses do not request priority from the junction J1; the lower curve show buses travel time between the two bus stops BR1_A2 and BR1_A3 when buses are asking for priority at junction J2. We can note that buses travel time improves and becomes very regular when bus priority is taken into account.

Figure 4 gives the results of the two strategies for very heavy traffic conditions. Figure 4.a shows the recorded delays for buses with the two control strategies and figure 4.b shows the same kinds of curves for private vehicles.

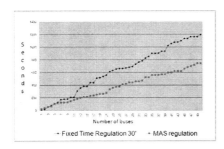

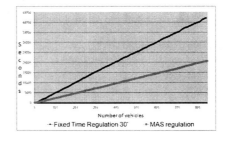

Fig. 4a Buses cumulated delays

Fig. 4b Private vehicles cumulated delays

These delays correspond to the sum of time lost by all buses (resp. vehicles) during stops at traffic lights. As shown on figures 4, the MAS strategy improves both traffic of buses and traffic of private vehicles. As we can see, there is a decrease of 38% on lost time spent by buses on traffic light; for the private vehicles, we got a decrease of about 51%. These results can be explained because giving priority to buses decreases traffic jams thus improving private vehicles traffic. However, more work should be done, especially studying Pareto front.

4 Conclusion

In this paper, we have developed a bimodal traffic control strategy based on a multi-agent system. It takes into account two transportation modes: public transportation i.e. buses and private vehicles. The originality of this strategy is the application of the new information and agent technologies. The entities representing the urban network can communicate among themselves and negotiate in order to solve traffic regulating problems. First, we have shown that classical methods of control systems of traffic regulation present several weaknesses: at a macroscopic level, they do not take into account mixed traffic and do not allow for the regulation of intervals between buses. Furthermore, computations at a microscopic level are time-consuming, especially for regulating large networks. Secondly, we presented the multi-agent strategy that computes traffic signals plans based on the actual traffic situation and on the priority needed by the buses. The priority is given to those buses that do not deteriorate the intervals between the vehicles on the same route. Thirdly, we ran a simulation prototype on the JADE platform. A comparison between buses travel time with and without bus priority shows the capacity of the priority method we developed to improve both the travel time and the regularity of buses. Our results also show that this bimodal MAS strategy improves conditions of global traffic and reduces bus delays. Additional work however is needed: a more realistic network should be defined in the simulation run and more validation and more testing should be undertaken with the definition of several indicators.

References

1. Balbo, F., Pinson, S.: Using intelligent agents for Transportation Regulation Support System design. Transportation Research Part C: Emerging Technologies 18(1), 140–156 (2010)
2. Balbo, F., Pinson, S.: Dynamic modeling of a disturbance in a multi-agent system for traffic regulation. Decision Support Systems 41(1), 131–146 (2010)
3. Ana, L.: Opportunities for multiagent systems and multiagent reinforcement learning in traffic control. Autonomous Agents and Multi-Agent Systems 18(3), 342–375 (2009)
4. Bhouri, N.: Constrained Optimal Control strategy for multimodal urban traffic network. In: IFAC Workshop on Control Applications of Optimization (CAO 2009), Finland (2009)

5. Bhouri, N., Lotito, P.: An intermodal traffic control strategy for private vehicle and public transport. In: 10th Euro Working Group on Transportation, Poznan, Poland (2005)
6. Bhouri, N., Haciane, S., Balbo, F.: A Multi-Agent System to Regulate Urban Traffic: Private Vehicles and Public Transport. In: 13th IEEE–ITSC, Portugal, pp. 1575–1581 (2010)
7. Cazenave, T., Balbo, F., Pinson, S.: Monte-Carlo Bus Regulation. In: 12th Int. IEEE ITSC 2009, St. Louis, MO, USA, pp. 340–345 (2009)
8. De Oliveira, D., Bazzan, A.L., Lesser, V.: Using Cooperative Mediation to Coordinate Traffic Lights: a case study. In: AAMAS 2005, New York, NY, USA, pp. 463–470 (2005)
9. Ferreira, E.D., Subrahmanian, E., Manstetten, D.: Intelligent agents in decentralized traffic control. In: IEEE Intelligent Transp. Systems Conf. IEEE-ITSC 2001, Oakland (CA), USA, August 25-29, pp. 705–709 (2001)
10. France, J., Ghorbani, A.: A multiagent system for optimizing urban traffic. In: Proceedings of the IEEE/WIC Inter. Conf. on IAT, pp. 411–414. IEEE Computer Society, Washington, DC (2003)
11. Kachroudi, S., Bhouri, N.: A multimodal traffic responsive strategy using particle swarm optimization. In: 12th IFAC Symposium, Redondo Beach, California, USA (2009)
12. Mailler, R., Lesser, V.: Solving distributed constraint optimization problems using cooperative mediation. In: AAMAS 2004, pp. 438–445. IEEE Computer Society, Los Alamitos (2004)
13. Mizuno, K., Fukui, Y., Nishihara, S.: Urban Traffic Signal Control Based on Distributed Constraint Satisfaction. In: Proceedings of the 41st International Conference on System Sciences, Hawaii, p. 65 (2008)
14. Roozemond, D.A.: Using intelligent agents for pro-active, real-time urban intersection control. European Journal for Operational Research 131, 293–301 (2001)
15. Vasirani, M., Ossowski, S.: A market-inspired approach to reservation-based urban road traffic management. In: AAMAS 2009, vol. 1, pp. 617–624 (2009)

Application of Holonic Approach for Transportation Modelling and Optimising

Jarosław Koźlak, Sebastian Pisarski, and Małgorzata Żabińska

Abstract. The goal of this work is to design an environment which makes it possible to build systems allowing flexible composition. Such systems may be used for modelling an organisation with a high level of complexity, with functionalities on different levels, realising heterogeneous activities. We use concepts and assumptions offered by multi–agent and holonic approaches. Theoretic representation of the abstract holon structure is given and a model for building a realisation for a domain, focusing especially on building and reorganising holons. The domain is the management of the transportation company and a delivery strategy for it. Selected results provided by the developed system are presented.

1 Introduction

Man-made systems (eg. organisations, firms, societies) very often have increasingly higher complexity levels and usually it is possible to decompose them into complex units with a certain well-defined functionality, which originate as a result of close collaboration of lower level unit groups. A multi-agent approach, especially the holonic approach, seems to be useful to describe such units.

As great significance is placed on economy in the transport sector, and that the related substantial cost, improvements of its organisation and limiting costs are of paramount importance, there have been many works devoted to problems of transport modeling and optimization. To enable solving problems close to real situations, it is necessary to take into consideration elements in the real world, which did not occur in the existing models and require the application of different algorithmic solutions. A very promising method to cope with this sort of complexity seems to be the holonic approach. The main goal is to create a holonic agent system solving the

Jarosław Koźlak · Sebastian Pisarski · Małgorzata Żabińska
Department of Computer Science, AGH University of Science and Technology,
AL. Mickiewicza 30, 30-059 Kraków
e-mail: kozlak@agh.edu.pl, seba_pis@interia.pl, zabinska@agh.edu.pl

Y. Demazeau et al. (Eds.): Adv. on Prac. Appl. of Agents and Mult. Sys., AISC 88, pp. 189–194.
springerlink.com   © Springer-Verlag Berlin Heidelberg 2011

transportation problem, independent - to a large degree – from a certain application domain, especially an algorithm of holon building.

2   State of the Art

A concept of holons was proposed by Koestler in [6]. Holons may be treated in two ways: as elements making a bigger whole, as well as elements which are built from smaller components. Holon components lose some of their autonomy when they connect into one holon of the higher level and make an object of new properties. In [3] there is a discussion of different types of management of holon groups (holarchy) dependent on autonomy of component holons. Starting from the highest stage of autonomy, they are: federation of autonomous agents (union of cooperating agents in a multi-agent system, which does not effect particularly autonomy of its members), moderated group (one agent becomes a representative head of a holon).

Created holon-based systems concern many domains, such as management of an automatic factory management of cooperative works controlling a group of robot-football players applying additionally techniques of immunology systems, composing of web services, or management of a firm realising transportation tasks [2].

Transportation problems such as PDPTW (Pickup and Delivery Problem with Time Windows) consists in serving a set of transport commissions by a fleet of vehicles. Each commission is described by a location of pick up and delivery and by the time windows defining the time range of due pick up and delivery. Moreover, the size of the transported load is connected with a commission and a vehicle is described by a capacity which defines the maximal acceptable size of transported goods. The quality of a solution is usually described by a number of used vehicles and a total distance. Due to the computational complexity of the problem defined in such a way, heuristic solutions are applied [8].

The other group of solutions comprises agent systems, for example the holonic system Teletruck [2]. They are based on allocation of commissions to vehicles by protocol Contract Net [9] and use of the optimising algorithm simulated trading [1].

3   Holonic System Model

3.1   Domain-Independent Holon Organisation

The main element of the system is an agent–holon. An abstract holon i is described as a following n–tuple $AH_i = (G_i, P_i, S_i, A_i, K_i, C_i, HO_i)$, where

- G_i – a goal function, described by an expression to be maximised or minimised,
- P_i – a plan, a sequence of actions leading to the realisation of the goal,
- S_i – a current state of the holon, e.g resources owned,
- A_i – set of attributes and their values, describing the features of the given entity,

- K_i – knowledge, a set of statements usually about the state and attributes of the other entities or previous and predicted state and attribute configurations concerning the given entity or other entities,
- C_i - a set of actions to be performed by every holon such as joining, splitting, sending a proposition for joining, analysis of possible reorganisation measures,
- HO_i – holon organisation.

Holon organisation HO_i describes relations of the given holon with other ones and preferences concerning construction of new holons:

$$HO_i = (HR_i, HC_i, HI_i, JP_i, SP_i) \qquad (1)$$

- HR_i – information about other holons how valuable they are as candidates for creating a super holon, and whether the holons are suitable to make a holon together,
- HC_i – holons which are included by a given holon,
- HI_i – super holon, in which a given holon is included,
- JP_i – joining protocol, the steps how the super–holons may be formed,
- SP_i - protocol of holon separation.

Description of the joining operation. The new goal function (G_k) is obtained which is calculated from the goal functions of constituent holons, for example, it may calculate a weighted sum of the input functions. After a joining operation, a new plan (P_k) is constructed, which evaluates the results of possible actions and their influence on the value of the G_k function. A new holon receives a new set of attributes A_k, which is obtained as a result of the execution of the given θ operations on the attributes of input holons.

3.2   Model for the Transportation Domain

This model consists of an environment (a directed graph, the nodes represent locations such as depot, pickup and delivery locations or intersections and the arcs represent roads), a special dispatcher agent responsible for sending transportation requests, a truck holon and a complex transportation unit holons which come into existence after joining trucks, trailer and drivers.

Transportation Unit – is a complex holon which consists of several sub–holons, usually a driver, a truck and a trailer.

$$AH = (G_{TU}, P_{TU}, K_{TU}, S_{TU}, C_{TU}, HO_{TU}), \qquad (2)$$

G_{TU} – the goal function, a weighted sum of the goal functions of included holons,
P_{TU} – a sequence of operations of movements, pickup and delivery operations, holon splitting and joining,
$S_{TU}^i = \{Loc_{TU}^i, AlReqs^i, HolStruc^i, Res_{DRA}^i, Res_{TRA}^i, Res_{TKA}^i\}$, where Loc_{TU}^i, – location, $AlReqs^i$ – allocated requests, $HolStruc^i$ – information about holon structure, reserved drivers (Res_{DR}^i), trucks (Res_{TR}^i) and trailers (Res_{TK}^i),

A_{TU} – attributes such as fuel consumption or velocity,
K_{TU} - agent's knowledge $K_{TU}^i = \{Reqs^i, Env^i\}$ comprising $Reqs$ - set of requests,
Env - information about environment (transport network),
C_{TU} - actions of the TU:

$$A_{TU} = \{ReqPart^i, Move^i, PickUp^i, Delv^i, HolReorg^i\}, \tag{3}$$

where: $ReqPart$ - participation in auctions of transport requests, $Move$ – transfer between locations, $PickUp$ - loading, $Delv$ - unloading, $HolReorg^i$ – holon reorganisation.

3.3   Dynamics and Optimization Techniques

Holon structures as well as routes constructions are based on guaranteeing the most advantageous values of the goal function. Holons join then, when it causes an increase of the goal function. Similarly, the division of a holon takes place, when interested holon-components may obtain a proper increase of their goal function values. There are two cases to be distinguished, i.e. two work modes depending on a way of sending transportation requests (one by one or all of them together). The first mode is to send subsequent single commissions by Dispatcher to agent components. Then negotiations concern a single commission, and on the basis of their result, holons are created. Thus, the result depends on the order of commissions, whereas holons created as a reply to a single commission are not optimal. The second mode is the case when Dispatcher sends a group of commissions, a negotiation is carried out and as a result, a list of offers to create new holons originates.

It is possible to consider four types of negotiation procedures between agents:

- building of preference lists of the most promising partners for holon creation. On the basis of the knowledge about upcoming requests or their groups and anticipation of cost of realisation for existing or created configurations, the agent creates a list of potential partners, who enable the best service of these requests. The first in this list are configurations, which serve the most of commissions at the lowest cost.
- holon creation. Agents merge into holons when all interested agents are high enough in their preference lists.
- holon reorganisation. Holons try to replace the component holons such as it is profitable from the point of view of their cost function. The following negotiations consist of creation of preference lists and performance of an appropriately modified process of holon creation.
- route optimisation without reorganisation of holons. The optimization of routes is based on simulated trading [1]. The agents try to reject these commissions which substantially worsen their goal function. These agents, who may realise them with the lower cost, accept them and commissions are assigned to them for execution. Agents may try to reject each commission in a row, which increases cost in a given range, or the given number of the most disadvantageous commissions.

4 Realisation

The realised system is based on the architecture of its previous version, presented in [7, 4]. The system is made up of several modules: its main part makes the computational module, the simulation controller, the simulation visualiser, the transportation network generator, and the test generator. The agent platform JADE [5] was used.

5 Results

The performed tests use holon construction and optimisation algorithms for each incoming request separately or analyse a package of requests together.

As the consequence four modes of algorithm functions are presented ion figures: NPC – handling single requests, preference for the cost functions, NPNC – handling single requests, preference for the number of used vehicles, PC – handling a package of requests, preference for the cost function, PNC – handling a package of requests, preferences for the number of used vehicles. The optimisation algorithm may be executed after assigning the requests one time, two times or a maximum number of times (subsequently for every request in the holon with decreasing cost contribution).

The results presented in fig. 1 shows that the quality of solutions is usually better with the higher number of the execution of the optimisation algorithm, but there are also some random deviations of this rule. The worst results considering the total costs and number of vehicles were obtained for NPC configuration.

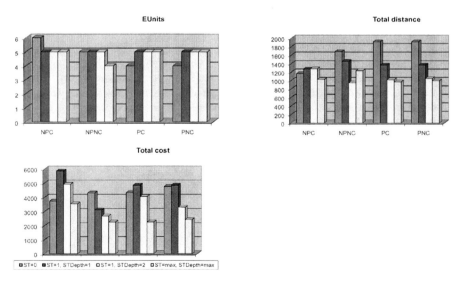

Fig. 1 Results (costs, vehicles, distance) for cluster configuration of requests, classical configuration of holons

6 Conclusion

A model of a flexible system for realisation of the transportation requests, based on the holonic approach, was presented. The features of the system, a holon organisation, composition algorithms and some selected results were described.

Future work will focus on extending the flexibility of the system thanks to the introduction of different kinds of holon construction and organisation algorithms. We will also introduce the handling of selected critical situations.

Acknowledgements. We would like to thank everybody who worked on the system, especially students from Dept. of Comp. Sci. AGH-UST M. Konieczny, M. Gołacki, M. Sieniek, P. Gurgul, P. Wójcik and M. Kuźniar. The work has been financed by a grant from the Polish Ministry of Science and Higher Education no. N N516 366236.

References

1. Bachem, A., Hochstattler, W., Malich, M.: The simulated trading heuristic for solving vehicle routing problems. Discrete Applied Mathematics 65, 47–72 (1996)
2. Burckert, H.J., Fischer, K., Vierke, G.: Transportation scheduling with holonic MAS - the TELETRUCK approach. In: Third International Conference on Practical Applications of Intelligent Agents and Multiagents, PAAM 1998 (1998)
3. Gerber, C., Siekmann, J., Vierke, G.: Holonic multi-agent systems. Tech. Rep. RR-99-03. Deutsches Forschungszentrum für Künstliche Intelligenz GmbH (1999)
4. Gołacki, M., Koźlak, J., Żabińska, M.: Holonic-based environment for solving transportation problems. In: Mařík, V., Strasser, T., Zoitl, A. (eds.) HoloMAS 2009. LNCS, vol. 5696, pp. 193–202. Springer, Heidelberg (2009)
5. Java Agent DEvelopment Framework, http://jade.tilab.com/
6. Koestler, A.: The Ghost in the Machine. Arkana Books (1989)
7. Konieczny, M., Koźlak, J., Żabińska, M.: Multi-agent crisis management in transport domain. In: Allen, G., Nabrzyski, J., Seidel, E., van Albada, G.D., Dongarra, J., Sloot, P.M.A. (eds.) ICCS 2009. LNCS, vol. 5545, pp. 855–864. Springer, Heidelberg (2009)
8. Li, H., Lim, A.: A Metaheuristic for the Pickup and Delivery Problem with Time Windows. In: Proceedings of 13th IEEE International Conference on Tools with Artificial Intelligence (ICTAI 2001), Dallas, USA (2001)
9. Smith, R.G.: The contract net protocol: high-level communication and control in a distributed problem solver. IEEE Transactions on Computer, 1104–1113 (1980)

A Multiagent System Approach to Grocery Shopping

Hongying Du and Michael N. Huhns

Abstract. We present an approach to social grocery shopping based on customers trading information about item prices and quantities in order for the customers to find the lowest prices for the goods they purchase and the most convenient plan for buying them. Because collecting and reporting prices is tedious, agents representing customers are needed to make this approach practical. Agents also have the potential to learn which other agents can be trusted. We use a realistic shopping list based on the U.S. Consumer Price Index in order to guarantee the realism of our results. By visiting actual grocery markets and comparing prices, we have discovered that the total cost of a list of groceries can vary by 13%. We have also discovered that by shopping optimally, that is, buying each item from the cheapest store, the result can be a savings of 16% over shopping at the store with the lowest total cost. To shop at minimum cost requires customers' agents to report prices to each other. If they do, each customer is likely to achieve at least a 10% savings. However, what if the reported prices are inaccurate? Would customers be worse off than if they just shopped randomly? We investigated the robustness of our multiagent shopping system in the presence of errors in reported prices. From this, we determine the potential savings an average customer might obtain.

1 Introduction

Aided by information systems for analyzing customer buying data, supermarket chains continually alter the prices of items to maximize their profits. They do this by, in essence, experimenting on their customers. For example, the price of an item might be raised at one store until customers stop buying it. This maximum price is then used at all of the stores in the chain. Customers however, do not have any comparable information systems that might aid them in price comparisons and are often at the mercy of the stores. Most stores do not post their prices online, so that customers have to visit each store to find the prices of groceries, which makes comparison shopping prohibitive.

Hongying Du · Michael N. Huhns
Department of Computer Science and Engineering, University of South Carolina,
Columbia, SC, 29208 USA

Y. Demazeau et al. (Eds.): Adv. on Prac. Appl. of Agents and Mult. Sys., AISC 88, pp. 195–200.
springerlink.com © Springer-Verlag Berlin Heidelberg 2011

Imagine an online system where customers could post the prices paid for groceries and where a prospective shopper could enter a grocery list and obtain a pointer to the store with the lowest total price. This would enable comparison shopping for groceries and would render the customer-to-store interactions fairer. It would also encourage stores to offer their true prices to avoid driving away potential customers. However, the effort required from the customers would be substantial. To make the effort reasonable and manageable, each customer could benefit from an agent that represented his/her interests and interacted with the agents of the other customers and, possibly, with store agents.

However, there is an expense in implementing and operating such a system. Moreover, its success is dependent on prices entered by other customers, on the availability of goods, and on prices that stores might change to yield an advantage for them to the disadvantage of customers. Hence, it is subject to errors and manipulation. To be feasible, the potential cost savings must substantially exceed the expense and effort of its implementation.

In this paper, we investigate the efficacy of a customer-oriented comparison-shopping system for groceries and the trade-offs in an implementation of it. Our approach is to use real data, normalize it according to typical customer actions, and simulate a system of stores and customers. We introduce both random and systematic (manipulation) errors into our simulation in order to evaluate its robustness. We provide a customer with the best combination of price and quality for a list of products available at different stores and recommend which store or stores would be optimal for shopping.

2 Background

Price comparison services (also known as comparison shopping services) allow people to query a product's prices at online stores. The services list the product's prices in all of the stores and sort the prices to provide customers with support for their online shopping. An intelligent software agent to implement comparison shopping is called a *shopbot* [1]. The first well-known shopbot, BargainFinder [2], provided comparison shopping for music CDs. It searched eight online music stores and displayed all prices on a webpage. Customers gained obvious benefit from BargainFinder and it has been used widely. Current shopbots have greater functionality than before by including information about shipping expenses, taxes, vendors' rates, and product reviews. The app RedLaser [3] accepts the barcode of a product from an iPhone's camera, searches many online stores, and shows their prices on the phone.

There are typically three steps for a shopbot to deal with data. First, it retrieves data from online stores or other shopbots, possibly by using an extraction method as in [4]. Second, the data is processed according to a user's command. Last, the results are shown to the user on a webpage. Other researchers are developing algorithms to improve the behavior of shopbots [5, 6] and make them more robust to changes in the stores' websites, such as by using Semantic Web concepts [7].

3 Analysis

There are a number of variables in grocery shopping. Our simulation of it uses five parameters: customer input, customer location, store location, item price, and item quantity. Customer input is a shopping list of the items a customer wants to buy and their quantities. Store location and customer location are used to calculate the fuel cost when driving to and from the stores. Item prices are those either reported by customers or by stores. We assume the quantity of a specific item in a store is either zero or infinity.

Our algorithm begins with a customer's shopping list of items and quantities. If the customer just goes to the stores with the lowest price for each item, the customer might need to go to many stores and spend more on fuel. So we search in all the stores and find the lowest and second lowest prices of each item on the list. If the prices of an item in several stores are the same, we consider the item with best quality first. We considered all possible combinations of the two prices and calculate the total cost as the sum of grocery cost and fuel cost. When calculating the fuel cost, we assume the customer's path is to go to the nearest store he needs to go to where he has not already shopped until he gets all the items. For comparison, we also calculate the cost if the customer chooses to go to stores using three other strategies: (1) choose one store randomly and buy all the items at that store, (2) go to the nearest store, or (3) randomly go to one of the five nearest stores. Then we calculate the ratio of the total grocery and fuel cost of these three methods over that of our optimal multiagent method.

Last, we evaluate robustness. There are two ways price information might be erroneous. First, if we rely on the stores themselves to report the prices, they might claim their prices are lower than they actually charge. Second, customers' agents can report prices by querying the RFID tags of the items the customers bought, but the agents might make mistakes when they acquire/report prices. We investigated both in our simulations.

4 Grocery Shopping Simulation

We used the Netlogo platform [12] for our grocery shopping simulation, which we separate into two phases. In the first phase, we simulate shopping according to fictitious prices and qualities generated randomly and examine the ratio of the cost of other methods over that of our method and evaluate the influence of different values for the parameters. For each combination of parameter values, we ran the simulation 100 times and used the mean of the 100 results. To consider deception, we assume that the deceptive stores say their prices are 10% lower than the real prices and the percentage of deceptive stores are 25%, 50% and 75% separately to see how the results change.

In the second phase, we use real prices collected from South Carolina stores to see whether there is a significant difference between using fictitious prices and real prices in the simulations. With real prices, the store location, item price, and item number are fixed. We assume here that an item's qualities are the same in all stores. However, quality information can be used in practice by allowing customers to rate it. For the customer input, we constructed a shopping list according to the U.S. Consumer Price Index (CPI). The CPI measures a price change for a constant market basket of goods and services from one period to the next within the same area (city, region, or nation) [13]. Along with the CPI, the relative importance of the items in the market basket is published. We created a realistic shopping list by selecting an item from each category according to its relative importance [14]. Since there are many categories, we selected a representative list of 33 items. For these, we collected item prices from 5 different stores. Table 1 shows a few of them. The complete price list can be found at [15]. We compared the savings when a customer went to two stores to buy goods than when the customer went to just one store. To measure robustness, we checked the results when there was a 10% possibility that the customers reported each digit of the prices wrong. When a desired item isn't available in a store, such as apples not available in store 4, we assume that the customer will go to another store to buy it.

Table 1 The shopping list

Item	Walmart store 0	Publix store 1	Food Lion store 2	BI-LO store 3	Target store 4
Tropicana: orange juice, 64oz	2.92	3.79	2.97	3.69	2.99
Simply Orange juice, 1.75l	3	3.79	2.99	3	2.99
Corona extra: 12oz*6	8.47	8.29	7.99	8.29	6.5
Budlight: 12oz*6	6.97	6.49	5.99	6.99	5.25
Totino's: pepperoni pizza, 10.2oz	1.25	1.49	1.67	1.67	1.2

5   Results and Discussion

In our simulation, we assumed there were 12 stores and 30 kinds of items. Given 10 items a customer wants to buy, we ran the simulation 100 times for a random change in one of the simulation parameters and calculated the mean, as shown below. We assumed the parameters are independent, so our simulation varied them one-at-a-time. The values in the table are the ratios of the total grocery and fuel cost using an alternative method over our optimal multiagent method.

The simulation results show that our approach to deciding at which stores to shop can save 21% or more, except when customers change their shopping lists. Our approach is better than the other 3 methods for all cases. When a shopping list changes, the savings are lower, possibly because the randomly generated customer input may contain fewer items. The consideration of fuel decreased the savings by only 2.4%, so we did not consider fuel cost for our results using real price data.

Table 2 Simulation results using randomly generated price data

	Mean Ratio of Shopping Method to Optimal Multiagent Method		
Simulation Parameters	Choose Store Randomly	Choose Nearest Store	Choose 1 Store Randomly from 5 Nearest
Vary Customer Location	1.2328	1.2365	1.2178
Vary Store Location	1.2351	1.2325	1.2269
Vary Item Price	1.2150	1.2180	1.2225
Vary Number of Items	1.2637	1.3317	1.2911
Vary Shopping List	1.1732	1.1080	1.1573

What if 25%, 50%, 75% stores are deceptive by claiming that their price is 10% lower than the real price, thereby luring customers to shop at the wrong stores? Our simulation chose deceptive stores randomly. The results were that a customer would save 5.2% less when 25% of the stores are deceptive and 9.1% less with 75% deceptive stores.

Using the real price data we collected (Table 1), the total cost of the goods on the shopping list if a customer shops at just one store varied from 114.27 from store 0 to 129.52 at store 1.

The cost of buying each item at its lowest price is 98.44, which is more than 13% lower than going to one store, but a customer would have to go to four stores to get this lowest price. Because a customer might not want to go to more than two stores, we tried all combinations of two stores and calculated the cost. The lowest cost of 106.58 is 6.7% lower than going to just one store.

What if the customers reported the price data wrong? We simulated this situation by giving each digit of a price a 9% possibility to change to another digit, each with a 1% possibility. When the price information is wrong, the only thing changed are the stores the customer would go to, because the customer would still pay the real price at the store. We ran the simulation 500 times and found there is a 2.2% possibility that the customer would go to another store due to wrong prices, rather than going to the store with the lowest price. The average cost is 114.57, which is very close to the optimum 114.27. For the results with two stores, there is a 37% possibility that a customer would go to stores other than the best combination of two stores. Though the possibility is significant, the average cost is 107.04, which is very close to 106.58, the lowest price possible for two stores. So on average, a customer can still save 6.3% by going to two stores compared to just one store, even if the prices are incorrect.

6 Conclusion

A societal grocery shopping system as described in this paper would be useful and practical, because it helps customers obtain a savings of 21% or more according to our simulation. Even with deceptive pricing by stores or incorrect price data reported by other customers, it will still produce savings. We considered five parameters that characterize real shopping experiences: customer location, store location, item price, item number, and customer input. We varied them in our simulation to explore this five-dimensional space and produced results consisting of the average savings achieved by customers. To validate our results further, we also used real price data in a simplified version of our simulation containing fewer stores and shopping at just two of them. The results indicate an average savings of 6.7% by choosing the best two stores. Even with incorrect price data, customers can still save 6.3% on average. An implementation of our approach would require a multiagent social infrastructure where agents representing customers could report prices they discovered and use prices reported by others. Based on both simulated and real data experiments, and the expected costs of such an infrastructure, our system would be useful and cost-effective in practice.

References

1. Clark, D.: Shopbots become agents for business change. IEEE Computer 33(2), 18–21 (2000), doi:10.1109/MC.2000.820034
2. Krulwich, B.: The BargainFinder agent: Comparison price shopping on the Internet. In: Williams, J. (ed.) Agents, Bots and Other Internet Beasties, SAMS.NET (1996)
3. http://www.redlaser.com/
4. Yang, J., Kim, T.H., Choi, J.: An Interface Agent for Wrapper-Based Information Extraction. In: Barley, M.W., Kasabov, N. (eds.) PRIMA 2004. LNCS (LNAI), vol. 3371, pp. 291–302. Springer, Heidelberg (2005), doi:10.1007/978-3-540-32128-6_22
5. Tang, Z., Smith, M.D., Montgomery, A.: The impact of shopbot use on prices and price dispersion: evidence from online book retailing. International Journal of Industrial Organization 28(6), 579–590 (2010), 10.1016/j.ijindorg.2010.03.014
6. Greenwald, A.R., Kephart, J.O.: Shopbots and Pricebots. In: Agent Mediated Electronic Commerce II (AMEC 2009), pp. 1–23. Springer, Heidelberg (2000), doi:10.1007/10720026_1
7. Lee, H.K., Yu, Y.H., Ghose, S., Jo, G.-S.: Comparison shopping systems based on semantic web – A case study of purchasing cameras. In: Li, M., Sun, X.-H., Deng, Q.n., Ni, J. (eds.) GCC 2003. LNCS, vol. 3032, pp. 139–146. Springer, Heidelberg (2004)
8. Kawa, A.: Simulation of dynamic supply chain configuration based on software agents and graph theory. In: Omatu, S., Rocha, M.P., Bravo, J., Fernández, F., Corchado, E., Bustillo, A., Corchado, J.M. (eds.) IWANN 2009. LNCS, vol. 5518, pp. 346–349. Springer, Heidelberg (2009), doi:10.1007/978-3-642-02481-8_49
9. BLS Information, Glossary, U.S. Bureau of Labor Statistics Division of Information Services. http://www.bls.gov/bls/glossary.htm#C (accessed October 4, 2010)
10. (2007-2008 Weights) Relative importance of components in the Consumer Price Indexes: U.S. city average, U.S. Bureau of Labor Statistics Division of Information Services. http://www.bls.gov/cpi/#tables (accessed October 4, 2010)
11. http://www.cse.sc.edu/~huhns/ShoppingList.html

Artificial Snow Optimization in Winter Sport Destinations Using a Multi-agent Simulation

M. Revilloud, J.-C. Loubier, M. Doctor, M. Kanevski, V. Timonin, and M. Schumacher

Abstract. This paper presents the Juste-Neige system for predicting the snow height on the ski runs of a resort using a multi-agent simulation software. Its aim is to facilitate snow cover management in order to i) reduce the production cost of artificial snow and to improve the profit margin for the companies managing the ski resorts; and ii) to reduce the water and energy consumption, and thus to reduce the environmental impact, by producing only the snow needed for a good skiing experience. The software provides maps with the predicted snow heights for up to 13 days. On these maps, the areas most exposed to snow erosion are highlighted. The software proceeds in three steps: i) interpolation of snow height measurements with a neural network; ii) local meteorological forecasts for every ski resort; iii) simulation of the impact caused by skiers using a multi-agent system. The software has been evaluated in the Swiss ski resort of Verbier and provides useful predictions.

Keywords: Multi-agent simulation, artificial snow optimization.

1 Introduction

Since the early 2000's, the use of artificial snow to guarantee the economic viability of ski resorts has been confronted with the paradigms of climate change and sustainable development [11]. Publications on climate change show that this situation will not improve [4]. Currently, no other economic model seems to be attractive enough to replace the ski economy. The Juste-Neige project is a scientific and commercial attempt to contribute to a more sustainable management of the resources needed for mass skiing. It was developed for the company GeoSnow (www.geosnow.ch), in cooperation with the Swiss cableway companies of Verbier, Champéry, Zermatt and Saas-Grund. To achieve this, Juste-Neige predicts the snow height on the ski runs of a resort using a multi-agent simulation software. Its aim is to facilitate snow cover management in order to i) reduce the production

M. Revilloud · M. Doctor · M. Schumacher
University of Applied Sciences Western Switzerland (HES-SO), Switzerland

J.-C. Loubier · M. Kanevski · V. Timonin
University of Lausanne, Switzerland

Y. Demazeau et al. (Eds.): Adv. on Prac. Appl. of Agents and Mult. Sys., AISC 88, pp. 201–210.
springerlink.com   © Springer-Verlag Berlin Heidelberg 2011

cost of artificial snow and to improve the profit margin for the companies managing the ski resorts; and ii) to reduce the water and energy consumption, and thus to reduce the environmental impact, by producing only the snow needed for a good skiing experience.

The result of the project is a tool that, based on meteorological factors and the usage intensity of the ski runs, provides maps with the predicted snow heights for up to 13 days. On these maps, the areas most exposed to snow erosion are highlighted. The persons responsible for the ski runs can then adapt their snow management. The software proceeds in three steps: i) interpolation of snow height measurements with a neural network; ii) local meteorological forecasts for every ski resort; iii) simulation of the impact caused by skiers using a multi-agent system. The software has been evaluated in the Swiss ski resort of Verbier for two seasons and provides useful predictions for the management of the ski runs.

Related literature on ski resort models and the prediction on snow height cover several aspects. The first topic concerns the study of ski resorts infrastructures [1], which discusses mainly graph aspects, without simulations. The second aspect concerns sociological and behavioral descriptions of skiers [13]. There is then an important amount of work on economical studies on ski resorts [6]. Among those works, [9] presents a very interesting analysis of management decisions on effectiveness of ski resorts. To achieve this, the authors have realized a simulation with skiers moving from cableways to cableways, however without simulating skier movements on slopes and therefore without simulating the impact on snow. Finally, several work concentrate on past snow height measures [2] and snow behavior in itself [8] thus mainly from a physical point of view. The Juste-Neige project mainly innovates in binding research on the infrastructure (graph aspect), skier behaviors with their simulation on the graph, and their impact on the snow cover of all slopes of the resort. Thus, the main novelty concerns the interaction between the snow cover and its usage.

This paper presents the software in general and the multi-agent simulation in particular. It is subdivided as follows: Sec. 2 presents the general aspects of the Juste-Neige software, the required data and the steps involved. Sec. 3 presents the multi-agent simulation of the skiers and of the impact calculation. In Sec. 4, a validation of the software is discussed, followed by a conclusion.

2   General Description of the Juste-Neige Software

The Juste-Neige tool presents many similarities with geographic information systems (GIS). The users can move, zoom and filter information in the cartographic environment. It manages two types of spatial representations: i) bitmap representation (for surfaces), based on a value matrix; and ii) vector representation (for cableways, runs, etc.), based on geometric primitives (point, line, circle, polygon), each with a certain number of attributes (position, color, filling).

To make predictions, the following *data* is needed for every resort: i) a map of the ski runs in a database format; ii) a digital terrain model (DTM) with a resolution of at least 5 meters (a DTM is described as a grid with squares of a specific length resolution); iii) map of the cableways in a vector format; iv) concavity map showing the density of the concave areas (derived from the DTM); v) meteoro-logical data (e.g. fresh snow in cm during a 24-hour span, rain-snow altitude, altitude where the temperature is -10°C; vi) a grid of snow height measurements; vii) dates of school holidays; viii) daily number of cableway passengers of previ-ous years (Skidata).

Juste-Neige carries out a simulation in the following *three steps* (the results of each step are used for the next step):

Automatic analysis of snow height. The goal of the first simulation step is the *spatial interpolation* of unknown points, which is a problem that can be found in many applications (e.g. [5]). The problem is the following: using measurements from the monitoring network of one or several variables, the goal is to predict a variable of interest at the places where there are no measurements. If this predic-tion is carried out on a grid, then a map can be produced. Several methods from Machine Learning can be used. Juste-Neige uses general regression neural net-works (GRNN), because of some useful features: automatic tuning of parameters (using leave-one-out cross-validation technique) and its ability to define so-called "validity domain", i.e. domain where reliable predictions can be performed. The input of the algorithm is an incomplete map of snow heights that was produced by georadar on the snow grooming machines (snow grooming machines cannot cover the whole ski resort every day). The details of the applied GRNN have been pub-lished in [12] and are out of scope of this paper, which concentrates on the multi-agent simulation.

Meteorological simulation. The map resulting from GRNN is then integrated into the simulation model of meteorological factors, that uses an algorithm acting like the influence of the weather on the snow height variation. The definition of this algorithm is based on the measurements of fresh snow and of snow heights made in Verbier in 2009. The data of local weather stations was used and corrected according to the daily snow reports. Assuming that the snow heights are known for the day in question (day D), the variation of the snow height Dh 24 hours later is calculated, i.e. for day D+1 (at the same time as on day D and at approx. 7.30 am the next day). To calculate the variation Dh for more than one day, this process is used iteratively up to 13 days. The result will be a new map of the different snow heights.

Multi-agent simulation. The impact of the skiers is caused by the friction of their skis on the snow. To produce a map that takes into account every individual skier's friction, we need a way to simulate every single skier individually, and its singular impact on the ski runs. From a methodological point of view, a model such as dif-ferential equations cannot calculate this kind of impact. This is however possible using an agent-based simulation of the skiers that will make it possible to study the erosion impact of individual skiers. This is what the 3rd step of Juste-Neige does: the skiers in a resort and their impact on the snow are simulated. Every impact de-pends on several parameters such as the pressure exerted on the floor by the skier, the contact surface, the type of material, and the intrinsic qualities of the snow cover. The agents' movement can be seen even during the simulation. The result of the simulation of the flow of skiers can be finally seen on a map of the resort. The remainder of this paper explains how the multi-agent simulation is carried out.

3 Impact Caused by Skiers (Multi-agent Simulation)

The multi-agent simulation used to study the skiers' impact can be subdivided into four steps: i) estimation of the number of skiers in every cableway; ii) simulation of the flow of skiers to determine their distribution in the resort; iii) cinematic simulation of their behavior for their distribution on the ski runs; iv) simulation of their impact on the snow cover. The aim of the first two steps is to simulate the daily load of the network (cableways and ski runs). Through statistical reasoning, we can estimate the number of agents per cableway by analogy with similar situations from previous years for which the daily number of skiers is provided by the cableway companies thanks to the Skidata system. Using probabilities based on the number of skiers at the bottom of the cableway and on a field study, we can simulate the flow of skiers from the top of the cableway stations to the bottom of another cableway station. A cinematic simulation allows us to simulate the skiers' movements and to coherently animate them on the ski runs. It allows estimating the skiers' distribution on the ski runs. Combined with a snow erosion law, we get the result of the skiers' impact on the ski run. In the remainder of this section, the stages of the multi-agent simulations are explained.

Our multi-agent simulations use two kinds of data: *The number of users of the different cableways in every resort,* obtained through the Skidata system (http://www.Skidata.com/), gives us information about the number of passenger transports for every cableway on every day of the past season. Using it, we can estimate the number of skiers for every day of the simulation.

The simulation then uses the results of a *survey* on different criteria (visitor influx, weather, snow quality, etc.) that may influence the probability of the skiers' choice of a ski run [3]. This survey has shown that among the 14 factors that influence the skiers' behavior, the snow quality and the weather have a significant influence on the choice of the ski run difficulty and provides the percentage of skiers who choose a blue, red, black or yellow run (difficulty levels). Those factors were integrated into the software as a set of rules.

3.1 Definition of Standard Days

A simulation of 13 days in the future is required. For every day, the number of skiers and their distribution in the resort must be estimated. Based on the Skidata database, the past day most similar to the simulation day has to be determined first. Past visitor traffic data of the cableways is used to determine the skiers' distribution in the resort.

Using statistical data analysis methods, it is possible to create representative groups of a situation. These groups are called *standard days* (e.g. *holiday, weekend, cloudy*). By defining a method for finding the standard day that is the closest possible to the day to be simulated, the number of skiers most likely to be present during the simulation of the day in question can be determined. In order to choose standard days, we carried out a multivariable statistical analysis to determine influential factors and confirm their interrelations and to eliminate redundant variables. The result is the following: i) holiday, weekend or weekday; ii) sunny, cloudy, with or without rain or snow; iii) temperature at midday at an average altitude of the resort. In a second phase, we used the factor scores of correspondence analysis to construct a robust classification system of the standard days, using a 3-step process: i) **Hierarchical ascendant classification** (HAC) according to individual scores on the factorial axes, in order to discover the optimal structuring of the data set. However, as the individuals are sometimes incorrectly classified, we used the k-means method to construct the effective classification: ii) **K-means classification**, that guarantees a good classification of the individuals in the different groups with the number of clusters provided by the HAC of the previous step; and iii) **Discriminant factor analysis** of the k-means classification to evaluate the robustness of the classification of standard days, using the confusion matrix between the a priori classified individuals (k-means method) and the a posteriori classification (discriminant factor analysis).

Finally, we have subdivided the statistical groups into four subgroups: holidays and weekends, holidays and weekdays, non-holiday weekends and non-holiday weekdays. This considerably reduces ambiguities and allows us to define the characterization rules for the standard days. As a final result, a standard day was attributed to every day to be simulated. The visitor information for this standard day (available through Skidata) was used to distribute the number of skiers to simulate and to calculate the probabilities of their trajectories.

3.2 Simulation of the Flow of Skiers

The number of skiers on every cableway is defined based on the Skidata data for the chosen standard day. We therefore know how many skiers leave the point of departure (cableway top station) and how many arrive at the point of arrival (cableway bottom station), as the number of arriving and departing skiers is the same. In the example in Figure 1, 600 + 1500 skiers arrive at the point of arrival A1 and 1500 + 300 leave from the point of departure D2.

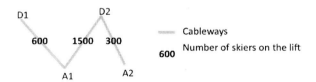

Fig. 1 Flow of skiers on cableways

To simulate the flow of skiers in our resort, we first have to model the resort as a directed graph. In this graph, each ski run section is a directed arc and each intersection is a node (cf. Figure 2). The processing speed can be optimized by deleting the points, which are not intersections from the ski runs, and by keeping only decision points. The agents will move along these arcs to their point of arrival.

Fig. 2 Creation of the flow graph

Once the graph has been designed and the number of skiers per cableway is known, the skiers must know their point of arrival at the bottom of the ski runs. This destination must comply with the constraint concerning the number of skiers (i.e. the number of skiers at the bottom and the top of the cableway stations given by Skidata) and the probabilities of the choice of a ski run based on the surveys.

From its starting point (at the top of the slope), a skier has many different possible destinations. We need therefore to calculate in advance all ***probabilities to arrive to all destinations***. For that, we use an iterative probability weighting algorithm that bases on Skidata counts of the cableways. Intuitively, the algorithm considers the chances of a skier as higher to go where the greatest number of skiers has arrived (given by the numbers in the cableways). For the four resorts in study, the algorithm converges in approximately 100 iterations. We do not present here details of the algorithm because of lack of space.

With this method, we can thus find the point of departure and arrival of all skiers of our simulation, using only the Skidata visitor information. We can now determine the ***trajectory of every skier to reach the point of arrival*** (the simulation of their movements will be made in the cinematic simulation section below). To determine a skier's route between the point of departure and the point of arrival, he must be moved along the graph. To avoid random movements or lost skiers, the nodes that are most likely to be passed are referenced in an efficient data structure. Only the nodes that allow the skier to go from the point of departure to the point of arrival are taken into consideration. The flow of skiers is then moved from referenced node to referenced node until a decision point is reached (point where the skiers have to decide on a route). At this point, the flow is divided according to the ski run probabilities provided by the survey. Finally, we obtain a number of skiers for every route and the number of skiers who have passed through every arc of our graph (cf. Figure 3).

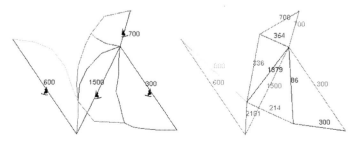

Fig. 3 Result of the distribution of the skiers

3.3 Cinematic Simulation

With the cinematic simulation, we assimilate the skiers with agents. We use a physically inspired cinematic simulation, similar to Craig Reynolds' algorithm [10] for the simulation of bird flocking. This type of simulation is based on the fact that the displacement of a body is nothing else than the sum of the forces that act on it. In our case, a skier's behavior can be assimilated to a body influenced by its attraction to the subsequent point on the ski run, the oscillation around the axis of the ski run and a random vector ensuring a homogeneous distribution. In our simulation, we made the assumption that all skiers remain on the runs. In reality, a small amount of skiers may choose sometimes to go outside the runs.

After a short estimation based on a worst case scenario for a ski resort such as Verbier, we envisage to simulate up to 100,000 agents per simulation day, i.e. 1,300,000 for 13 days. This means though that we have to use a simplified skier behavior model and that we need to considerably optimize the corresponding algorithm.

Action plan. To link the flow simulation and the cinematic simulation, we have introduced an "action plan". The action plan is an ordered list of points that represent the route of a skier. At the end of the flow simulation, we design an action plan for every route that has been found. The arcs are successively decomposed into ski run points. This action plan is then used to create the corresponding number of skiers.

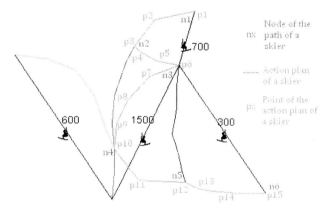

Fig. 4 Action plan

In the example of Figure 4, the flow simulation has resulted in the estimation that x skiers have taken the following route: n1 => n2 => n3 => n4 => n5 => n6. The route is decomposed into points on the ski run (p1, p2 ... p15), which form the action plan. Hence, x skiers are constructed with this action plan.

Modeling of the skiers' behavior. Next, the skiers' behavior has to be modeled and implemented. The aim is not to create a realistic skier behavior – which would be far too complex to model and to simulate – but to create a model that allows a distribution of realistic routes of the skiers on the ski run. The behavior is thus relatively simple: it does not take into account the skiers' individual features (weight, level of expertise) or the events that could change their actions (collision). These simplifications allow us to disregard the time factor: all agents leave at the same time. The thrust is therefore to move the agents from point to point with a trajectory that is similar to the one of a real skier. We model this behavior using three influence vectors:

- *Directional vector*: It is recalculated at every time step and directed towards the next target point (i.e. the next point to reach in the action plan). Its norm is unitary in order to preserve the mathematical properties of the oscillation vector, cf. below.
- *Oscillation vector*: It is characterized by a sinusoidal curve whose attributes are described hereafter: i) the axis is the straight line that goes through the last node and the following node; ii) the amplitude of the oscillation is a random value between 0.4 and 0.9 times the width of the ski run; iii) the period is a random value between 20 and 40 meters; as the directional vector is unitary, the period is not disturbed. The attributes of the sinusoidal curve evolve progressively in order to avoid discontinuity problems with the trajectories.
- *Random vector*: Allows a homogeneous distribution of the skiers. Its norm is unitary so not to distort the skier's trajectory too much.

The results for one skier (cf. Figure 5a) and for 1,000 skiers (cf. Figure 5b) show a satisfactory distribution.

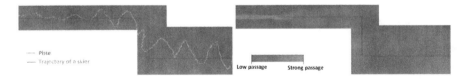

Fig. 5 a) Trajectory of a skier and b) distribution of skiers on the ski run

3.4 Law of Skiers' Impact on Snow Cover

Now, the impact of the skiers' passage on the snow has to be determined. The characteristics of the ski run (incline, hardness, concavity) at the moment of passage have to be taken into account. We define an impact law that quantifies the erosion of the snow cover for the skiers who have passed through. We base on the assumption that there exists an average coefficient of the pressure factor of a skier according to his weight and level, the material and the contact area with the ground. This enables us to ignore the particular features of the skier. This estimate is justified for two reasons: i) the relative homogeneity of the types of skiers in a resort; ii) a sufficient number of skiers to envisage an average coefficient for all of these phenomena.

The analysis of the snow heights in the different resorts has shown that the impact of the skiers mainly depends on three parameters:

- *Slope of the ski run:* The flatter the ski run, the less the skiers oscillate and the less they erode snow. This is expressed by a slope coefficient linked to the difference in altitude between two nodes of the run, divided by the distance between them.
- *Concavity of the ski run:* The snow has a tendency to accumulate in the concave parts of the ski run. We have therefore designed a concavity map based on the digital terrain model, which has a concavity coefficient in every cell of the raster.
- *Hardness of the ski run:* The warmer the snow cover, the bigger the skiers' impact. We have therefore established a hardness equation based on the distribution of the temperatures measured at 7 am during the entire winter season. This equation defines a hardness coefficient.

The product of the coefficients of the ski run, the concavity and the hardness gives an overall form of the skiers' impact, weighted by parameterizable coefficients that integrate the average coefficient of the pressure factor.

4 Validation and Conclusion

The tests carried out for a 13 days prediction have shown that the Digital Terrain Model's (DTM) resolution plays an important role for the quality of the simulation. Long-term simulations have to be carried out on DTMs with high resolutions for the estimations being usable for the decision-makers. The results of the test station in Saas-Grund have shown that with a DTM resolution of 5 m, the errors exceed 1 m, which is inacceptable for the ski run managers. However, with a resolution of 1 m, the estimation can considerably be improved: the maximum error is approx. 90 cm and 85% of the errors are inferior to 75 cm. This result is extremely interesting given the fact that it is only based on one set of values taken 13 days prior. This shows that the Juste-Neige software can produce quite efficient estimations if the DTM is relatively precise. The error histogram shows a tendency towards overestimation though, which means results that can be compared to a long-term meteorological prediction. Currently, developments are being made to integrate this information into the predictions by calculating a quality index, which will be introduced into the outputs.

Version 1.0 of the Juste-Neige software is operational and has been deployed in the four participating ski resorts. Its commercialization is underway. The multi-agent simulation method has proved to be adequate for predicting the skiers' impact. The projections on the potential for improvement are important. In Zermatt, for instance, the production of 1 m^3 of artificial snow costs 2.50 CHF. With 1,300,000 m^3 of snow produced, this corresponds to yearly production costs of 3.2 million CHF. In addition, the surface where artificial snow is being used corresponds to approx. 250 ha. Thanks to this software, the production of artificial snow could be reduced by 10% within the next five years. Consequently, the snow production and management cost will be reduced by 320,000 CHF per year. Currently, 1,300,000 m³ of snow are produced, which requires 565,000 m³ of water. With an economy of 10%, 56'500 m³ water could be saved in Zermatt. New developments, such as the optimization of the flow of skiers, are planned. In the long term, we can expect a complete environment for the sustainable management of the ski industry.

Bibliography

[1] Brissaud, I.: Fractal distribution of Ski lifts and Ski slopes. Example of the Serre Chevalier resort, Cybergeo. European Journal of Geography, Systèmes, Modélisation, Géostatistiques, article 366 (2007)

[2] Brown, R.D., Braaten, R.O.: Spatial and Temporal Variability of Canadian Monthly Snow Depths, 1946—1995. Atmosphere-Ocean 36(1), 37–54 (1998)

[3] Dubosson, Y.: Etude sur les critères influençant le choix d'une piste de ski sur le domaine skiable de Verbier, Ecole Suisse de Tourisme, Sierre, 53 p (2009)

[4] Elsasser, H., Bürki, R.: Climate change as a threat to tourism in the Alps. Clim. Res. 20, 253–257 (2002)

[5] Kanevski, M., Pozdnoukhov, A., Timonin, V.: Machine learning for spatial environmental data, Theory, applications and software. EPFL and CRC Press (2009)

[6] Knaffou, R.: Les Stations intégrées de sport d'hiver des Alpes françaises, Masson (1978)

[7] OcCC. Klimaänderung und die Schweiz 2050. Erwartete Auswirkung auf Umwelt, Gesellschaft und Wirtschaft, Bern, p.168 S (2007)

[8] Pinzer, B.R., Schneebeli, M.: Snow metamorphism under alternating temperature gradients: Morphology and recrystallization in surface snow, Geophysical Research Letters (2009)

[9] Pullman, M.E., Thompson, G.M.: Evaluating capacity- and demand-management decisions at a ski resort: this model reveals some effective and not-so-effective way. Cornell Hotel & Restaurant Administration Quarterly (2002)

[10] Reynolds, C.W.: Flocks, Herds and schools: A distributed behavioral model. Computer Graphics 21(4), 25–34 (1987)

[11] Scott, D., McBoyle, G., Mills, B.: Climate change and the skiing industry in southern Ontario (Canada): exploring the importance of snowmaking as a technical adaptation. Clim. Res. 23, 171–181 (2003)

[12] Timonin, V., Kanevski, M., Loubier, J.-C., Doctor, M.: Analysis and modeling of snow depth patterns in Alpine ski resorts. In: Proceedings of the 16th European Colloquium on Theoretical and Quantitative Geography, Maynooth, Ireland, September 4 - 8 (2009)

[13] Vermeir, K.: le risque sur les domaines skiables alpins. Analyse des représentations sociales des pratiquants. Thèse de géographie (2008),
http://tel.archives-ouvertes.fr/tel-00322735/fr/
oai:tel.archives-ouvertes.fr:tel-00322735

Middle Age Social Networks: A Dynamic Organizational Study

L. Lacomme, V. Camps, Y. Demazeau, F. Hautefeuille, and B. Jouve

Abstract. This article describes the main components of the multiagent model and the tool we are designing in order to analyze the structure and the dynamics of an historical social network from the Middle Age. This social network has been built through the discovery of many notarized documents and first studied according to a mathematical approach. Using a customizable multiagent system to simulate the network's evolution and Markov chain calculus to adjust behavioral parameters, we propose a method to assess behaviors' importance and effects on the global network's evolution. From a historical point of view, this multiagent model also aims at explaining this network's dynamics during some specific periods such as the Hundred Year's War where numerous notarized documents disappeared.

1 Introduction

This study began in 2004 and has been involving historians, mathematicians as well as computer scientists with the aim of reconstructing social networks in

L. Lacomme
Université de Grenoble
Laboratoire d'Informatique de Grenoble
e-mail: `Laurent.Lacomme@imag.fr`

V. Camps
Institut de Recherche en Informatique de Toulouse – UPS Toulouse III
e-mail: `camps@irit.fr`

Y. Demazeau
CNRS
Laboratoire d'Informatique de Grenoble
e-mail: `Yves.Demazeau@imag.fr`

F. Hautefeuille
Laboratoire TRACES, Université Toulouse II Le Mirail
e-mail: `florent.hautefeuille@univ-tlse2.fr`

B. Jouve
Institut de Mathématiques de Toulouse - Université Toulouse II
e-mail: `jouve@univ-tlse2.fr`

Y. Demazeau et al. (Eds.): Adv. on Prac. Appl. of Agents and Mult. Sys., AISC 88, pp. 211–216.
springerlink.com
© Springer-Verlag Berlin Heidelberg 2011

medieval peasant society and highlighting possible different structuring levels [8]. The study is based on exceptionally abundant notarized documents from the Middle Ages covering a small geographic region of southwestern France (more than 8000 contracts for most agricultural and marginally wills, marriage contracts or feudal homages). These documents were collected over three centuries (1250-1550) and archived for most of them in the departmental archives of the Lot. The working hypothesis is that this material allows highlighting relational phenomena between different social strata of rural society medieval peasants and lords, peasants and church, farmhouses and peasant farmers from the villages... or just neighbors. It thus allows the construction of a graph that describe the organization of the society with the simple idea where a node represents an individual named in some manuscript and where a link between two nodes means that two people are listed in the same transaction. This literature does not focus specifically on the ruling classes (clergy, nobility). The peasant world accounts for 90-95% of the population and has left very few writings. This makes difficult to penetrate the private sphere and to try to write a history of peasant societies of the Middle Ages. A database was created, containing 8725 individuals identified during 250 years and situated on parcels of land. The population base is estimated around 3000 people maximum (1330) and half at minimum (1420).

The particularity of this study is the existence of partial social network over a long period of time, but which is barely incomplete and uncertain. Our objectives are to study this network as well as its dynamics, by attempting to recreate its structure and properties through an agent-based simulation. To achieve this, we try to model the composition of the system and then to evaluate different possible local mechanisms to make the network evolve. This involves the study of dynamic organizations, which is a known subject within the multiagent domain [6].

A state of the art of the mathematical analysis of social networks is given after this introduction. We describe in the third section the theoretical choices we made about the problem modeling, and the resulting multiagent model. Then we describe in the fourth section the approach we took in our work over the historical problem and our further expectations. Finally, we conclude on the tool, the historical perspectives and the perspectives over our multiagent approach and model.

2 State of the Art in the Analysis of This Social Historical Network

A relational graph is built from the database in order to represent the peasant social network: each node of the graph corresponds to a peasant named in the transactions. Two nodes are linked together by an edge if they appear in a same contract (rule 1), or if they appear in two different contracts separated by less than 15 years and on which they are related to the same lord or to the same notary (rule 2). Because of this second rule, some main lords have been removed from the network to avoid the creation of large complete subgraphs that show the predominance of a lord in the area. Two separate studies were carried out and are still in progress: one on a small area of 25 km² (2462 nodes for the period 1240-1340) [2, 3, 4, 5, 9, 13] and the other on an expanded territory of 150 km² (3853 nodes for the same period) [10]. These studies are based on two parameters, the node degree (number of neighbors) and the betweenness (frequency of the shortest paths between any two nodes of the graph in which this node occurs), whose distributions are typical of small world networks. We observe that the resulting social network has a low global connectivity and a high local connectivity.

For the small area, we have shown the existence of a hierarchy in the organization of the network. This is measured by a scaling relation between the clustering coefficient and the close to the one that Barabasi proposed as an indication of a hierarchy [1]. Nodes of a *rich-club* (set of the highest degree nodes strongly interlinked) and small communities composed of highly interconnected nodes are linked by the way of high betweenness nodes (called *relay individuals*).

For the large area, we have removed all nodes whose degree was higher than the highest degree reached by a peasant and then we have constructed the graph following the rules 1 and 2. This global graph does not reveal a rich-club but it reveals a correlation between the peasant social network's structure and the geography of the area (studied in [11]).

Structural elements of the society can be observed from these first analyses, but several questions remain. The analysis has indeed been achieved by focusing on the links between individuals and we are just starting to explore other information stored in the database (acts type, relevant geographic areas...). One of the issues is the importance of the relationship that seems to exist between the structure of peasant social networks we have revealed and the geography. Moreover, in the studies we have made of these networks, we did not consider the time and have thus removed all dynamics that might help to explain their high structuring.

3 The Multiagent System Approach

On one hand, the information given by the database is local: each individual has characteristics, such as its profession and social status, and each transaction is linked to specific individuals: farmers, merchants, notaries and lords. On the other hand, the network properties provided by mathematical analysis are global properties: degrees, centrality, distributions, etc. Then, when studying the network dynamically, we have to make assumptions and choices over local properties and to observe and compare global properties. Through multiagent paradigm, we are able to elicit local parameters and to measure global effects on the structures of the system. We could use other methods such as cellular automata modeling, but the incompleteness of the considered data drives us toward a more flexible method than these. The method we consider for exploring and analyzing the real network is to recreate a similar virtual network based on customizable agents and to compare global structures and properties of this virtual network with the real historical network. We first specify a system in which parameters can be adjusted to match the real network. Its structure clearly results from an emergent process based on local rules belonging to individuals. Through guess and evaluation we can give to agents some rules that seem to fit individuals' normal behavior, then measure the properties of the resulting virtual network and compare them to the real network's properties. We then adjust local behavioral models to match the real network's

properties through a simulation-adaptation cycle. Then, we might be able to analyze the resulting rules and make historical sense of them.

3.1 The Proposed Model

We describe the virtual system according to VOWELS dimensions [7]: agents (A),
environment (E), interactions (I) and organization (O). We rapidly discuss the A E
I and then focus on the O.

The real network evolves according to decisions made by individuals. Each individual interacts with other and these interactions are recorded as transactions
(basic items of the database). Hence, in the virtual system, each agent (A) represents a unique individual. Because archives are not very precise in time, we can,
without bias, propose a synchronous system. The time unit corresponds to one
year in the real network, because it is the smaller significant time step in the database. If we consider that life expectancy was about 50 years, each individual could
have a social life during 30 years, so each agent has a life time of 30 time units.
The virtual environment (E) represents the set of the external forces influencing
the population evolution, such as the external mortality factor, which is variable in
time because of events such as wars or diseases. It also includes localization for
each agent, based on three levels of information: parcels, place names and par-
ishes. Information about each agent's localization can be incomplete or uncertain.
During its life, each agent acts accordingly to its behavior specifications by interacting with others. Each interaction (I) between agents is a social interaction, such
as a sale or an inheritance. Each interaction between agents symbolizes a transaction existing between individuals, as described in the historical database.

At an organizational level (O), we describe types of agents, relations between
agents and organizational mechanisms. Firstly, agents are typed according to their
static characteristics, such as their job and social condition. Secondly there is a
relation between two agents if they know each other and are likely to formally
interact (if they do so, they will then appear in the same transaction, which is ma-
terialized in the multiagent system as an interaction). Each relation is typed according to the kind of interaction it can involve in the database: sales, weddings,
inheritances, etc. The mechanisms are the central point of the problem exploration.
They describe how each agent makes its neighborhood evolve. For example, a
farmer should learn to know every merchant that settle in the same village. In order to explore the network structure generation, we define generic mechanisms.
Then we try to express how these mechanisms can alter the global network properties to verify if these mechanisms are correct or not, and how they have to be adjusted. In order to minimize the required number of simulations and to simplify
the parameters' adjustment, we intend to specify the organizational mechanisms as
Markov chains [12]. While this is a loss of information because it goes from exact
mechanisms to probabilistic ones, it will allow us to seek indications about global
properties by calculating steady state distributions of neighborhoods. These steady
state distributions, although theoretical give information about the neighborhood
composition, and, by a combination with the transition matrix of the chain, information about the neighborhood dynamics. All this information can be used to
guess and try parameters adjustments before running the simulations.

3.2 The Objectives and Methods

The multiagent simulation has to answer to at least three main objectives:

- **Properties conservation:** On a global level, we observed that some global properties exist in the system, such as a *rich club* and *relay individuals* (see section 2). The question raised by the existence of such elements is if they are an intrinsic property of the network or if their existence is contingent. We try to determine if elementary behaviors of individuals are intrinsically the cause of such properties. We choose some individual behavior and make them vary (*e.g.* by changing their probability of occurrence) we can determine if a behavior (a mechanism in the multiagent system) is related to a property of the network.
- **Self-organization:** Another way to proceed is to create a blank system, where agents correspond to the individuals of a given date in the real system, but without any relation. We let the system evolve, given a specification of which social behaviors – organizational mechanisms, in terms of agents – to apply. We then compare the final state of the system to the real one at the chosen date.
- **Family exploration:** Another objective of the work is to better understand the families' evolution in terms of size, influence and situation, for instance. We propose to simulate behavior of families inside the system, given adequate individual behaviors. We measure their evolution given specific parameters, both internal – chosen jobs for the children, localization, etc. – and external – war, mortality, etc. By adjusting the parameters and mechanisms on well-known families by comparing their simulation to their real evolution, we can explore the evolution of families on which we lack information.

4 Conclusion and Perspectives

In order to analyze an historical social network, we have developed the first steps of a particular tool. This tool includes a generic multiagent system model, which has an organizational customizable part, as well as a way to experiment with this model. Markov calculus and simulation are combined to build a faster way to explore various reorganization rules and methods and to define the most appropriate – appropriateness being here the resemblance with the real historical network following given criteria and properties –. The tool is adequate to the application, because it can be used for computational prediction through multiagent systems. The approach follows an iterative method that includes prediction through Markovian model, optimization of the parameters and verification by simulation. This can be applied to different parts of the analysis: understanding the evolution of families through times, verifying the intrinsic nature of properties of the network and filling blanks in badly known periods such as the war.

As a first perspective, our goal is to use the tool to better understand the historical network, and to be able to find historical understanding of the network's evolution. We propose to define, compare and analyze the organizational mechanisms used in the system, and make sense from them. This may lead to the discovery of the impact of each mechanism on the network structure, and so to the understanding of important social relationships to the global historical network.

This tool is not only appropriate for analyzing this particular social network; it also defines a more general approach that could be applied to the analysis of any kind of incomplete social networks where nodes or links are missing. Through simulation and comparison, we propose to study the impact of various organizational and behavioral parameters on the global network configuration and properties. This will permit to study the network both statically and dynamically.

References

1. Barabási, A.L., Oltvai, Z.N., Wuchty, S.: Characteristics of biological networks. In: Lect. Notes Phys., vol. 650, pp. 443–457 (2004)
2. Boulet, R., Hautefeuille, F., Jouve, B., Kuntz, P., Le Goffic, B., Picarougne, F., Villa, N.: Sur l'analyse de réseaux de sociabilité de la société paysanne médiévale. Méthodes Computationnelles pour Modèles et Apprentissages en Sciences Humaines et Sociales, Brest (2007)
3. Boulet, R., Jouve, B.: Partitionnement d'un réseau de sociabilité à fort coefficient de clustering. Revue des Nouvelles Technologies de l'Information (9), 569–574 (2007)
4. Boulet, R., Jouve, B., Rossi, F., Villa, N.: Batch kernel SOM and related Laplacian methods for social network analysis. Neurocomputing 71(7–9), 1257–1273 (2008)
5. Boulet, R.: Comparaison de graphes, application aux réseaux de sociabilités paysans du moyen âge. PhD Thesis in mathematics, Université de Toulouse (2008)
6. Costa, A.R., Demazeau, Y.: Towards a formal model of multi-agent systems with dynamic organizations. In: Proceedings of the Int. Conf. on Multi-Agent Systems, Kyoto, Japan, p. 431. MIT Press publisher, Cambridge (1996)
7. Demazeau, Y.: From Cognitive Interactions to Collective Behaviour in Agent-Based Systems. In: 1st European Conference on Cognitive Science, Saint-Malo, pp. 117–132 (1995)
8. Hautefeuille, F., Jouve, B.: Analyse d'une base de données de contrats agraires pour comprendre l'espace et la société paysanne du Moyen-Age. In: Séminaire Grands Réseaux d'Interactions (invited speaker). LIAFA, Paris (2004)
9. Hautefeuille, F.: L'extraordinaire ascension d'une famille de marchands de Castelnau-Montratier (46): Les Trapas (1250-1350), dans Minorités juives, pouvoirs, littérature politique en péninsule ibérique, France et Italie au Moyen Age, études offertes à Béatrice Leroy, édition Atlantica, pp. 51-64 (2006)
10. Hautefeuille, F., Jouve, B.: Modélisation des réseaux sociaux des communautés paysannes de fin du Moyen Age. In: Population and Language Change Int. Work.p Lacito, Villejuif (2010)
11. Laurent, T., Villa-Vialaneix, N.: Analysis of the influence of a network on the values of its nodes: the use of spatial indexes. In: Proceedings of MARAMI 2010, Toulouse (2010)
12. Tatlor, H.M., Karlin, S.: Introduction to stochastic modeling, 3rd edn (1998)
13. Villa, N., Rossi, F., Trong, Q.D.: Mining a medieval social network by kernel SOM and related methods. In: Proceedings of MASHS 2008 Modèles et Apprentissages en Sciences Humaines et Sociales, Créteil (2008)

From an Individual Perspective to a Team Perspective in Agent-Based Negotiation

Víctor Sánchez-Anguix, Vicente Julián, and Ana García-Fornes

Abstract. Agent-based negotiation teams are groups of two or more interdependent agents with their own and possibly conflicting goals that join together as a single negotiating party because they share a common goal that is related to the negotiation. In this paper we present a general workflow of tasks for agents that participate in a negotiation team. Furthermore, we focus on the study of intra-team organizations, which define the roles played by team members, what decisions are taken by the team, and how and when they are taken, and the negotiation strategy carried out with the opponent. The performance of the proposed organizations is tested in environmental scenarios where team diversity is different.

1 Introduction

Automated negotiation is considered as one of the key technologies in the family of technologies which is nowadays known as *Agreement Technologies* [5]. From the first works in the area until the present, a vast amount of literature has provided solutions to different negotiation scenarios: bilateral negotiations [3], multi-party negotiation [2], mechanisms for complex utility functions [6], and so forth.

From the point of view of agent-research, the negotiation processes which have been studied involve parties which are formed by single individuals. However, most parties in a real world negotiation process are formed by more than a single individual. They are a negotiation team. The reasons to send a team to represent a negotiation party are twofold. First, each party member may represent different interests inside a organization which need to be taken into account. Second, due to an inherent complexity of the domain, collaboration among individuals with different expertise is needed. In any case, team members are usually aligned with a common

Víctor Sánchez-Anguix · Vicente Julián · Ana García-Fornes
Universidad Politécnica de Valencia, Departamento de Sistemas Informáticos
y Computación, Camí de Vera s/n, 46022, Valencia, Spain
e-mail: {sanguix,vinglada,agarcia}@dsic.upv.es

Y. Demazeau et al. (Eds.): Adv. on Prac. Appl. of Agents and Mult. Sys., AISC 88, pp. 217–223.
springerlink.com   © Springer-Verlag Berlin Heidelberg 2011

and shared goal which drives them to collaborate as a single party during the negotiation process. These situations can also be found in a direct/indirect manner in complex distributed systems such as open and dynamic multi-agent systems. For instance, it can be found in e-commerce systems, and decision-making processes involving agent organizations. Our goal is providing computational solutions for negotiations which involve negotiation teams as one of the parties.

In this paper, firstly, we introduce a general work-flow of tasks for agents that participate in a *negotiation team* [7]. In Section 3, we focus in one of the phases of this workflow: the intra-team organization. More specifically, we propose several preliminary organizations for a negotiation team which negotiates with an opponent following an alternating bilateral negotiation protocol. We decided to study this phase first due to the fact that it directly impacts team dynamics during the negotiation process and the final outcome obtained by the team. In Section 4, experiments carried out to show how these organizations behave in different negotiation environments are described and analyzed. Finally, this present work is compared with related work, and some conclusions and future lines of work are given.

2   General Workflow of Tasks for Agent-Based Negotiation Teams

In [7], we proposed a general workflow of tasks for agents that participate in a negotiation team. Next, we briefly describe each of its phases.

- Identify Negotiation: The agent should reason whether it is convenient/necessary to form a negotiation team or not, which agents can be considered as potential teammates and which agents can be considered as opponents.
- Team Formation: In some scenarios, team composition is static and is linked with the problem instance. However, in other scenarios, the team must be formed and teammates must be selected from a pool of prospective teammates.
- Plan Negotiation Issues: Teams members should state which negotiation attributes they consider relevant to the negotiation process.
- Agree Negotiation Issues: The team should interact with the opponent in order to the set of attributes which are relevant for teammates and the opponent.
- Plan External Negotiation Protocol: Team members should reason about which external protocols are more convenient for the current negotiation scenario.
- Agree External Negotiation Protocol: The team should agree an external negotiation protocol.
- Plan Intra-Team Organization: This phase governs team dynamics during the negotiation process. The coordination issues that should be settled are: (i) roles (A flat structure or an organization based on roles may be employed); (ii) intra-team strategy (Team-mates have to identify which decisions may be taken during the negotiation process, and how and when this decisions are taken); (iii) initial negotiation strategy (The team has to decide a proper negotiation strategy to carry with the opponent).

- Negotiation & Adaptation: If the dynamics of the negotiation process do not lead towards an expected result, team members may decide to re-plan some internal aspects in the team.

3   A First Approach into a Negotiation Team Model

In this paper, we present a negotiation model where one of the parties is a negotiation team and the other party is an individual agent. When we present the negotiation team model, we will not describe a negotiation strategy, but a Intra-team Organization, which is one of the phases that we presented in Section 2. This phase directly affects team dynamics during the negotiation process and, thus, the performance of the team. Both parties employ the alternating offer bilateral protocol to interact with the other party. Some of the general assumptions of our model are: (i) The negotiation team has been already formed. Team membership remains static during the negotiation process. The team is formed by $A = \{a_1, a_2, ..., a_M\}$ agents; (ii) Agents do not know other agents' preferences and negotiation strategies; (iii) The domain is composed of n real valued attributes whose domain ranges from 0 to 1; (iv) Agents have a private reservation utility RU which represents the utility obtained by the agent if the negotiation process fails; (v) As for agent preferences, they employ linear utility functions to represent their preferences:

$$U(X) = w_1 V_1(x_1) + w_2 V_2(x_2) + ... + w_n V_n(x_n) \tag{1}$$

where X is a n-attributes offer, x_i is the value of the i-th attribute, $V_i(.)$ is a linear function that transforms the attribute value to $[0,1]$, and w_i is the weight or importance that is given by the agent to the i-th attribute. Weights given by the opponent to attributes may also be different. Agents do not know the form of other agents' utility functions, even if they are teammates. For each attribute, $V_i(.)$ is common for all of the teammates and the reverse function for the opponent. Therefore, if the value preferred by a team member for attribute i is 1, then the value preferred by the opponent for that attribute is 0.

3.1   Opponent Negotiation Strategy

For the opponent, the fact that he is negotiating with a team is transparent. Therefore he negotiates employing the same strategies that he would employ in a bilateral negotiation process. The negotiation strategy $s_{op}(.)$ employed by the opponent is a time-based concession strategy which can be defined as follows:

$$s_{op}(t) = 1 - (1 - RU_{op})(\frac{t}{T_{op}})^{\frac{1}{\beta_{op}}} \tag{2}$$

where t is the current negotiation round, RU_{op} is the opponent reservation utility, t is the current round, T_{op} is his private deadline, and β_{op} determines how concessions are made towards the reservation utility. At each negotiation round t, the opponent will propose an offer to the team whose utility (from the point of view of the opponent) is equal to $s_{op}(t)$. As for the acceptance criteria employed by the opponent, he accepts the offer sent by the team $X_{A \to op}^t$ at round t if the utility it reports is greater than or equal to the utility of the offer that will be proposed by him. This acceptance mechanism $ac_{op}(.)$ can be formalized as follows:

$$ac_{op}(X_{A \to op}^t) = \begin{cases} accept & \text{if } s_{op}(t+1) \leq U_{op}(X_{A \to op}^t) \\ reject & \text{otherwise} \end{cases} \tag{3}$$

3.2 Intra-Team Organization

In this subsection, we describe several intra-team organizations that may be employed by a team of agents in this scenario. We propose two different intra-team organization paradigms: a representative approach, and a voting-based approach. Next, we describe both approaches in more detail.

3.2.1 Representative-Based Intra-Team Organization

Roles
One of the teammates assumes the role of representative a_{re}. This agent is entitled with the tasks of team decision-making and communicating with the opponent. In these experiments, we assumed that the representative is chosen randomly.

Negotiation Strategy
The representative employs a time-based strategy as the one presented in Eq. 2. The value of the strategy parameter β_A and the private deadline T_A have been agreed upon by the team. The representative negotiates using his own utility function $U_{a_{re}}$.

Intra-Team Strategy
The decisions that must be taken each round are the offer to be sent to the opponent, and whether the opponent offer is accepted or not. The offer sent to the opponent follows the utility defined by the strategy $s_A(.)$. The acceptance criterion employed by the representative $ac_A(.)$ is analogous to the one employed by the opponent and presented in Eq. 3.

$$s_A(t) = 1 - (1 - RU_{a_{re}})(\frac{t}{T_A})^{(\frac{1}{\beta_A})} \tag{4} \qquad ac_A(X) = \begin{cases} accept & \text{if } s_A(t+1) \leq U_{a_{re}}(X) \\ reject & \text{otherwise} \end{cases} \tag{5}$$

3.2.2 Voting-Based Intra-Team Organization

Roles
Agents adopt a flat structure and use voting processes in decision-making tasks. A trusted mediator governs the voting process and communicates with the opponent.

Negotiation Strategy
Teammates use a time-based concession strategy where β_A and T_A have been previously agreed upon. However, each agent employs his own strategy $s_{a_i}(.)$ and his own utility function $U_{a_i}(.)$ in the decision-making processes (i.e., voting).

$$s_{a_i}(t) = 1 - (1 - RU_{a_i})(\frac{t}{T_A})^{(\frac{1}{\beta_A})} \tag{6}$$

Intra-Team Strategy
Each agents participates in the team decision-making process by means of voting processes.

- Offer proposal: The trusted mediator opens an anonymous process where each team member a_i can propose up to one offer $X^t_{a_i \to A}$ to the team. This offer has a utility equal to $s_{a_i}(t)$ for the proposing agent a_i. Once all of the offers XT^t have been received, the mediator makes them public and an anonymous voting process starts. Each teammate a_i classifies each offer $X^t \in XT^t$ as acceptable if $U_{a_i}(X^t) \geq s_{a_i}(t)$. The trusted mediator sums up acceptability votes and selects with the higher number of votes. This offer is sent to the opponent.
- Offer acceptance: The mediator makes public the offer sent by the opponent $X^t_{op \to A}$ and opens an anonymous voting process. The offer sent by the opponent is acceptable for a_i if $U_{a_i}(X^t_{op \to A}) \geq s_{a_i}(t+1)$. The mediator accepts the opponent offer if the majority of the team members have accepted the offer ($\frac{A}{2} + 1$).

4 Experiments

4.1 Experimental Settings

We decided to check the performance of both intra-team organization paradigms in different environments. Environments differ in team preference diversity: very similar teammates or very dissimilar teammates. Next, we detail how these environmental scenarios were generated:

- 25 different linear utility functions were randomly generated. These utility functions represented the preferences of potential team members for n=4 negotiation attributes. Team size was set to M=4 members. The dissimilarity between two utility functions (teammates) was measured taking a sample of 1000 offers and averaging the absolute difference in terms of the utility reported by the offer. The team dissimilarity (td) is calculated averaging the dissimilarity among pair of teammates. For all of the possible teams, we measured their dissimilarity and calculated the dissimilarity mean $\bar{td}$ and standard deviation σ. We considered $td \geq \bar{td} + 1.5\sigma$ as very dissimilar teams and $td \leq \bar{td} - 1.5\sigma$ as very similar teams. For each type of team, 100 random negotiation teams were selected for the tests.
- T_{op}, T_A are selected randomly from a uniform distribution U[5,10].
- β_{op}, β_A are randomly set from a uniform distribution U[0.4,0.99] (boulware).
- RU is randomly chosen from a uniform distribution U[0,0.25].

In each environmental scenario, we measure the performance of the proposed organizations (average utility for team members, and the number of negotiation rounds).

4.2 Results

The results for the experiments can be observed in Fig. 1. On the one hand, the results show that when teammates are very similar, the representative approach es capable of obtaining similar results to the voting approach in terms of average utility. However, the representative approach takes less number of rounds to finish than the voting approach. On the other hand, when teammates are very dissimilar, we can observer that the voting approach obtains significantly better results in terms of utility than the representative approach (Representative Confidence Interval 95%=[0.44-0.45], Voting Confidence Interval 95%=[0.57-0.58]). In this case, the number of rounds taken by the representative approach is also lower. These results imply that there is not an intra-team organization which is universally better than the other for every possible scenario. Thus, a more extended study of the performance of several organizations in different scenarios would allow negotiation teams to select the most adequate mechanism according to the current situation. This kind of study is pointed out as future work.

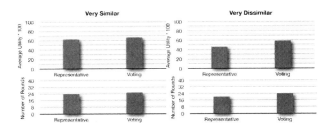

Fig. 1 Results for the experiments carried out. It show the average utility and the number of rounds for both methods in both scenarios

5   Related Work and Conclusions

As far as we are concerned, the topic of negotiation teams has not been thoroughly studied in agent literature, which has mainly focused on negotiation processes where parties represent individuals [2, 3, 6]. However, there are some topics which are closely related. Customer coalitions are groups of self-interested agents who join together in order to get volume discounts from sellers [4]. Customer coalitions usually consider scenarios where there is a single attribute that is equally important for every buyer. Negotiation teams also face the problem of multi-attribute tasks, where teammates may have different opinions about the different negotiation issues. Another topic that is closely related is multi-agent teams [1]. However, as far as we know negotiation teams have not been considered.

In this paper we have presented the novel topic of agent-based negotiation teams. First, we have presented a general workflow of tasks for agents that participate in negotiation team. After that, we have focused on studying two different intra-team organizations: one based on a representative, and the other based on voting processes. Results have shown that neither of both intra-team organizations is universally better than the other for every possible scenario. This points out to the fact that the performance of the different approaches may be affected by environmental settings. It may prove useful for agents to know which strategies work better in different scenarios in order to select the most adequate option. Thus, this fact leaves rooms for further research in providing agents with the proper knowledge to select the most adequate organization according to environmental settings.

Acknowledgments

This work is supported by TIN2008-04446, PROMETEO/2008/051, TIN2009-13839-C03-01, CSD2007-00022 of the Spanish government, and FPU grant AP2008-00600 awarded to Víctor Sánchez-Anguix.

References

1. Cohen, P., Levesque, H., Smith, I.: On team formation. In: Contemporary Action Theory. Synthesis, pp. 87–114. Kluwer Academic Publishers, Dordrecht (1999)
2. Ehtamo, H., Kettunen, E., Hamalainen, R.P.: Searching for joint gains in multi-party negotiations. European Journal of Operational Research 130(1), 54–69 (2001)
3. Faratin, P., Sierra, C., Jennings, N.R.: Negotiation decision functions for autonomous agents. Int. Journal of Robotics and Autonomous Systems 24(3-4), 159–182 (1998)
4. Li, C., Rajan, U., Chawla, S., Sycara, K.: Mechanisms for coalition formation and cost sharing in an electronic marketplace. In: ICEC 2003, pp. 68–77 (2003)
5. Luck, M., McBurney, P.: Computing as Interaction: Agent and Agreement Technologies. In: Proc. IEEE Conf. Distributed Human-Machine Systems, pp. 1–6 (2008)
6. Sánchez-Anguix, V., Valero, S., Julián, V., Botti, V., García-Fornes, A.: Genetic-aided multi-issue bilateral bargaining for complex utility functions. In: AAMAS 2010, pp. 1601–1602 (2010)
7. Sánchez-Anguix, V., Julian, V., Botti, V., Garcá-Fornes, A.: Towards agent-based negotiation teams. In: Group Decision and Negotiation 2010, pp. 328–331 (2010)

An Exploratory Analysis for Designing Robust Interactions in Multi-Agent Systems

Celia Gutiérrez and Iván García-Magariño

Abstract. In large multi-agent system (MAS), the definition of protocols between pairs of agents is not enough for guaranteeing the absence of undesirable communication organizations and the presence of desirable ones. This paper presents a technique and a tool for designing robust MAS communication architectures. In particular, in the proposed technique, designers are not only recommended to design the desired communication protocols, but also the undesired ones and the organization structures. The proposed tool performs data mining and on-line analytical processing (OLAP) on the logs of execution of MAS that use the Foundation for Intelligent Physical Agents (FIPA) protocol, in order to debug MAS communication architectures. The experimental results advocate that this approach is useful for MAS debugging and testing.

1 Introduction

Executions of concurrent systems become a complex task when the number of involved participants grows [1]. This is the case of MAS, where complex problems require a considerable number of agents that interact, giving place to executions difficult to be tracked and debugged by designers. Research on MAS has acquired more interest on debugging tools that provide aid about the executions, serving in two ways: testing that desired patterns appear, but also detecting the presence of undesirable patterns. Our work departs from precedent works, designing the patterns in a formal way, and using data mining and OLAP tasks to evaluate them from the logs of MAS executions. This work is an extension of [2], which uses sequential patterns to extract execution patterns from MAS execution logs, though the present work introduces a technique using more varied methods (e.g. OLAP),

Celia Gutiérrez
Departamento Ingeniería de Software e Inteligencia Artificial, Facultad de Informática, Universidad Complutense de Madrid, Jose Garcia Santesmases, s/n, 28040, Madrid, Spain
e-mail: `cegutier@fdi.ucm.es`

Iván García-Magariño
Departamento de Ciencias e Ingeniería, Facultad de Enseñanzas Técnicas. Universidad a Distancia de Madrid, c/ Camino de la Fonda 20, 28400 Collado Villalba Madrid, Spain
e-mail: `ivan.garcia-magarino@udima.es`

Y. Demazeau et al. (Eds.): Adv. on Prac. Appl. of Agents and Mult. Sys., AISC 88, pp. 225–230.
springerlink.com

and the aim is more complete, because it detects the desired and undesirable patterns from a previous design.

The proposed technique assists designers in defining robust communications. Desirable and undesirable patterns are provided in a formal way. These behaviors can be specific of the MAS (e.g. two types of agents that should not directly exchange information) or more general (e.g. deadlocks and redundancies). In fact, this paper presents a corpus of general behaviors that are usually undesirable. Designers must customize the patterns to the specific case, maybe adding or removing some ones. Then, the tool support guarantees the detection of the presence of undesirable behaviors and non presence of desirable ones. The tool support belongs to the OLAP (to identify undesirable relationships between agents) and the data mining classes (to detect sequential patterns). In brief, the structure of the document is the following: section 2 introduces the technique aspects for the communication agents; section 3 introduces the experimental results based on the development of the Cinema MAS, and describes the related work and conclusions.

2   Technique for Defining Robust Communications

The presented technique for defining robust communications is composed of two stages:

    1)   Stage 1: Definition of the general framework.
This stage is performed once for all MAS analysis, and is composed of the following steps:
-   Definition of a formal language for patterns. A simple but complete formal language is defined with the variables and operators necessary to define the regular expressions.
-   Definition of the regular expressions that recognize the desired and undesired patterns. These behaviors are general of all MAS.
-   Detection of the regular expressions in the MAS executions with a proper tool.

    2)   Stage 2: Customization of the environment.
This stage is executed for each MAS iteratively, and consequently can be repeated as many times as necessary for each MAS. Section 3 describes the details about an example of this customization. This stage contains the following steps:
-   Extraction of execution logs in the proper format.
-   Customization of the patterns defined in stage 1 with the concrete MAS, specifying the concrete agents and protocols.
-   Development of the MAS, detecting the desirable and undesirable patterns in the testing phase.
-   Interpretation of the detected patterns.

2.1 Definition of a Formal Language for Patterns and Associations

In the current approach, a MAS designer defines the desired and the undesired communication patterns. The first step is to define and implement them by the use of regular expressions. These patterns may involve one conversation, or may involve all the conversations of an execution. In this case, the interest is the discovery of patterns that appear a minimum of times.

For this purpose, a formal language, R, is necessary to define the patterns and associations. R must include a set of variables and operators, among the most important elements, to define the patterns and associations:

- $A, B, \ldots$ are the variables used for the sets of agents that identify the types of agents in the associations.
- [] are association operators for two types of agents or a type of agent with a concrete agent in a whole execution.
- If a is an agent of A type; b is an agent of B type, etc., $a, b, \ldots$ are the variables used in the sequences of the regular expressions.
- * is the operator used for any number of appearance of agents in a conversation, including zero. It also specifies that the precedent subsequence of agents appears more than once (for defining cycles).
- + is the operator used for at least one appearance of an agent of any type.
- () are association operators for a sequence of agents in a conversation.
- n is the threshold to determine bias between agents. In subsection 2.2 and section 3, this will be further explained.

2.2 Definition of the Patterns

In the proposed approach, designers define both the desirable and undesirable patterns for each MAS. In this manner, designers can easily debug and test their systems. This section proposes some undesirable and desirable patterns that are usually common in most developments.

Some common undesirable patterns using R notation are the following:

1. Deadlocks between a and b: *(ab)*. This situation occurs when each agent needs a progress of the other agent to progress. In this situation, none of these agents can progress.
2. Agents (agent a) at an inanition state, that they do not represent a final state, when the initial state is represented by another agent (agent b): $b*a$
3. Redundancies, or sequences among agents, that may be resumed in a single transition. These sequences may involve agents of any type: *$a+b$*, where ab is the desirable shortest subsequence.
4. Bias for an agent a toward the same agent b: in a whole execution, where n is the support of b on a: $(ab)n$
5. Unexpected association of two types of agents, A and B, in the whole execution: [AB]
6. Bias of association of agents of A type towards b agent, in a whole execution: [Ab]n

The desirable patterns are formulated as a desired sequence composed by at least two agents. These sequences depend on the specific conversation, like:

1. Expected sequence of agents of different types: $a \ b \ *$

2.3 Detection of the Patterns with a Suitable Tool

There are two types of patterns: sequences and associations. For the first one, discovery of sequential patterns is the most suitable data mining task for that purpose. The WUM [3] tool, which is already used in [2], provides a framework to extract them by means of the query language MINT.

As the purpose of WUM is to extract sequential patterns on logs of web server sessions, the MAS logs need to be readapted to the format of to web server logs, in the same way as in [2]. The current framework assimilates a web server to a MAS, and the visited web pages to message recipient agents. The framework keeps the timestamp, so WUM can organize the sequence of agents that participate in different conversations. With this information, it is easy to test the appearance of subsequences, and also the number of times and the order.

For the association patterns, OLAP techniques are suitable, because it provides the discovery of associations among n dimensions at different levels, which are collected in table 1:

Table 1 Dimensions and levels for OLAP operations

Dimension	Level
Conversation	1 (Conversation Identifier)
Conversation Initiator	2(Agent type; agent identifier)
Message receiver	2(Agent type; agent identifier)

OLAP operations count the number of occurrences (or intersections) among the three dimensions in exchanged messages of a whole execution. The level serves to identify the intersections at a more detailed level (*drill* operation) or at a higher level (*roll* operation). At the same time the designer can select the information using projections on the dimensions (*slice and dice* operation). Results are shown both graphically and numerically.

The presented technique and framework can be easily adapted to any MAS that follows the FIPA protocol. This adaptation only needs that the execution logs are provided in the correct format.

3 Experimental Results and Conclusions

The Cinema MAS [4] has been selected for the experiments, because one of its versions is complex enough with a great number of concurrent conversations.

Designers must decide the values of the A, B, a, b and n variables defined in the prior section, depending on the objective of the conversations. They can also customize the place of the subsequence. These values can vary for each pattern. In the Cinema case, the following values are chosen:

Undesirable patterns:

1. Deadlocks between any *InterfaceAgent* and any *BuyerAgent*. This transition must happen once.

2. Agents at an inanition state: any *SellerAgent* for the conversations initiated by a *BuyerAgent* or an *InterfaceAgent*.
3. Redundancies: the transition between the *BuyerAgent* and a *SellerAgent*
must be direct, without intermediaries.
4. Bias between a *BuyerAgent* toward the same *SellerAgent* (n=1).
5. Unexpected associations:
    a. *BuyerAgent* with *InterfaceAgent*.
    b. *InterfaceAgent* with *SellerAgent*.
6. Bias of the *BuyerAgents* towards certain *SellerAgent* (n=max
(count(SellerAgent)).

Desirable patterns:

1. Expected sequences:
    a. An *InterfaceAgent* must be followed by any *BuyerAgent* (2 times, one
    for the broadcasting), and then followed by the *InterfaceAgent*.
    b. A *BuyerAgent* must be followed by a *SellerAgent* (2 times, one for the
    broadcasting).

The logs have been imported in the right format as in [2]. The sequential patterns
have been built and the queries and OLAP have been customized. According to
this, table 2 summarizes the obtained results, detailing the identification and interpretation of the patterns that have been detected.

Table 2 Results and interpretation for the patterns

Type of pattern	Pattern-number	Result	Interpretation
Undesiderable	1	Not found	
	2	Not found	
	3	Not found	
	4	Not found	
	5a	Conversation ID 4.	Misused communication: revise conversation.
	5b	Not found	
	6	SellerAgent2_2: 789 times	2.1% deviation of the average, negligible
Desiderable	1a	Found	Expected result
	1b	Found	Expected result

There are different approaches that study the patters of execution in MAS such
as the Activity Theory (AT) [5]. We have followed a similar approach to formu-
late the patterns in a formal way, which is also used in these works to recognize
the system analysis diagrams. However, we use regular expressions. ACLAnalyser
tool is used for debugging MAS by means of clusters [6]. Our work specially resembles that work though our work also performs other data mining and OLAP

tasks. When patterns affect to the relationships among agents, we perform a technique that is similar to the inter-agent debugging of Sudeikat J et al [7]. This work resembles ours because it has the purpose of detecting message exchange patterns, though they do it validating the semantic of the agents by means of assertions.

Our work has been influenced by the mentioned ones, trying to grasp the most relevant aspects from them. The novelty of our work relies on a new technique for defining robust communications, in which both desired and undesired communication structures are defined. In addition, our work provides a corpus of general patterns that are usually undesirable in MAS developments. If the tool support detects any undesirable pattern, the designer is automatically warned. In this manner, the technique and tool support facilitate the debugging and testing.

Acknowledgements. This work has been developed with support of the grants TIN2008-06464-C03-01 by Spanish Council for Science and Innovation, and Banco Santander - UCM GR58/08.

References

1. Lijun, W., Jinshu, S., Kaile, S., Xiangyu, L., Zhihua, Y.: A concurrent dynamic logic of knowledge, belief and certainty for multi-agent systems. Knowledge-Based Systems 23(2), 162–168 (2010)
2. Gutierrez, C., Garcia-Magariño, I.: Extraction of excution patterns in multi-Agent Systems. IEEE Latin American Transactions 8(3), 311–317 (2010),
   http://ewh.ieee.org/reg/9/etrans/ieee/issues/vol8/
   vol8issue3June2010/8TLA3_15Gutierrez.pdf
   (accessed September 30, 2010) (in Spanish)
3. Spiliopoulou M, Faulstich LC, WUM 7.0 Beta,
   http://ka.rsten-winkler.de/hypknowsys/wum/download/
   WUM.v70.zip
   (accessed September 30, 2010)
4. García-Magariño, I., Gómez-Sanz, J.J., Fuentes-Fernández, R.: An evaluation framework for MAS modeling languages based on metamodel metrics. In: Luck, M., Gomez-Sanz, J.J. (eds.) AOSE 2008. LNCS, vol. 5386, pp. 101–115. Springer, Heidelberg (2009)
5. Fuentes, R., Gómez-Sanz, J.J., Pavón, J.: Activity theory for the analysis and design of multi-agent systems. In: Giorgini, P., Müller, J.P., Odell, J.J. (eds.) AOSE 2003. LNCS, vol. 2935, pp. 110–122. Springer, Heidelberg (2004)
6. Botia, J.A., Hernansaez, J.M., Skarmeta, F.G.: On the Application of Clustering Techniques to Support Debugging Large-Scale Multi-Agent Systems. In: Proceedings of Programming Multi-Agent System 2006, pp. 217–227 (2007)
7. Sudeikat, J., et al.: Validation of BDI Agents. In: Proceedings of the 4th international conference on Programming multi-agent systems, pp. 185–200 (2006)

Comparison of DCSP Algorithms:
A Case Study for Multi-agent Exploration

Pierre Monier, Arnaud Doniec, Sylvain Piechowiak, and René Mandiau

Abstract. Distributed Constraint Satisfaction Problems (DCSP) provide a formalism for representing, in a simple and efficient way, distributed problems. Different agents collaborate with each other in order to find a global solution subject to constraints. In general, researchers evaluate DCSP algorithms on graph coloring problems or uniform random binary DCSP. In this paper, we propose to compare DCSP algorithms on a multi-agent exploration problem. This problem is a particular case of multi-agent coordination problems and the evaluation of DCSP algorithms on this problem is difficult. In order to measure different aspects of this multi-agent exploration problem, we use six different criteria: classical criteria from DCSP benchmarks and specific criteria for this problem.

1 Introduction

In a multi-agent context, important issues concern both the study of coordination between agents and the reasoning performed by these agents. A lot of coordination problems are studied in Multi-Agent Systems such as foraging problem [9] or sensorDCSP [4]. In this paper, we focus on a multi-agent exploration problem [2]. A fleet of situated agents [8] must coordinate in order to explore an unknown

Pierre Monier · Arnaud Doniec · Sylvain Piechowiak · René Mandiau
Univ Lille Nord de France, F-59000 Lille, France
e-mail:{pierre.monier,sylvain.piechowiak,
        rene.mandiau}@univ-valenciennes.fr

Pierre Monier · Sylvain Piechowiak · René Mandiau
UVHC, LAMIH, F-59313 Valenciennes, France

Pierre Monier · Sylvain Piechowiak · René Mandiau
CNRS, UMR FRE 3304, F-59313 Valenciennes, France

Arnaud Doniec
Ecole des mines de douai, F-59500 Douai, France
e-mail: arnaud.doniec@mines-douai.fr

Y. Demazeau et al. (Eds.): Adv. on Prac. Appl. of Agents and Mult. Sys., AISC 88, pp. 231–236.
springerlink.com   © Springer-Verlag Berlin Heidelberg 2011

environment over connectivity constraints. To solve this problem, we assume that the reasoning is defined by a set of relations between variables and we use the Distributed Constraint Satisfaction Problem (DCSP) formalism [4, 11].

However, the use of DCSP for real multi-agent problems is not common. The evaluation of DCSP algorithms are usually based on "artificial" metrics like cycles or number of exchanged messages. Unfortunately these metrics do not always reflect the efficiency of the algorithms for real-world problems. Solving a real-world problem can bring up some issues which are not considered with classical benchmarks (graph coloring problems [7] or uniform binary random DCSPs [3]).

In this article, we have used different DCSP algorithms to solve a multi-agent exploration problem and we focus on the evaluation of theses algorithms. We analyze and compare different aspects of the multi-agent exploration problem using six different criteria from classical benchmark and real-world criteria.

Section 2 introduces generalities about DCSP. Section 3 briefly explains the multi-agent exploration problem. Experimental results are presented and discussed in section 4. Concluding remarks are provided in section 5.

2 Preliminaries

A Distributed Constraint Satisfaction Problem (DCSP) is a tuple (X, D, C, A). $X = \{X_1, X_2, \ldots, X_N\}$ is a finite set of N variables. A domain D_i is associated to each variable X_i and denotes the finite set of values for the variable. $C = \{C_1, C_2, \ldots, C_e\}$ is a finite set of constraints. $A = \{A_1, A_2, \ldots, A_m\}$ is a finite set of m agents. All agents encapsulate mutually exclusive subsets of X. A priority order (denoted as $\succ$) is assigned to each agent in order to avoid infinite loop problems.

Three DCSP algorithms will be compared on a multi-agent exploration problem:

- *ABT* [11] (for Asynchronous Backtracking) is often used as the reference algorithm by the community. It uses nogoods, i.e. a list of couples (*variable*, *value*) to determine the context of a backtrack message. In addition, each agent saves received nogoods in order to avoid submitting redundant information.
- *AWC* [11] (for Asynchronous Weak Commitment) is based on the principle of *ABT* combined with a dynamic priority for the agents.
- *DBS* [5, 6] (for Distributed Backtracking with Sessions) does not use a list of couples (*variable*, *value*) to manage the message context but only an integer called *session*.

3 Context of Exploration by Situated Agents

The exploration of an unknown environment under communication constraint is known as a difficult problem [10] in multi-agent systems. In such an application, we consider agents with communication abilities (usually based on Mobile Ad-hoc NETwork - MANET) and sensing abilities like laser, sonar or camera. These two abilities are subjected to hardware limitations. For instance, two agents can

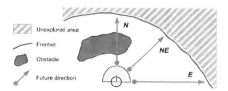

Fig. 1 Frontier based principle exploration

communicate only if they share a communication area (for example a wifi range). Based on the same idea, an agent can only sense away objects of its own environment: its perception is limited to a sensor range.

The agents move toward the frontier between open space and unexplored area [2]. In addition, agents have to collaborate to fan out on the environment and to keep in touch with each other (requirement to be able to exchange partial maps during the resolution and to maintain a communication link with a control center). These requirements can be integrated into a coordination scheme as constraints of a DCSP. Each agent has one variable to instantiate. This variable represents the next direction the agent will explore. The values of each domain are the 8 cardinal directions: $\{N, NE, E, SE, S, SW, W, NW\}$. The assignment of each variable has to satisfy two constraints [2]: (1) according to a direction, the future position of an agent does not have to break the connectivity of the network; (2) according to a direction, the future position of an agent does not have to create overlapping with the sensor range of its neighbors.

The need for each agent to come closer to the frontier at each movement can be expressed as a particular ordering of the eight directions present in the domains. This ordering can also take into account the potential obstacle. In figure 1, the shortest point of the frontier is situated at north. However, this point can not be reached since an obstacle is present on the agent's path. The two other directions, namely north-east and east both allow to reach a point of the frontier. But clearly, the frontier will be more quickly reached following the north-east direction. This means that the value NE will be before E and before N in the domain of the agent.

The exploration algorithm consists in repeating 3 steps: (1) Perception: each agent updates its knowledge about the ad-hoc network and updates its map with the last known position of its neighbors and the frontier; (2) Decision: constraints of the DCSP are computed. Each agent orders the values of its domain taking into account to the position of the frontier and the obstacle then the DCSP resolution begins; (3) Action: each agent operates its movement based on its future direction resulting from the resolution of the DCSP.

The mission is over when there is no more area to explore. Before starting the exploration, agents are laid out to random positions respecting the following rule: an agent must have at most two neighbors. Thanks to this rules, a wifi path exists between each pair of agents at the beginning of the exploration. The agents will maintain these path throughout all the exploration.

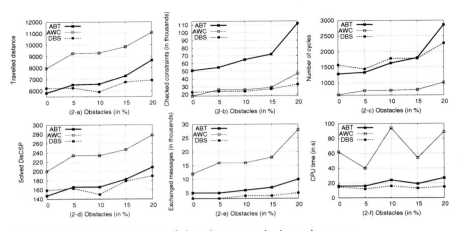

Fig. 2 Range of the percentage of obstacles present in the environment

4   Evaluation

This section presents the results obtained by *ABT*, *AWC* and *DBS* algorithms on the multi-agent exploration problem. Experiments were performed using a 2.8 Ghz dual-core computer having 3 Go of RAM on the multi-agent platform NetLogo. Our environment is closed and modeled as a 100x100 grid. We decided to use six criteria [5] to evaluate *ABT*, *AWC* and *DBS* algorithms: the length of the travelled path, the number of checked constraints, the number of cycles [11], the CPU time, the number of exchanged messages and the number of DCSP solved. These criteria are on x-axis for the figure 2 and each point represents an average of 10 explorations.

Figure 2 shows results where the agents explore an environnent containing obstacles represented by a square of 10 x 10 cells. The number of obstacles ranges from 0 % to 20 % of the environment. We note that the more percentage of obstacles increases, the less the agents are effective even if the number of empty cells to explore is lower. We observe that, using *AWC*, the agents are less effective to circle obstacles. The difference between *AWC* and others algorithms becomes larger when the number of obstacles increases. We see that agents in *AWC* travel a higher distance (fig. 2-a) in a higher CPU time (fig. 2-f) to explore the environment than for *ABT* and *DBS* which are very close. This is due to the dynamic priority in *AWC*. At each step of the simulation, the highest priority agent, which impose its favourite direction to lower priority agents (see [5]) changes. We see (fig. 2-c) that *AWC* has the lowest number of cycles but each cycle is more costly in *AWC* than for the other algorithms. The number of checked constraints in high for *ABT* (fig. 2-b) because of the creation of minimal nogoods [11]. We see (fig. 2-d) that for the three algorithms, the travelled distance is proportional to the number of solved DisCSP.

Others experiments were done varying the percentage of area to explore (from 40% to 100%). We have plot, for the 6 criteria, results with the surface to explore on x-axis. The curves are steepest when the surface to explore ranges from 40% to 60% than from 60% to 80%. This means that it requires more effort for agents to explore the environment from 40% to 60% rather than from 60% to 80%. Similarly, we observe that the curve increases strongly in the last percentages (nearly 100%). This is due to the exploration of scattered cells in the environment. When the agents need to explore more than 40% of the environment, *ABT* and *DBS* (which obtained very close results) are better than *AWC* because of *AWC*'s dynamic priority.

We done others experiments varying the wifi range of each agent from 20 to 50 (by default, we put it arbitrarily to 30). Recall that the wifi range represents the maximal distance beyond which the agents can not communicate. The more the wifi range increases, the more the agents are effective to explore the environment. For each of 6 criteria, with the *AWC* algorithm, the values decrease strongly from 20 to 30, moderate from 30 to 40 and less from 40 to 50. For algorithms *ABT* and *DBS*, the trends of the curves are slightly different: the curve decreases from 20 to 30-40 and then stagnates. A higher wifi range (> 50) would only have a very small impact on the multi-agent system performances. The lowest the wifi range is, the hardest to solve the DCSP is, due to the strong decrease of solutions for the DCSP. For these hard DCSP containing few solutions, we see that *AWC* is not very effective.

We have seen in figure 2 that *AWC* is not adapted because the dynamic priority penalizes the multi-agent coordination. Experiments done about the percentage of the unknown environnement to explore shows that the effort required to explore the environment is not the same at the beginning or at the end of the exploration. It is in the interval [60%-80%] that the rate (exploration efforts / gains obtained by exploring the cells) is the best. Experiments done about the variation of the wifi range shows that the more the wifi range is important, the less it requires effort to explore the environment. Others experiments were done and explained in [5] such as range of the numbers of agents or explored area by the highest priority agent.

AWC is known to be better than *ABT* for solving DCSP where the number of satisfiable instances is important. We can see that it is not the case here because the dynamic priority order for agents is not adapted to this multi-agent exploration problem (see [5]). Finally considering all obtained results, *DBS* which is very close to *ABT*, seems to be the best alternative among the three considered algorithms.

5   Conclusion

We have focused the article on the evaluation of three DCSP algorithms on a multi-agent exploration problem. The similarities and differences between the algorithms *ABT*, *AWC* and *DBS* were specified. Then, we detailed the multi-agent exploration problem and the choices we made. Unlike traditional DCSP benchmarks such as graph coloring problems or uniform random binary DCSP, we have seen that the use of a dynamic priority among agents is unfavorable to our problem. The results show that *DBS* is more effective than *ABT* and *ABT* is more effective than *AWC* for our problem. So, *DBS* seems to be the best choice.

Our results, obtained in simulation, will be confirmed, in a future work, with real experimentations on physical robots. Such experimentations adresse some problems. Just to name a few, we have to ensure that the localization of each robot is known at each decision step, the cardinal domain has to be express according the cinematic of each robot (range of the angular speed). Moreover, it will be interesting to evaluate the impact of algorithms used to detect the end of the DCSP search [1]. Maybe such algorithms could induce some latencies during the exploration.

Acknowledgment

This research work was financed by the european community, la Délégation Régionale la Recherche et la Technologie, le Ministère de l'Education Nationale, de la Recherche et de la Technologie, la région Nord Pas de Calais, le Centre National de la Recherche Scientifique, et l'Agence Nationale de la Recherche : projet AMORCES.

References

1. Chandy, K.M., Lamport, L.: Distributed snapshots: Determining global states of a distributed system. ACM Transactions on Computer Systems 3, 63–75 (1985)
2. Doniec, A., Bouraqadi, N., Defoort, M., Le, V.T., Stinckwich, S.: Distributed constraint reasoning applied to multi-robot exploration. In: 21st International Conference on Tools with Artificial Intelligence (ICTAI), New Jersey, USA (2009)
3. Ezzahir, R., Bessière, C., Wahbi, M., Benelallam, I., Bouyakhf, E.H.: Asynchronous inter-level forward-checking for discsps. In: Gent, I.P. (ed.) CP 2009. LNCS, vol. 5732, pp. 304–318. Springer, Heidelberg (2009)
4. Fernández, C., Béjar, R., Krishnamachari, B., Gomes, C.: Communication and computation in distributed CSP algorithms. In: Van Hentenryck, P. (ed.) CP 2002. LNCS, vol. 2470, p. 664. Springer, Heidelberg (2002)
5. Monier, P., Doniec, A., Piechowiak, S., Mandiau, R.: Metrics for the evaluation of discsp: Some experiments of multi-robot exploration. In: IEEE/WIC/ACM International Conference on Web Intelligence and Intelligent Agent Technology, Toronto, Canada, pp. 370–373 (2010)
6. Monier, P., Piechowiak, S., Mandiau, R.: A complete algorithm for discsp: Distributed backtracking with sessions (dbs). In: Workshop OptMas, Autonomous and Multi-Agent Systems (AAMAS 2009), Budapest, Hungary, Mai (2009)
7. Omomowo, B., Arana, I., Ahriz, H.: DynABT: Dynamic asynchronous backtracking for dynamic disCSPs. In: Dochev, D., Pistore, M., Traverso, P. (eds.) AIMSA 2008. LNCS (LNAI), vol. 5253, pp. 285–296. Springer, Heidelberg (2008)
8. Pepin, N., Simonin, O., Charpillet, F.: Intelligent tiles: Putting situated multi-agents models in real world. In: International Conference on Agents and Artificial Intelligence (2009)
9. Sempere, M., Aznar, F., Pujol, M., Rizo, R.: On cooperative swarm foraging for simple, non explicitly connected, agents. In: Practical Applications of Agents and Multi-Agent Systems, pp. 581–589. Springer, Salamanca (2010)
10. Vazquez, J., Malcolm, C.: Distribute multirobot exploration maintaining a mobile network. In: 2nd international IEEE Conference on Intelligent Systems (2004)
11. Yokoo, M.: Distributed Constraint Satisfaction: Foundation of Cooperation in Multi-agent Systems. Springer, Heidelberg (2000)

Heterogeneous Multiagent Architecture for Dynamic Triage of Victims in Emergency Scenarios

Estanislao Mercadal, Sergi Robles, Ramon Martí, Cormac J. Sreenan, and Joan Borrell

Abstract. This paper introduces a double multiagent architecture allowing the triage of victims in emergency scenarios and the automatic update of their medical condition. Gathering updated information about the medical condition of victims is critical for designing an optimal evacuation strategy that minimizes the number of casualties in the aftermath of an emergency. The proposed scheme, currently under development, combines Wireless Sensor Networks, an Electronic Triage Tag and a double multiagent system (Agilla-JADE) to achieve a low cost, no infrastructure-based, efficient system.

1 Introduction

The first moments of a mass casualty incident aftermath are decisive to save the highest number of lives. Only a reduced number of resources are often available at this point, making it essential to coordinate efforts and evacuate and give urgent treatment to the most seriously injured yet curable victims. Thus, the field work of trained personnel (doctors, nurses, paramedical) triaging victims accordingly to their medical status is of paramount importance. The triage information is later used to decide who are the victims to be moved first. For many years, this has been achieved by using a cardboard triage tag (like the one depicted in Fig. 1) which includes an easily identifiable color code indicating the severity of the injuries, besides other basic medical information. Triage personnel follows a standard expedited triage method like MTS [10] or START [17].

Estanislao Mercadal · Sergi Robles · Ramon Martí · Joan Borrell
Department of Information and Communications Engineering,
Universitat Autònoma de Barcelona,
08193 Cerdanyola del Vallès, Spain
e-mail: {emercadal,srobles,rmarti,jborrell}@deic.uab.cat

Cormac J. Sreenan
Departament of Computer Science, University College Cork, Cork, Ireland
e-mail: cjs@cs.ucc.ie

Y. Demazeau et al. (Eds.): Adv. on Prac. Appl. of Agents and Mult. Sys., AISC 88, pp. 237–246.

It is obvious that these scenarios can take a clear advantage of Information and Communication Technologies to improve efficiency, although some strong limitations have to be taken into account. Emergency personnel, for instance, have to act quickly and will refuse to interact with complex systems or fill in forms *in situ*. Communications, on the other hand, should not rely on any local infrastructure. Existing infrastructure (wired or not) can be damaged or out of order, and setting up a new one can be very expensive in terms of money and/or time.

Mobile Agent based Electronic Triage Tag (MAETT) [11] is an interesting agent-based application showing how mobile agents [3] can significantly help in this type of scenarios while keeping the budget low. A touch screen handheld device, equipped with a GPS receiver, wireless network interface and a JADE [1] mobile agent platform, is the key element of the system. When a victim is evaluated in the triage stage, a mobile agent is created holding the START color code, GPS position and other information. The traditional cardboard triage tag is also placed on the victim. This agent will make its way to the emergency coordination center, leaping forward from device to device. It may be stored in a handheld device for a while, being physically carried by a member of the emergency personnel, waiting until a candidate device is at reach. The network created in this fashion goes beyond *ad hoc* or MANET possibilities, for no concomitant communication links are required from the source device to the destination. The routing protocol uses the estimated time to return to the coordination center of the handheld device bearer.

MAETT works well, and makes information about victims available to the coordination center at a low cost. This information is later used to plan the evacuation of the victims. Unfortunately, changes in victims' medical conditions, which have a great impact on the subsequent rescue planning, are never conveyed to the coordination center. The solution to this is not trivial; the new information is not triggered by the field personnel but by the victims. The chances for personnel of finding a victim are much higher than the other way around.

Our final goal is to devise a new dynamic electronic triage tag system where a variation in the medical situation of victims can be easily communicated to the coordination center. This dynamism will be achieved by placing a wireless node equipped with medical sensors in every victim, in addition to the triage tag in the MAETT scenario. These nodes run an Agilla [5] mobile agent platform and can communicate with each other if they are near by forming a Wireless Sensor Network (WSN) [4]. The nodes monitor the victim's vital signs, and communicate any significant change to the emergency personnel nearby, should they are in range. The nodes in the WSN are very limited in resources, and cannot be compared to the devices carried by the emergency personnel. Thus, communication strategies, carried data and code differ completely from the network created in MAETT, and an interface is required to connect both.

To improve the probability of sending information from a node in the WSN to a handheld device, an efficient and adaptable mechanism based on mobile Agilla agents and genetic algorithms is currently under development, to explore all the nodes of the WSN and to fuse all the information about changes on victims' status. As a result, any node in the WSN will be able to communicate these changes as a whole to any JADE platform (triage or rescue member) that moves into the wireless coverage area of any node in the WSN. JADE mobile agents will act as data mules [16] for the independent WSNs created in the emergency area, and will carry to the coordination center the initial classification of the victims and any significant subsequent change in their medical status.

2   Background

2.1   MAETT

MAETT (Mobile Agent Electronic Triage Tag) [11] is a system providing early resource allocation during emergencies when no network infrastructure is available.

The foundation of the system is mobile agent technology [3], which allows information to be directly transported from terminal to neighbor terminal regardless of the status of the rest of the network at that particular time. Handheld devices run an execution environment for agents, the platform, where mobile agents can be created, executed and forwarded to other terminals. Are the agents themselves who decide the route to follow depending upon the available information on the neighbors.

The main actors of the system are the victims, the triage personnel, and the rescue teams (see scenario in Fig. 1). Victims are supposed to be scattered over an arbitrarily large area of emergency. The triage personnel scour all this area looking for victims and triage them according to standard methods. The result of this triage is written in a physical tag and placed visibly on the victim. Finally, the rescue teams collect all the victims, prioritizing depending on triage results. The Emergency Coordination Center (ECC) coordinates all actions. Triage personnel, rescue teams and the ECC have wireless devices with a JADE mobile agent platform and a GPS receiver.

Triage personnel leave the ECC, and have an estimation on when they will get back (TTR). When a victim is found, they use the standard START method [17] and place a cardboard triage tag (see Fig. 1) on the neck of the victim with their evaluation written on it. The tag has an integrated RFID. At the same time, an agent is created containing the information in the tag, plus the GPS position of the victim and the RFId of the tag. All this information will be used later in the ECC to optimize the route of the rescue teams. This agent is transmitted to neighbor devices only if the bearer has a lesser TTR. This is to make sure that moving the information is never going to make it arrive later. Consequently, all handheld devices carried by triage personnel are used to create agents with information about found victims, and also to forward agents corresponding to other victims. When agents arrive to the ECC, the ECC send the rescue teams with a detailed schedule of the route based on the GPS position of victims as well as their medical condition.

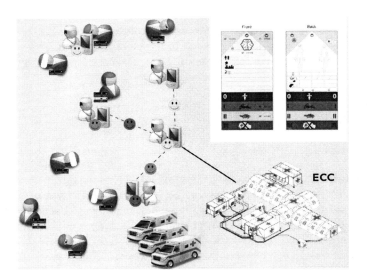

Fig. 1 MAETT triaging scenario showing also the classical cardboard triage tags

2.2 Wireless Sensor Networks and Agilla Mobile Agent Middleware

A Wireless Sensor Network (WSN) [4] is a specific type of ad hoc networks, built using wireless radio communication. It consists of sensor nodes collecting particular measures, i.e. sensor readings, and processing elements, which collect these measures for further processing. Usually sensor networks are strongly resource-restricted in terms of communication, processing and storage capabilities, and in terms of available energy. For the most part, all sensor nodes deliver their data to a base station or *sink* of data. This *sink* can be part of the network, or be external, accessed trough a gateway to other networks.

Application examples of WSNs [4, 5] are increasing, and include, among others, emergency operations, habitat monitoring, precision agriculture, home automation and health care or logistics. Initial applications within WSNs were static, i.e. all the nodes run statically installed software loaded prior to their deployment. Nowadays, diferent systems have been developed to allow more adaptable WSN applications (see [5] for an excellent survey of such systems), from mechanisms that allow a total or partial reprogrammation of the nodes, to middlewares that allow the execution of several mobile agents inside each node, allowing application self-adaptability. Among the different mobile agent middlewares proposed for WSNs, Agilla [5] is the first one deployed inside real WSNs.

Agilla provides a programming model in which applications consist of evolving communities of agents that share a WSN. Agents can dynamically enter and exit the WSN, can autonomously clone and migrate themselves in response to environmental changes, and can mantain a global coordination. Agilla was implemented on top of TinyOS WSN operating system [6], and experimentally evaluated on several real WSNs, for instance those consisting of TelosB [14] nodes.

3   Heterogeneous Multiagent Architecture to Provide Dynamism to MAETT

Despite being two different agent technologies, JADE and Agilla can coexist and share information to build a more complex agent system. Albeit agents themselves can not migrate transparently between the two platforms, little changes in Agilla's code make this cooperation possible.

In our solution, we use Agilla to continuously monitor Mass Casualty Incident (MCI) victims inside WSNs and JADE to carry the monitored data to the Emergency Coordination Center (ECC), introducing dynamism to MAETT. We take advantage of the communication between both technologies to share victim information and route details, thus improving the efficiency of the triaging system.

We extend MAETT scheme (Fig. 1) by adding to each victim a wireless human body monitoring device, for example that of Fig. 2(a), a TelosB compatible node manufactured by Maxfor (http://maxfor.co.kr), thus creating a WSN among neighboring victims, which we can use to dynamically update the medical status of every victim.

To communicate the WSN and the handheld devices of triage or rescue members we also need these members carry a WSN node (Fig. 2(b)) attached to their handheld device, also a TelosB compatible node from Maxfor.

3.1   WSN Set Up and Operation

In the first run of the triage personnel, every victim receives the classical triage tag, and it is also equipped with a wireless body monitoring sensor node. Inside this node there is an Agilla monitor agent with the specific data of the patient (Victim ID and Medical Condition). Each of this monitor agents continuously supervise the associated victim health constants mantaining an up-to-date log with his/her medical condition. In turn, the wireless body monitoring sensor node is identified by a two dimensional coordinate (WId, AId), the first one (WId) indicates its WSN and the second one (AId) identifies the node into the network.

As every nearby victim is both paper tagged and electronically tagged, neighboring wireless nodes belonging to the same triage member wirelessly connect creating a growing WSN of victims.

When every victim in the vicinity is tagged and every body sensor node monitors its own victim, the triage team member ends the creation of his WSN. First of all, the handheld device starts the calculation of a genetic algorithm following the appproach of [12, 18] (see Section 3.2) to compute a good enough itinerary to roam the newly created WSN to fetch the updates of every victim's medical condition. A roaming agent with this computed itinerary is then injected inside the last node of the WSN. This roaming agent gets and sets the differences in the state on every node.

When a new rescue or triage team member approaches the WSN, and their handheld device contacts a node in the WSN, the sensor node attached to the handheld device automatically identifies itself as the *sink* node. Then a mobile agent containing the information of every victim is sent to the node on the handheld device to flush all the gathered data. As every node of the WSN has the most up-to-date state as possible, any member of the WSN in direct contact with the *sink* node can indistinctly route the information. The handheld device then creates a JADE mobile agent containing the new status of the victims, and routes it to the ECC, as was done in MAETT.

(a) Watch type body monitoring device

(b) A sensor node connected to a handheld

Fig. 2 The different devices of the scenario.

To be able to use any node of the WSN to flush the status of every victim to the handheld device, every node must mantain a record of every other node status. This status is identified for each victim by her concrete medical condition, the ID of the victim, and a modification flag. The roaming mobile agent continuously updates this information mantaining an up-to-date record of every victim.

3.2 Roaming the Whole WSN

The problem of visiting every node avoiding repetition and optimizing distance is deeply studied in graph theory, also known as the Travelling Salesman Problem, and is known to be NP-Complete [13] even in the Euclidean plane.

The case of covering all the nodes of a WSN can be seen as a particular case of the Travelling Salesman Problem, with the particularity of pretending to also optimize the energy consumption, and is also proven to be NP-Complete [18] [12].

Genetic algorithms are being used to solve NP-Complete problems since the early 90's [7] and have been proved useful to efficiently solve the concrete particular of the Travelling Salesman Problem [2] [15] [8].

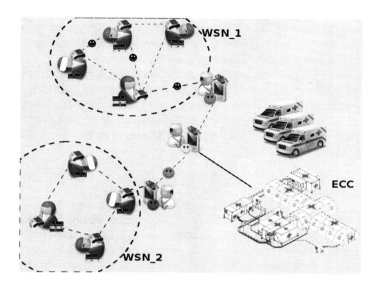

Fig. 3 Dynamic MAETT triaging scenario

In our handheld device, with its restrictions, both of time and computation power, is not feasible to calculate the optimal route for a problem of this kind. Thus, we use a highly studied genetic algorithm approximation [12], and which is proved to offer good enough solutions. Albeit not being the optimal solution, it returns a route which is satisfactory both in computation time and roaming time.

The chosen genetic algorithm starts creating an initial population of random paths that cover the network. Then, from that set of paths only those whose number of hops to cover the network is smaller or equal to the time remaining to meet the time deadline (maximum number of nodes per path) are selected. After that, a new generation is generated by selecting x individuals to be crossed. A subset of the paths of two individuals are crossed to create a new individual. Finally, all solutions are evaluated and sorted from the minimum to the maximum. We are currently adapting this genetic algorithm not to calculate the minimum route to a node, but to determine an efficient way to cover the whole WSN.

Additionally, if a problem in the communication between WSN nodes occur, e.g. due to a malfunctioning or broken node, the roaming agent has the ability to reroute itself and find an alternative path to cover the whole WSN. In our case the mobile agent tries to move to the next available node in the WSN. If it is not reachable tries to route the network backwards.

Note that the computation of the genetic algorithm in the handheld device is possible due to the following facts:

1. The triage personnel saves the topology of the WSN while tagging victims. When finished the device calculates a genetic algorithm adapted to that particular topology, and injects an agent with the final algorithm into the WSN. The Agilla agent

then covers the whole WSN collecting and informing changes in victim's state, thus writing the information of every victim in every node. The agent keeps doing the same route keeping the victim's medical condition up-to-date.

2. The number of nodes is limited. Both to speed up the route calculation process and to lighten the medical personnel bags. Notice that every health monitoring sensor weights about 70g including the required two 1.5V batteries. Assuming that the triaging personnel carries the handheld device and the paper tags, they may agree to carry the extra weight of 20 to 25 sensor nodes, that is, the weight of a small laptop (i.e. 1.5Kg).

3.3   WSN Maintenance

Another addition that boosts the dynamism of our triaging method is the possibility to modify an already deployed WSN, such that a victim can be added or removed, while being added to or removed from the itinerary of the roaming agent. Using JADE agents and their routes to the ECC, we have an excellent infrastructure to attain this dynamism.

In every triage, the handheld device of the triage personnel is also loaded with the position of every node of the WSN, thus ending up with a full view of the topology of the network. After the computation of the genetic algorithm this topology is also loaded into a JADE agent and routed to the ECC.

On arrival to the ECC, the JADE agent flushes the WSN topology, which is shared to every other handheld device deployed in the emergency. Hence if some other triaging member has to tag or remove a victim in another's WSN, he just removes or adds this new victim to the corresponding topology in the handheld device, and calculates a new itinerary (using our genetic algorithm) with the modification.

Of course, just as in the first calculation, the new topology is routed, along with the medical data of the victims in the WSN, with a JADE agent to the ECC.

3.4   The WSN – Handheld Device Interface

To properly set the interface we had to modify Agilla to run side-by-side with JADE. Doing so Agilla can obtain the instance of the running platform and generate an agent there. Then, we modified Agilla's AgentInjector to switch on JADE when invoked, this way we are sure that from that moment on, every recieved agent will be able to generate a JADE agent. The agent which migrates or sends data to the handheld device, the one to be encapsulated into a JADE agent, needs a special portion of code permitting the communication with the Injector at the other side of the USB interface. This code moves or sends the data to a special address pertaining to the Injector in the handheld device.

Our test platform (Nokia N810 with the Maemo OS (Diablo 5.2008.43.7)) did not come with the needed drivers for the mote to work. Thus, we had to recompile the stock kernel adding the appropiate drivers, and load them into the device. Moreover, albeit the device supports *host mode usb*, not every usb cable is prepared to notify the host device of this requirement, thus, we had to explicitly tell our test platform to switch itself to *host mode*. Once our test platform was plugged and the drivers loaded, the system recognized a new usb device, but complained about not being able to access it. The problem was related to the power the device was recieving not being sufficent to mantain it operating. We had to add a rule to the *udev* daemon of the handheld device indicating that this particular device needs more power than the allocated by default.

4 Conclusion

The triage stage during the aftermath of an emergency is critical to minimize the number of casualties. There are many systems in the bibliography introducing Information and Communication Technologies to achieve this, most times at a high cost or requiring a non-practical interaction with the system that field personnel refuse to perform. This paper presents a multiagent architecture allowing the triage of victims in emergency scenarios and the automatic update of their medical condition at a low cost.

In our approach, neighbor victims are automatically grouped together by placing a wireless sensor node in each of them monitoring their status. This groups form a Wireless Sensor Network. Changes are shared within the WSN using Agilla mobile agents and a gathering algorithm designed using genetic algorithms. Now, changes can be communicated to any member of the emergency personnel by any node of the group, increasing in this way the probabilities of sending this information. From there, a JADE mobile agent will carry the changes to the coordination center where they will update old information.

Currently we are working on the implementation, and simulating different routing strategies using *TOSSIM* and *TinyViz* [9] to determine the number of agents needed to cover the WSN as fast as possible and therefore propagating the changes in victim's state rapidly.

Acknowledgements

This work has been funded by the Spanish Ministry of Science and Innovation through the project TIN2010-15764.

References

1. Bellifemine, F.L., Caire, G., Greenwood, D.: Developing multi-agent systems with JADE. Wiley Inc., Chichester (2007)
2. Braun, H.: On solving travelling salesman problems by genetic algorithms. In: Schwefel, H.P., Männer, R. (eds.) Parallel Problem Solving from Nature. LNCS, vol. 496, pp. 129–133. Springer, Heidelberg (1991)
3. Cucurull, J., Ametller, J., Martí, R.: Agent mobility. In: Developing multi-agent systems with JADE, pp. 115–130. Wiley Inc., Chichester (2007)

4. Dressler, F.: Self-Organization in Sensor and Actor Networks. Wiley Inc., Chichester (2007)

5. Fok, C.-L., Roman, G.-C., Lu, C.: Agilla: A mobile agent middleware for self-adaptive wireless sensor networks. ACM Trans. Auton. Adapt. Syst. 4(3), 1–26 (2009)

6. Hill, J., Szewczyk, R., Woo, A., Hollar, S., Culler, D., Pister, K.: System architecture directions for networked sensors. In: Proceedings of the ninth international conference on Architectural support for programming languages and operating systems, pp. 93–104. ACM, New York (2000)

7. De Jong, K.A., Spears, W.M.: Using genetic algorithms to solve NP-complete problems (1989)

8. Larrañaga, P., Kuijpers, C.M.H., Murga, R.H., Inza, I., Dizdarevic, S.: Genetic algorithms for the travelling salesman problem: A review of representations and operators. Artificial Intelligence Review 13, 129–170 (1999), doi:10.1023/A:1006529012972

9. Levis, P., Lee, N., Welsh, M., Culler, D.: TOSSIM: accurate and scalable simulation of entire tinyos applications. In: SenSys 2003: Proceedings of the 1st international conference on Embedded networked sensor systems, pp. 126–137. ACM, New York (2003)

10. Mackway-Jones, K. (ed.): Emergency triage, 2nd edn. Wiley Inc., Chichester (2006)

11. Martí, R., Robles, S., Martín-Campillo, A., Cucurull, J.: Providing early resource allocation during emergencies: The mobile triage tag. Journal of Network and Computer Applications 32, 1167–1182 (2009)

12. Massaguer, D.: Multi mobile agent deployment in wireless sensor networks. Master's thesis, University of California, Irvine (2005)

13. Papadimitriou, C.H.: The Euclidean travelling salesman problem is NP-complete. Theoretical Computer Science 4(3), 237–244 (1977)

14. Polastre, J., Szewczyk, R., Culler, D.: Telos: Enabling ultra-low power wireless research. In: IPSN 2005 Proceedings of the 4th international symposium on Information processing in sensor networks, pp. 364–369. ACM and IEEE, New York (2005)

15. Potvin, J.-Y.: Genetic algorithms for the traveling salesman problem. Annals of Operations Research 63, 337–370 (1996), doi:10.1007/BF02125403

16. Shah, R.C., Roy, S., Jain, S., Brunette, W.: Data mules: Modeling a three-tier architecture for sparse sensor networks. In: Proc. Sensor Network Protocols and Applications (SNPA), pp. 30–41. IEEE, Los Alamitos (2003)

17. Super, G.: START: a triage training module. Hoag Memorial Hospital Presbyterian, Newport Beach, California (1984)

18. Wu, Q., Rao, N.S.V., Barhen, J., Iyengar, S.S., Vaishnavi, V.K., Qi, H., Chakrabarty, K.: On computing mobile agent routes for data fusion in distributed sensor networks. IEEE Trans. on Knowl. and Data Eng. 16(6), 740–753 (2004)

A Decision Support System for Hospital Emergency Departments Built Using Agent-Based Techniques

Manel Taboada, Eduardo Cabrera, and Emilio Luque

Abstract. This paper presents the results of an ongoing project that is being carried out by the Research Group in Individual Oriented Modelling (IoM) of the University Autonoma of Barcelona (UAB) with the participation of Hospital Emergency Department (ED) Staff Teams. Its general objective is creating a simulator that, used as decision support system (DSS), aids the heads of the ED to make the best informed decisions possible. The defined ED model is a pure Agent-Based Model, formed entirely of the rules governing the behaviour of the individual agents which populate the system. The actions of agents and the communication between them are represented using Moore state machines extended to include probabilistic transitions. The model also includes the environment in which agents move and interact. With the aim of verifying the proposed model an initial simulation has been created using NetLogo.

Keywords: Healthcare operational management, agent-based modeling, individual oriented simulation, emergency department, decision support systems.

1 Introduction

Hospital Emergency Departments (ED) may well be one of the most complex and fluid healthcare systems that exists, consuming a large portion of economic budgets for health services. However, patients often feel neglected and that the service is saturated. The simulation of complex systems is of considerable importance and is used in a broad spectrum of fields such as engineering, biology, economy and health care. There are no standard models to describe these complex systems, but they may share many common traits. Agent-Based Modelling (ABM) is an efficient and well

Manel Taboada
Tomas Cerda Computing Science School
e-mail: manel.taboada@eug.es

Eduardo Cabrera · Emilio Luque
Computer Architecture and Operating Systems Department (CAOS) -
Universidad Autonoma of Barcelona (UAB), Spain
e-mail: ecabrera@caos.uab.es, emilio.luque@uab.es

Y. Demazeau et al. (Eds.): Adv. on Prac. Appl. of Agents and Mult. Sys., AISC 88, pp. 247–253.

utilised technique that has many advantages, amongst others its increased detail in experiments based on simulation, a transparent learning process, and the ability to control and modify individual behaviour.

This paper presents the results of an ongoing project whose general objective is to create a simulator that, used as decision support system (DSS), aids the heads of the ED to make the best informed decisions possible. The construction of the tool is carrying out following an iterative and spiral process. Each cycle involves 5 phases (system analysis; model design; simulator implementation; simulator execution and results analysis; simulator validation). Once a cycle is completed, based on the conclusions obtained during the analysis and validation phase, the model is updated and a new cycle is carried out. The process will be repeated until the objectives are achieved. Through the first cycle it has started the design of an Agent-Based Model for ED, in which all rules within the model concern the agents; no higher level behavior is modeled. The System behavior emerges as a result of local level actions and interactions. This model describes the complex dynamics found in an ED, representing each individual and system as an individual agent. State machines are used to represent both, the actions of each agent and the communication between agents. Such communication is modelled as the inputs that agents receive and the outputs they produce. In order to control the agent interaction, the physical environment in which these agents interact has also to be modelled.

The remainder of this article is organised as follows; section 2 describes the related work. The proposed ED model is detailed in section 3, while the results of an initial simulation are given in section 4. Section 5 closes with future work and conclusions.

2   Related and Previous Work

The modelling and simulation of ED sits at the intersection of a number of distinct fields. Agent-based techniques have been used in the modelling of healthcare operational management, but there are few pure agent-based models to be found in the literature that have been rigorously validated against their real world counterparts. Economics, biology, and social sciences are the three fields in which agent-based models are most utilised [1]. Modelling techniques using agents can bring the most benefit when applied to human systems where agents exhibit complex and stochastic behaviour, the interaction between agents are heterogeneous and complex, and agent positions are not fixed [2]. In the particular case of social sciences ABM are used in situations where human behaviour cannot be predicted using classical methods such as qualitative or statistical analysis [3]. Human behaviour is also modelled with ABM in the fields of psychology [4] and epidemiology [5] amongst others.

Agent technology is a useful tool when applied to healthcare applications. Previous works modelling healthcare systems have focused on patient scheduling under variable pathways and stochastic process durations, the selection of an optimal mix for patient admission in order to optimise resource usage, and patient throughput [1]. Work has been performed using differing degrees of agent-based modelling for evaluating patient waiting times under the effects of different ED physician staffing schedules [6] or patient diversion strategies [7]. Having into account the general objectives of the project presented in this paper, the potential improvements in relation to the previous work can be summarized in: 1) Generality of the model: what means the possibility of applying the model and the tool in different Hospital ED, after a tuning process of the configuration parameters that will be carried out through parallel simulations and similarity search techniques; 2) the variety of patients of this model is higher and closer to the reality than the models that have been found in the previous works. In addition, this model includes ED Staff of different kinds (expertise), taking into account how these differences affect the length of stay of patients in the emergency department.

3 Emergency Department Model

The ED model defined in this work is a pure Agent-Based Model, formed entirely of the rules governing the behaviour of the individual agents which populate the system. Through the information obtained during the interviews carried out with ED staff at the Hospital of Mataro and the Hospital of Sabadell, two kinds of agents have been identified: active and passive agents. The active agents represent people and other entities that act upon their own initiative (patients, companions of patients, admission staff, sanitarian technicians, triage and emergency nurses, staff emergency doctors, specialists, and social workers). The passive agents represent systems that are solely reactive, such as the loudspeaker system, patient information system, pneumatic pipes, and central diagnostic services (radiology service and laboratories).

This section is dedicated to describe the various components of the general model in detail. Section 3.1 explains the manner in which active agents are modelled. Passive agents are discussed in section 3.2. The communication model is defined in section 3.3. Finally in section 3.4 the details of the environment where the agents interact are outlined.

3.1 Active Agents

Active agents are described through Moore state machines. Each state has an output, and transitions between states are specified by the input. The current state of an active agent is represented by a collection of "state variables", known as the state vector (T). Each unique combination of values for these variables defines a distinct state. A basic set of state variables and some of their possible values has been defined through the round of interviews performed (see table 1). The variable "personal details" collects the patient's information which is important in relation with his stay in the ED. The "physical condition" and "symptoms" contain that information that will let the ED Staff to identify the patient's priority level. The "communication skills" and the "level of experience" of the ED staff will influence in the time required for completing the process. Both the variables and their possible values should be improved in the future work.

Table 1 Initial selection of state variables and their values

Variables	Values
Name/identifier <id>	Unique per agent
Location <location>	Entrance; admissions; waiting room; triage box; Consultancy Room; Treatment Box; ...
Action	Idle; requesting information from <id>; giving information to <id>; searching; moving to <location>; ...
Physical condition	Hemodynamic-constant; Bartel index (sample)
Symptoms	Healthy; cardiac/respiratory arrest; severe/moderate trauma; headache; vomiting; diarrhoea; (sample)...
Communication skills	Low; medium; high
Level of experience (Staff)	None; Low; medium; high

Upon each time step the state machine moves to the next state. This may be another state or the same one it was in before the transition. The next state the machine takes depends on the input received. The input may be more accurately described as an input vector (I) that contains a number of input variables, each one of which may take a number of different values. As this is a Moore machine, the output depends only on the state, so each state has its own output, although various states may have outputs that are identical. Again, the output is more accurately described as an output vector (O), a collection of output variables, each with a number of defined possible values. Thus transitions between states are dependent on the current state at time t (St) and the input at time t (It). In dynamic and complex systems such as ED exists the necessity for a model not to be entirely deterministic. In these cases the state machine can be modelled with more than one possible next state, giving an input combination to the current state. The specific transition is chosen randomly at the time of the transition, based on weights that provide a means for specifying transitions that are more or less likely for a given individual. In these cases the state transition table is defined with probabilities on the "effect" of the input. The state machine can be represented as a "transition state table", as shown in figure 1 (a), where each row represents a unique state input combination, showing the output and the state in the next time step, but also in form of diagram (shown in figure 1 (b)).

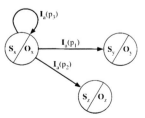

Current state/Output	Input	Next state/Output
--	--	--
S_x / O_x	$I_a(p_1)$	S_y / O_y
S_x / O_x	$I_a(p_2)$	S_z / O_z
S_x / O_x	$I_a(p_3)$	S_x / O_x
--	--	--

Fig. 1 (a) Probabilistic state transition table; (b) Probabilistic state transition graph

3.2 Passive Agents

The state machine of the passive agents will be a simple system for interacting with active agents. The model is not, however, purely a state machine. In order to represent data storage or other systems that may have a very large number of combinational states a simple memory model will be used. In some cases passive agent may (although it is not necessary) have a simple record based memory system, allowing it to store and repeat information provided by active agents.

3.3 Communication Model

The interaction between agents is carried out through communication. Such communication is modelled as the Input that agents receive and the Outputs that they produce. The communication model represents three basic types of communication: 1) **1-to-1**, between two individuals (the message has a single source and a single destination, as happens between admission staff and patient, during the admission process); 2) **1-to-n**, representing an individual addressing to a group (like a doctor giving information to patient and nurses during the diagnostic process); 3) and **1-to-location**, when an individual speaks to all occupants of a specific area (for instance when any staff member uses the speaker system to address a message to all the people who are in a specific waiting room). Messages contain three parts; the "message source" (the individual who is communicating, speaking in many cases), the "message destination" (the individual to whomever is speaking to), and thirdly the "content" (what is being said). These three parts form the message tuple (<src>, <dst>, <content>).

3.4 Environment

All actions and interactions modelled take place within certain locations, collectively known as the environment. Although the environment itself can be defined in two different levels (at a low or high granularity scale) depending on the positional precision required for the model, in the specific case of the ED it is enough use a low granularity positioning scale, although it is important to represent distances between distinct areas, to correctly model travel times form one area to another. Figure 2 shows a representation of topographical distribution of the ED.

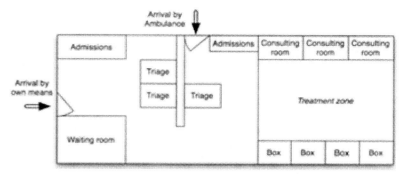

Fig. 2 Simplified Emergency Department layout.

4 Initial Simulation

With the aim of verifying the model designed in the first cycle, an initial simulation has been created using the agent-based simulation environment NetLogo, a high level platform particularly well suited for modelling complex systems developing over time [8]. Such simulation initially uses a simplified set of patient attributes and a less complicated patient flow in order to make a preliminary demonstration of how accurate the simulation can be produced using only reduced parameters. Four primary areas have been considered: admissions, triage (3 boxes), waiting rooms (one for patients before triage, and the second for patients who have passed the triage process, and are waiting for treatment), and diagnosis and treatment area (that include four boxes). The types of active agents represented in this simulation are patients, admission staff, triage nurses, and doctors. In the case of the ED Staff, two distinct levels of experience have been considered (low and high). Less experienced staff will need more time to carry out his part of the process than the most experienced one. In this model patients are shown following the same path through the ED, even though in reality they are treated differently depending on the level of severity of their condition. The time spent at diagnosis stage may also represent laboratory tests, which are not shown explicitly. Finally user can define easily both, the number of each type of ED Staff and their level of expertise using a "console of configuration".

Netlogo stores information about all what happens during the running, but in this initial experiment the report only includes information about the number of patients (the amount that have arrived to the ED during the running and the amount that have completed the process), and the time that each patient have spent in each one of the ED zones (absolute values, mean, minimum, maximum and standard error). In order to analyze the behaviour of the simulator in front of the variables that have influence in the ED, the simulation has been executed several times for the equivalent to one day of activity. The results obtained let to conclude that the gradual increase in the number of ED staff and the improving of their experience cause a gradual growth in the number of patients treated and a reduction of the time spent in the ED. There is further validation to be performed, but these initial experiments show very promising results about its usefulness to support the decision making process.

5 Future Work and Conclusions

A concrete example of Agent-Based Model for Hospital Emergency Departments has been presented, which represents a hospital ED following system analysis performed at a number of different hospitals, under the advice of healthcare professionals with many years of experience. The model uses state machine based agents which act and communicate within a defined environment, providing the ability to study the dynamic of complex systems without the difficulty of obtaining exhaustive system descriptions required by other modelling paradigms.

An initial simulation has been created in order to verify the validity of the model. In initial stages future improvements of the model will be carried out adding gradually new agents and state variables. In the next stages parallel simulations with different parameters will be performed, and after comparing data from simulation and real system, adjustments in the model will be made in order to achieve both, a proper similarity level, and an enough predictive power of the simulator. Once the predictive ability of the simulator achieves the proper level, the tool could be used as DSS, in order to aid healthcare managers make the best informed decisions possible. The simulator will let to divine what will happen to the ED a whole if one or more changes are made to the parameters that define it.

The distribution pattern of patients' arrival to the ED varies depending on time, day of week and season. For this reason is desirable to run simulations for an annual period of time. In addition, as a result of the potential number of individuals, and also of states in the state machine of each individual, a great amount of values should be computed. Considering also the parallel simulations that will have to be performed during the tuning process, High Performance Computing will have to be used.

Acknowledgements. Supported by the MICINN Spain under contract TIN2007-64974.

References

1. Hutzschenreuter, A.K., et al.: Agent-based patient admission scheduling in hospitals. In: Proceedings of the 7th Int. Joint Conf. on Autonomous Agents and Multiagent systems, pp. 45–52 (2008)
2. Bonabeau, E.: Agent-based modeling: Methods and techniques for simulating human systems. Proceedings of the National Academy of Sciences 99, 7280–7287 (2002)
3. Norling, E., Sonenberg, L., Rönnquist, R.: Enhancing multi-agent based simulation with human-like decision making strategies. In: Moss, S., Davidsson, P. (eds.) MABS 2000. LNCS (LNAI), vol. 1979, pp. 214–228. Springer, Heidelberg (2001)
4. Smith, E.R., Conrey, F.R.: Agent-based modelling: A new approach for theory building in social psychology. Pers. Soc. Psychol. Rev. 11(1), 87–104 (2007)
5. Epstein, J.M.: Modelling to contain pandemics. Nature 460, 687 (2009)
6. Jones, S.S., Evans, R.S.: An agent based simulation tool for scheduling emergency department physicians. In: AMIA Annual Symposium Proceedings, p. 338 (2008)
7. Laskowski, M., Mukhi, S.: Agent-based simulation of emergency departments with patient diversion. In: Weerasinghe, D. (ed.) eHealth 2008. Lecture Notes of the Institute for Computer Sciences, Social Informatics and Telecommunications Engineering, vol. 1, pp. 25–37. Springer, Heidelberg (2009)
8. Allan, R.: Survey of Agent Based Modelling and Simulation Tools. STFC Daresbury Laboratory – Warrington (2010)
   http://193.62.125.70/Complex/ABMS/ABMS.html
   (accessed December 2010)

An Agent-Based Dialog Simulation Technique to Develop and Evaluate Conversational Agents*

David Griol, Nayat Sánchez-Pi, Javier Carbó, and José M. Molina

Abstract. In this paper, we present an agent-based dialog simulation technique for learning new dialog strategies and evaluate conversational agents. Using this technique the effort necessary to acquire data required to train the dialog model and then explore new dialog strategies is considerably reduced. A set of measures has also been defined to evaluate the dialog strategy that is automatically learned and compare different dialog corpora. We have applied this technique to explore the space of possible dialog strategies and evaluate the dialogs acquired for a conversational agent that collects monitored data from patients suffering from diabetes.

1 Introduction

Conversational agents have became a strong alternative to provide computers with intelligent and natural communicative capabilities. A conversational agent is a software that accepts natural language as input and generates natural language as output, engaging in a conversation with the user. To successfully manage the interaction with the users, conversational agents usually carry out five main tasks: automatic speech recognition (ASR), natural language understanding (NLU), dialog management (DM), natural language generation (NLG) and text-to-speech synthesis (TTS).

The application of statistical approaches to the design of this kind of agents, specially regarding the dialog management process, has attracted increasing interest during the last decade [8]. Statistical models can be trained from real dialogs, modeling the variability in user behaviors. The final objective is to develop

---

David Griol · Nayat Sánchez-Pi · Javier Carbó · José M. Molina
Group of Applied Artificial Intelligence (GIAA),
Computer Science Department, Carlos III University of Madrid
e-mail: {david.griol,nayat.sanchez,
        javier.carbo,josemanuel.molina}@uc3m.es

* Funded by projects CICYT TIN2008-06742-C02-02/TSI, CICYT TEC2008-06732-C02-02/TEC, CAM CONTEXTS (S2009/TIC-1485), and DPS2008-07029-C02-02.

Y. Demazeau et al. (Eds.): Adv. on Prac. Appl. of Agents and Mult. Sys., AISC 88, pp. 255–264.
springerlink.com   © Springer-Verlag Berlin Heidelberg 2011

conversational agents that have a more robust behavior and are easier to adapt to different user profiles or tasks.

The success of these approaches depends on the quality of the data used to develop the dialog model. Considerable effort is necessary to acquire and label a corpus with the data necessary to train a good model. A technique that has currently attracted an increasing interest is based on the automatic generation of dialogs between the dialog manager and an additional module, called the user simulator, which represents user interactions with the conversational agent [7, 4]. This way, a very important application of the simulated dialogs is to support the automatic learning of optimal dialog strategies.

In this paper, we present an agent-based dialog simulation technique to automatically generate the data required to learn a new dialog model for a conversational agent. We have applied our technique to explore dialog strategies for the DI@L-log conversational agent, designed to collect monitored data from patients suffering from diabetes. In addition, a set of specific measures has been defined to evaluate the main characteristics of the acquired data and the new dialog strategy that can be learned from them. The results of the comparison of these measures for an initial corpus and a corpus acquired using the dialog simulation technique show how the quality of the dialog model is improved once the simulated dialogs are incorporated.

The remainder of the paper is organized as follows. Section 2 describes the proposed agent-based dialog generation technique and the measures used to evaluate the quality of dialogs acquired with different dialog strategies. Section 3 describes the DI@L-log conversational agent and the acquisition of a initial corpus for this task. Section 4 shows the results of the comparison of the measures for the two corpora acquired for the DI@L-log task. Finally, some conclusions and future work lines are described in Section 5.

2 Our Agent-Based Dialog Simulation Technique

Our proposed architecture to provide context-aware services by means of conversational agents is described in [3]. It consists of five different types of agents that cooperate to provide an adapted service. *User agents* are configured into mobile devices or PDAs. *Provider Agents* supply the different services in the system and are bound to *Conversational Agents* that provide the specific services. A *Facilitator Agent* links the different positions to the providers and services defined in the system. A *Positioning Agent* communicates with the ARUBA positioning system to extract and transmit positioning information to other agents in the system. Finally, a *Log Analyzer Agent* generates user profiles that are used by Conversational Agents to adapt their behaviour taking into account the preferences detected in the users' previous dialogs.

In this paper we focus on the simulation of the user and conversational agents to acquire a dialog corpus. In our dialog generation technique, both agents use a random selection of one of the possible responses defined for the semantics of the task (expressed in terms of user and system dialog acts). At the beginning of the simulation, the set of system responses is defined as equiprobable. When a success-ful dialog is simulated, the probabilities of the answers selected by the the conver-sational agent simulator during that dialog are incremented before beginning a new simulation.

An error simulation agent has been implemented to include semantic errors in the generation of dialogs. This module modifies the dialog acts by the user agent simulator once it has selected the information to be provided to the user. In addition, the error simulation module adds a confidence score to each concept and attribute in the semantic representation obtained from the user turn.

For the study presented in this paper, we have improved this agent using a model for introducing errors based on the method presented in [6]. The generation of confidence scores is carried out separately from the model employed for error gen-eration. This model is represented as a communication channel by means of a gen-erative probabilistic model $P(c, a_u | \tilde{a}_u)$, where a_u is the true incoming user dialog act $\tilde{a}_u$ is the recognized hypothesis, and c is the confidence score associated with this hypothesis.

The probability $P(\tilde{a}_u | a_u)$ is obtained by Maximum-Likelihood using the initial labeled corpus acquired with real users and considers the recognized sequence of words w_u and the actual sequence uttered by the user $\tilde{w}_u$. This probability is de-composed into a component that generates a word-level utterance from a given user dialog act, a model that simulates ASR confusions (learned from the reference tran-scriptions and the ASR outputs), and a component that models the semantic decod-ing process.

$$P(\tilde{a}_u | a_u) = \sum_{\tilde{w}_u} P(a_u | \tilde{w}_u) \sum_{w_u} P(\tilde{w}_u | w_u) P(w_u | a_u)$$

Confidence score generation is carried out by approximating $P(c | \tilde{a}_u, a_u)$ assuming that there are two distributions for c. These two distributions are handcrafted, gen-erating confidence scores for correct and incorrect hypotheses by sampling from the distributions found in the training data corresponding to our initial corpus.

$$P(c | a_w, \tilde{a}_u) = \begin{cases} P_{corr}(c) & if \quad \tilde{a}_u = a_u \\ P_{incorr}(c) & if \quad \tilde{a}_u \neq a_u \end{cases}$$

The conversational agent simulator considers that the dialog is unsuccessful when one of the following conditions takes place: i) the dialog exceeds a maximum num-ber of system turns slightly higher than the average number of turns of the dialogs acquired with real users; ii) the answer selected by the dialog manager in the conver-sational agent simulator corresponds to a query not made by the user simulator; iii) a query to the database generates an error because the user agent simulator has not provided the mandatory data needed to carry out the query; iv) the answer generator generates an error when the answer selected by the conversational agent simulator involves the use of a data item not provided by the user agent simulator.

A user request for closing the dialog is selected once the conversational agent simulator has provided the information defined in its objective(s). The dialogs that fulfill this condition before the maximum number of turns are considered successful.

Figure 1 shows the complete architecture for the proposed dialog simulation technique.

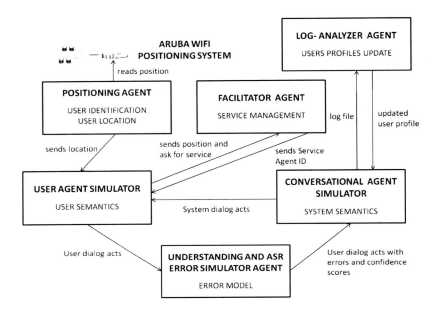

Fig. 1 Graphical scheme of the proposed agent-based dialog simulation technique

For the evaluation of the quality of the dialogs provided by a conversational agent, we have defined a set of quantitative evaluation measures based on prior work in the dialog literature [5, 1]. This set of proposed measures can be divided into two types:

- High-level dialog features: These features evaluate how long the dialogs last, how much information is transmitted in individual turns, and how active the dialog participants are.
- Dialog style/cooperativeness measures: These measures analyze the frequency of different speech acts and study what proportion of actions is goal-directed, what part is taken up by dialog formalities, etc.

Six high-level dialog features have been defined for the evaluation of the dialogs: the average number of turns per dialog, the percentage of different dialogs without considering the attribute values, the number of repetitions of the most seen dialog, the number of turns of the most seen dialog, the number of turns of the shortest dialog, and the number of turns of the longest dialog. Using these measures, we tried to evaluate the success of the simulated dialogs as well as its efficiency and variability with regard to the different services.

For dialog style features, we defined and counted a set of system/user dialog acts. On the system side, we have measured the confirmation of concepts and attributes, questions to require information, and system answers generated after a database query. On the user side, we have measured the percentage of turns in which the user carries out a request to the system, provide information, confirms a concept or attribute, Yes/No answers, and other answers not included in the previous categories.

3 Case Application: The DI@L-log Conversational Agent

DI@L-log is a spoken conversational agent which acts as a voice logbook to collect home monitored data from patients suffering from Type-2 diabetes [2]. The data collected by the system are the patient's weight, blood pressure (systolic and diastolic values) and sugar levels. The system validates and analyses the data, providing some immediate feedback to the patients regarding their current progress as well as communicating the results to doctors at the hospital who are able to review the patient's progress graphically and deal with any alerts generated by the system.

To apply our dialog simulation technique to this task, the definition of the semantics was carried out considering the information that is required to monitor the patients and inform them about their condition. Taking into account the requirements of the task, an initial strategy was designed for the DM. Figure 2 shows an example of a dialog acquired using this strategy.

S1: Welcome to the DI@L-log System. Please say or enter your four digit pin.
U1: 1356.
S2: Are you ready to give your measurements?
U2: Yes.
S3: Let's begin with your weight. How many stones?
U3: 12.
S4: I have understood 12 stones. Is it correct?
U4: Yes.
S5: What is your blood sugar?
U5: 11.
S6: I have understood 11. Is it correct?
U6: Yes.
S7: And finally, for your blood pressure you will be asked for the top number and then the lower number. What is the top number of your blood pressure?
U7: 160.
S8: I have understood 160. Is it correct?
U8: Yes.
S9: And what is the lower number?
U9: 86.
S10: I have understood 86. Is it correct?
U10: Yes.
S11: Your readings have been successfully updated. According to them, your weight is higher than last week, be very careful. Thanks for contacting DI@L-log. Goodbye.

Fig. 2 An example of a dialog for the DI@L-log task

As can be observed, three different phases are present in every dialog. Firstly, there is an identification phase in which the system asks the user about his login and password and then waits until the user says that he is ready to provide the control data (S1 and S2 system turns). Secondly, the system analyzes which data is required for the current user, taking into account that the weight and sugar values are mandatory and the blood control is only carried out for specific patients (S3 to S10 system turns). In this phase, the system requires the user to provide this data. Every item is confirmed after the user has provided its value. The user can only provide one item at a time. In the last phase, the system consults the information that the patient has provided during the current dialog and compares it with the data that is present in a database that contains the values that he provided in previous dialogs. By means of this comparison, the system is able to inform the user about his condition and provide him with instructions that take this into account (S11 system turn).

A corpus of 100 dialogs was acquired using this strategy. In order to learn statistical models, the dialogs of the corpus were labeled in terms of dialog acts. In the case of user turns, the dialog acts correspond to the classical frame representation of the meaning of the utterance. For the DI@L-log task, we defined three task-independent concepts (*Affirmation*, *Negation*, and *Not-Understood*) and four attributes (*Weight*, *Sugar*, *Systolic-Pressure*, and *Diastolic-Pressure*).

The labeling of the system turns is similar to the labeling defined for the user dialog acts. A total of twelve task-dependent concepts was defined, corresponding to the set of concepts used by the system to acquire each of the user variables (*Weight*, *Sugar*, *Systolic-Pressure*, and *Diastolic-Pressure*), concepts used to confirm the values provided by the user (*Confirmation-Weight*, *Confirmation-Sugar*, *Confirmation-Systolic*, and *Confirmation-Diastolic*), concepts used to inform the patient about his condition (*Inform*), and three task-independent concepts (*Not-Understood*, *Opening*, and *Closing*).

4 Evaluation Results

By employing the agent-based architecture proposed in Section 2 for dialog simulation, a set of 100,000 automatically labeled dialogs was acquired for the DI@-log task. Table 1 summarizes the statistics of the acquisition of this simulated corpus. A set of 11 different scenarios was defined to specify the objectives of the simulation, taking into account if the pressure values are necessary and different possibilities for the generation of errors and confidence measures. Given that the first and third phases of the dialog are mandatory and they have always the same structure, only the second phase in which the system collects the different values to monitor patients was taken into account for the simulation.

Table 1 Statistics of the corpus acquisition for the DI@L-log system

Simulated dialogs	100,000
Successful dialogs	27,521
Different dialogs	1,573

Figure 3 shows an example of a dialog from the acquired corpus. The objective defined for the dialog was to collect the weight, sugar and pressure values. The values defined in the scenario are 12, 11, 160, and 80 respectively. Confidence scores generated by the error simulator agent are shown between brackets. A sentence in natural language, that is equivalent to the labeled system turn selected by the conversational agent simulator, has been incorporated to clarify the explanation.

LOGIN PHASE	U4: (*Systolic*): 150 [0.1]
	S5: (*Confirmation-Systolic*)
S1: (*Weight*)	I have understood 150. Is it correct?
Tell me your weight in stones?	U5: (*Rejection*) [0.9]
U1: *Weight*: 12 [0.1]	S6: (*Systolic*)
S2: (*Confirmation-Weight*)	Tell me your systolic pressure.
I have understood 12 stones. Is it correct?	U6: *Systolic*: 160 [0.9]
U2: (*Acceptance*) [0.9]	S7: (*Diastolic*)
S3: (*Sugar*)	And what is the lower number?
What is your blood sugar?	U7: *Diastolic*: 80 [0.9]
U3: *Sugar*: 11 [0.9]	
S4: (*Systolic*)	DATA ANALYSIS - FINAL ANSWER
Tell me your blood systolic pressure.	

Fig. 3 A dialog extracted from the simulated corpus of the DI@L-log task

In this dialog, the conversational agent simulator begins asking the user about his weight. As a low confidence measure is introduced for the value provided by the user agent simulator in U1, the conversational agent simulator decides to confirm this value in S2. Then, this agent asks for the sugar value. The user agent simulator provides this value in U3 and a high confidence measure is assigned.

The conversational agent simulator asks for the systolic pressure in S4. An error is introduced in the value provided by the error simulator agent for this parameter (it changes 160 to 150) and a low confidence measure is assigned to this value. Then, the conversational agent simulator asks the user agent simulator to confirm this value. The user agent simulator rejects this value in U5 and the conversational agent simulator decides to ask for it again. Finally, the conversational agent simulator asks for the diastolic pressure. This value is correctly introduced by the user agent simulator and the error simulator agent also assigns a high confidence level. Then, the conversational agent simulator obtains the data required from the patient, next the third phase of the dialog carries out the analysis of the condition of the patient and finally it informs him.

4.1 High-Level Dialog Features

The first group of experiments covers the following statistical properties to evaluate the quality of the dialogs obtained using different dialog strategies: i) Dialog length, measured as the number of turns per task; number of turns of the shortest dialog; number of turns of the longest dialog; and number of turns of the most seen dialog; ii) Different dialogs in each corpus, measured as the percentage of different dialogs (different labeling and/or order of dialog acts) and the number of repetitions of the most observed dialog; iii) Turn length, measured as the number of actions per turn; iv) Participant activity, measured as the ratio between system and user actions per dialog. Table 2 shows the comparison of the different high-level measures for the initial corpus and the corpus acquired incorporating the successfully simulated dialogs.

Table 2 Results of the high-level dialog features defined for the comparison of the dialogs for the initial and final strategy

	Initial Strategy	Final Strategy
Average number of turns per dialog	12.9±2.3	7.4±1.6
Percentage of different dialogs	62.9%	78.3%
Repetitions of the most seen dialog	18	3
User turns of the most seen dialog	9	7
User turns of the shortest dialog	7	5
User turns of the longest dialog	13	9

The first improvement that can be observed is the reduction in the number of turns. This reduction can also be observed in the number of turns of the longest, shortest and most seen dialogs. These results show that improving the dialog strategy makes it possible to reduce the number of necessary system actions. The greater variability of the resulting dialogs can be observed in the higher percentage of different dialogs and less repetitions of the most seen dialog obtained with the final dialog strategy.

We have observed that there is also a slight increment in the mean values of the turn length for the dialogs acquired with the final strategy. These dialogs are statistically longer, as they showed 1.6 actions per user turn instead of the 1.3 actions observed in the initial dialogs. This is also due to the better selection of the system actions Regarding the dialog participant activity, Figure 5 shows the ratio of user versus system actions. Dialogs in the final corpus have a higher proportion of system actions because the systems needs to make a smaller number of confirmations.

4.2 Dialog Style and Cooperativeness

The experiments described in this section cover the following statistical properties: frequency of different user and system actions (dialog acts), and proportion of goal-directed actions (request and provide information) versus grounding actions (confirmations). We consider as well the remaining possible actions. The histograms in Figure 4 show the frequency of the most dominant user and system dialog acts in

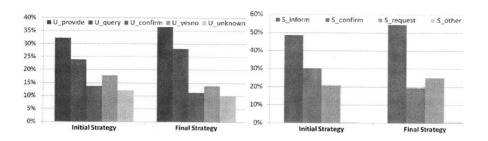

Fig. 4 Histogram of user dialog acts (left) and system dialog acts (right)

the initial and final strategy. In both cases, significant differences in the dialog acts distribution can be observed.

With regard to user actions, it can be observed that users need to employ less confirmation turns in the final strategy, which explains the higher proportion for the rest of user actions in this strategy. It also explains the lower proportion of yes/no actions in the final strategy, which are mainly used to confirm that the system's service has been correctly provided. With regard to the system actions, it can be observed a reduction in the number of system requests for data items. This explains a higher proportion of turns to inform and confirm data items in the dialogs of the final strategy. Finally, we grouped user and system actions into categories in order to compare turns to request and provide information (goal directed actions) versus turns to confirm data items and make other actions (grounding actions), as shown in Figure 5. This study also shows the better quality of the dialogs in the final strategy, given that the proportion of goal-directed actions is higher in these dialogs.

5   Conclusions

In this paper, we have described a technique for exploring dialog strategies in conversational agents. Our technique is based on an automatic dialog simulation technique to generate the data that is required to re-train a dialog model. The results of applying our technique to the DI@L-log system, which follows a very strict initial interaction flow, show that the proposed methodology can be used to automatically explore new enhanced strategies. Carrying out these tasks with a non-automatic approach would require a very high cost that sometimes is not affordable. As a future work, we are adapting a previously developed statistical dialog management technique to learn a dialog manager for this task and evaluate the complete agent-based architecture with real users.

References

1. Ai, H., Raux, A., Bohus, D., Eskenazi, M., Litman, D.: Comparing Spoken Dialog Corpora Collected with Recruited Subjects versus Real Users. In: Proc. of the 8th SIGdial Workshop on Discourse and Dialogue, Antwerp, Belgium, pp. 124–131 (2007)
2. Black, L., McTear, M.F., Black, N.D., Harper, R., Lemon, M.: Appraisal of a conversational artefact and its utility in remote patient monitoring. In: Proc. of the 18th IEEE Symposium CBMS 2005, Dublin, Ireland, pp. 506–508 (2005)
3. Griol, D., Sánchez-Pi, N., Carbó, J., Molina, J.: An Architecture to Provide Context-Aware Services by means of Conversational Agents. Advances in Intelligent and Soft Computing 79, 275–282 (2010)
4. Paek, T., Horvitz, E.: Conversation as action under uncertainty. In: Proc. of the 16th Conference on Uncertainty in Artificial Intelligence, San Francisco (USA), pp. 455–464 (2000)
5. Schatzmann, J., Georgila, K., Young, S.: Quantitative Evaluation of User Simulation Techniques for Spoken Dialogue Systems. In: Proc. of the 6th SIGdial Workshop on Discourse and Dialogue, Lisbon, Portugal, pp. 45–54 (2005)
6. Schatzmann, J., Thomson, B., Weilhammer, K., Ye, H., Young, S.: Agenda-Based User Simulation for Bootstrapping a POMDP Dialogue System. In: Proc. of Human Language Technologies HLT/NAACL 2007 Conference, Rochester, USA, pp. 149–152 (2007)
7. Schatzmann, J., Weilhammer, K., Stuttle, M., Young, S.: A Survey of Statistical User Simulation Techniques for Reinforcement-Learning of Dialogue Management Strategies. Knowledge Engineering Review 21(2), 97–126 (2006)
8. Young, S.: The Statistical Approach to the Design of Spoken Dialogue Systems. Tech. rep., CUED/F-INFENG/TR 433, Cambridge University Engineering Department, Cambridge, UK (2002)

Cloud Computing Service for Managing Large Medical Image Data-Sets Using Balanced Collaborative Agents

Raúl Alonso-Calvo, Jose Crespo, Victor Maojo, Alberto Muñoz,
Miguel García-Remesal, and David Pérez-Rey

Abstract. Managing large medical image collections is an increasingly demanding important issue in many hospitals and other medical settings. A huge amount of this information is daily generated, which requires robust and agile systems. In this paper we present a distributed multi-agent system capable of managing very large medical image datasets. In this approach, agents extract low-level information from images and store them in a data structure implemented in a relational database. The data structure can also store semantic information related to images and particular regions. A distinctive aspect of our work is that a single image can be divided so that the resultant sub-images can be stored and managed separately by different agents to improve performance in data accessing and processing. The system also offers the possibility of applying some region-based operations and filters on images, facilitating image classification. These operations can be performed directly on data structures in the database.

1 Introduction

Different medical imaging devices and techniques used in clinical routine in many settings, such as CT, MRI, PET, X-Rays, etc., generate a huge quantity of information every day. Moreover, continuous advances in the optical resolution of medical capturing instruments and sensors imply that image sizes are growing, introducing larger storage requirements.

Raúl Alonso-Calvo · Jose Crespo · Victor Maojo · Miguel García-Remesal · David Pérez-Rey
DLSIIS & DIA - Biomedical Informatics Group,
Facultad de Informática, Universidad Politécnica de Madrid, Campus de Montegancedo,
Avda Montepríncipe, S/N, 28660 Boadilla del Monte, Spain
e-mail: ralonso@infomed.dia.fi.upm.es

Alberto Muñoz
School of Medicine, Universidad Complutense de Madrid, Spain

Y. Demazeau et al. (Eds.): Adv. on Prac. Appl. of Agents and Mult. Sys., AISC 88, pp. 265–270.
springerlink.com   © Springer-Verlag Berlin Heidelberg 2011

Building effective storage and indexation systems for medical images poses a challenging problem. The standard approach Picture Archiving and Communication Systems (PACS) aims to address and solve important issues like robustness, scalability, interoperability, and simple access to data. For this purpose there are several projects that use image databases with Content-Based Image Retrieval (CBIR) techniques.

Different kinds of CBIR have been proposed. CBIR systems usually utilize low-level features (such as, for example, color histogram techniques and texture analysis) for performing image matching. In this group we can include well-known systems like QBIC [7], Virage [3], VisualSeek [11] and Blobworld [4]. However, medical images and their contents are usually linked to semantic information such as patient data, date, image type, and additional information (such as the body parts shown in the image). By using such textual information, software systems can refine content-based image searches.

In this context, storing and managing this medical imaging information becomes a crucial task for most medical organizations. Besides, integrating this multimedia information with other textual sources is useful for different aspects related to clinical workflows, such as diagnosis, therapeutic decisions, research and teaching.

In this paper we present a distributed multi-agent prototype system capable of storing and indexing very large image collections. This approach extends previous work on our group in ontology-based database integration of structured [2] and unstructured sources [10], [8] and in image processing [5], [6].

Distributing large image databases is commonly used for improving scalability and systems response. A distinctive aspect of our work with respect to other systems previously mentioned is that, in those other systems, data distribution is limited to spreading images to different servers, whereas in our system an image can be divided so that image parts can be stored in two or more different agents. This aims to improve performance.

This paper is organized as follows. In the next section we describe the methods used to create our multiagent system for storing large collections of images. Next, we present and discuss the results obtained for querying and retrieving information from stored images in the system. Finally, a conclusion section ends the paper.

2   Methods

2.1   Cloud Computing Service

As is well known, cloud computing [9] is intended to offer an interface to use virtual resources accessible on demand, and they can be controlled by the service owner. A goal of our work is to offer a cloud computing service for storing, analyzing and retrieving medical multimedia information. Institutions and medical professionals that do not have enough storage and computing resources could manage medical data through applications built on top of these types of services. This kind of Virtual PACS based on the Cloud is different to other previously reported systems.

The Cloud computing service is accessed by users having an authorized account. Images loaded by users are normally private, although they can be set as public if desired. By using Cloud services for accessing images, private PACS belonging to concrete medical organizations can be extended in a relatively easy manner.

2.2   Multi-Agent System Architecture

A system prototype for storing and managing images has been implemented following the agent-oriented paradigm. There are three agent roles in the system:

(i)   The Cloud Computing Service Access Point Agent is a unique agent in the system. This agent is an entry point for final registered users. It offers methods for (a) storing images in the system, (b) applying operations on stored images, and (c) querying for images in the system.

(ii)   The Resource Index Agent can be seen as a directory service. This agent contains information about location, current load, estimation function for performance prediction, and image processing status, of each working agent of the system.

(iii)   Working Agents are at the core of the system. They contain all the functionalities implemented in the system, namely extracting image low-level and semantic features, storing images and its associated information, retrieving them, and applying filters and operations.

Each working agent manages a relational database implementing the region-based datastructure created for storing images. Besides, they have methods implemented to be capable of:

- Dividing an input image. As result of this division a set of processed regions and two untreated subimages are obtained. Detailed explanation of the algorithm used can be found in our previous work [1].
- Extracting regions, low-level descriptors, relationships between regions, and textual semantic information (contained, if available, in DICOM headers) from the input image.
- Storing processed information in the graph-based data-structure. By inserting the data obtained in a relational database with the graph-based datastructure logical schema.
- Sending untreated parts of the input image, obtained in the division step, to other agents to be processed.
- Obtaining and updating information from the resource index agent for load balancing.
- Communicating to other agents to synchronize information of divided images.
- Applying filters and operations that have been implemented in the system. We use mainly region-based (or graph-based) morphological operations.

2.3   Managing Images in the Prototype: Agent Behaviors

We can distinguish three different behaviors of agents in the system depending on the kind of task that has to be performed: (i) processing and storing a new image, (ii) performing a filter or operation on an image, and (iii) querying the system for image contents (this query can be a query by image-content, or simply accessing to parts of a given image).

(i)   When a new image is introduced in the system to be stored, the input image is sent to a working agent that follows an algorithm with six main steps, depicted in Figure 1. Note: An input image is divided if its area is larger than a given threshold [1].

(ii) When a petition of applying a filter or operation on an image is received in the cloud computing service access point agent, it spread the message to those working agents that contains sub-images of the processed image. Then those working agents initialize the requested operation in their stored sub-images belonging to the selected image. These filters and operations are directly applied to the data-structure stored in the agent local database. Distributed adaptations of some most representative morphological region-based operations has been developed the system, namely erosion, dilation, and watershed.

(iii) When the system is queried by a final user for obtaining a part of the information stored (through the cloud computing service access point), the query is propagated to all agents of the system (as in project Ontofusion [2], [10]). Each agent returns information matching with the user query restrictions that are stored in its database. Note that results are not necessarily entire images. All results are returned by the cloud computing service access point agent to the final user.

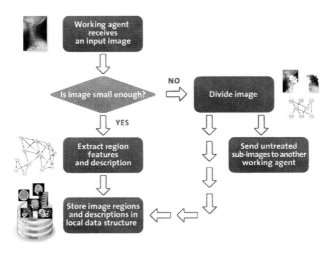

Fig. 1 Storing an image in the prototype

3 Evaluation and Discussion

We have conducted various image retrieval experiments using a set of images stored in the system. As illustrative cases that correspond to three selected images of different sizes, we have displayed in some figures execution times for three images: Image 1 is 1000x1000, Image 2 is 3000x3000, and Image 3 is 5000x5000. Results in seconds for some example queries in the system are shown when the number of agents (which are deployed in different servers) is increased from 1 to 3. Images are divided among the available agents. When more than one agent is involved, the execution time is the maximum of all agents.

Figure 2 shows the time reduction in a crop operation (where a 300x600 pixel rectangle is retrieved from the image, with the corresponding label of the region to which it belongs).

Finally, Figure 3 visualizes the execution time of a more complex image multi-criteria query. Particularly, the query has retrieved all regions having (a) an area less than a constant, (b) an intensity value greater than a threshold, and (c) both a major and a minor axis smaller than 90% the image size (such a multi-criteria query is often used to eliminate background regions).

As can be observed the time needed for executing these queries decrease when the number of agents in the system increases. The data division and distributed management among different agents, has decreased the time needed by database searches and operations.

Many of the existing database CBIR use distributed systems for storing images. This image distribution in different servers reduces the time needed for database queries. Besides this kind of parallelization, our system can also divide input images and store subimages in a distributed manner to further reduce the time for queries. This paper has focused on this second, novel type of reduction. Another important feature of our system is that operations are applied directly to databases.

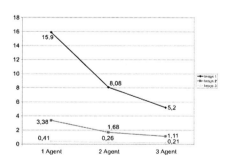

Fig. 2 Crop operation   **Fig. 3** Multi-criteria query

4 Conclusions and Future Work

In this paper we have presented a scalable multi-agent system prototype capable of storing very large image collections. The system offers methods for managing, analyzing and querying images by content through a cloud computing service. The system uses a distributed region-oriented data-structure that can be stored, in a distributed manner, in one or more databases. Some experimental results have been provided that show the performance increase in normally used operations when the number of agents is increased.

First, we plan to extend the image operators implemented by adding new ones, such as region merging. Another planned improvement is to create methods in the cloud computing service for offering intelligent streaming of image contents, which can be used to visualize complex medical image data.

Acknowledgements

This work has been supported in part by "Ministerio de Ciencia e Innovación" of Spain (Ref.: TIN2007-61768).

References

1. Alonso-Calvo, R., Crespo, J., García-Remesal, M., Anguita, A.: On distributing load in cloud computing: A real application for very-large image datasets. In: International Conference on Computational Science, vol. 1(1), pp. 2663–2671 (2010), doi:10.1016/j.procs.2010.04.300
2. Alonso-Calvo, R., Maojo, V., Billhardt, H., Martín-Sánchez, F., García-Remesal, M., Pérez-Rey, D.: An agent- and ontology-based system for integrating public gene, protein, and disease databases. Journal of Biomedical Informatics 40(1), 17–29 (2007)
3. Bach, J., et al.: Virage Image search engine: An open framework for image management. In: SPIE Storage and Retrieval for Image and Video Databases, vol. 2670, pp. 76–87 (1996)
4. Carson, C., Thomas, M., Belongie, S., Hellerstein, J.M., Malik, J.: Blobworld: A system for region-based image indexing and retrieval. In: Huijsmans, D.P., Smeulders, A.W.M. (eds.) VISUAL 1999. LNCS, vol. 1614, pp. 509–517. Springer, Heidelberg (1999)
5. Crespo, J., Serra, J., Schafer, R.W.: Theoretical aspects of morphological filters by reconstruction. Signal Process. 47(2), 201–225 (1995)
6. Crespo, J., Schafer, R.W., Serra, J., Gratin, C., Meyer, F.: The flat zone approach: a general low-level region merging segmentation method. Signal Process. 62(1), 37–60 (1997)
7. Flickner, M., et al.: Query by Image and Video Content: The QBIC System. IEEE Computer 28(9), 23–32 (1995)
8. Maojo, V., García-Remesal, M., Billhardt, H., Alonso-Calvo, R., Pérez-Rey, D., Martín-Sánchez, F.: Designing new methodologies for integrating biomedical information in clinical trials. Methods of Information in Medicine 45(2), 180–185 (2006)
9. NIST. definition of Cloud Computing, vol. 15 NIST (2010), http://csrc.nist.gov/groups/SNS/cloud-computing/ (cited April 2011)
10. Pérez-Rey, D., Maojo, V., García-Remesal, M., Alonso-Calvo, R., Billhardt, H., Martín-Sánchez, F., Sousa, A.: Ontology-based integration of genomic and clinical databases. Comput. Biol. Med. 36(7-8), 712–730 (2006)
11. Smith, J., Chang, S.: Blobworld: VisualSeek: A fully Automated content-based image query system. In: ACM International Conference on Multimedia, pp. 87–98 (1996)

Software Agents as Cloud Computing Services

Ignacio Lopez-Rodriguez and Mario Hernandez-Tejera

Abstract. The community devoted to the agency theory needs to create practical solutions capable of representing users in the virtual societies which are emerging as a result of Internet. Cloud Computing has precisely succeeded as a model able to bring software solutions to users in a practical and transparent manner. To foster the adoption of agent-based solutions by the users, this paper describes a model where software agents figure as a new Cloud Computing service which would represent clients in virtual environments. We discuss the challenges that entail the proposal, the technologies necessary for its implementation and finally we develop a proof of concept to confirm its viability.

1 Introduction

As a result of the success of Internet and the continuous progress of information technologies are emerging virtual societies from which users could gain many benefits through automating their participation. Intelligent agents are usually proposed as a feasible technology for providing autonomy and intelligence to this task. However, the community devoted to the agency theory has not managed to create models adapted to the actual user habits, who increasingly demand transparent and simple solutions which hide the technological details, as shows the recent success of the Cloud Computing model. With the aim of achieving these properties, this paper develops the Agency Services model in which software agents figure as another cloud service which companies and consumers can contract so they can participate in virtual societies without being taken up by the technological details and without sacrificing any of the advantages of intelligent agents.

Ignacio Lopez-Rodriguez · Mario Hernandez-Tejera
Institute for Intelligent Systems (SIANI), University of Las Palmas de Gran Canaria,
Las Palmas, Spain
e-mail: {ijlopez, fmhernandez}@siani.es

Y. Demazeau et al. (Eds.): Adv. on Prac. Appl. of Agents and Mult. Sys., AISC 88, pp. 271–276.
springerlink.com   © Springer-Verlag Berlin Heidelberg 2011

The remainder of this paper is organized as follows. Section 2 discusses the lack of success of intelligent agents in recent virtual societies. Section 3 describes the key concepts of the Cloud Computing paradigm. Section 4 introduces the Agency Services model. Section 6 concludes.

2   Searching for the Intelligent Agents

While there are many solutions in which software agents are run on the server side and middleware layers, the community has not managed to create successful agent-based models capable of representing users in the virtual societies that are emerging as a result of Internet. Some representative examples of these shortcomings are the auction portals, computational grids and sites for buying and selling. Most proposals in these areas focus on algorithm optimization and designing architectures which are impractical and even require users to implement their own agents by using development kits [7]. These proposals are not aligned with the type of software that currently predominates, which entail no additional user efforts. This condition is even more important in the case of multi-agent systems, since the problems involved are complex in nature, including tasks related to planning, cooperation and negotiation in distributed environments which are shared and cohabited. Aware of this complexity, recent research opts for installing an agent on the server of the business site [5]. However, this is not a valid approach for many contexts because agents could not be objective, it does not offer the possibility of monitoring local resources and interaction is always carried out as part of a web session.

For intelligent agents to provide real solutions for new virtual environments it is necessary to design new models in line with the simplicity demanded by the habits and expectations of current users.

3   Cloud Computing

The continuous development of processing and storage resources, the ubiquity of communications and lower costs of technology have meant that the traditional model based on direct software sales and the installation of infrastructures is being gradually replaced by a model based on the rental of resources to remote data centers that are accessed through standard service-oriented protocols. This new way to market and consume computing resources is known as Cloud Computing; a paradigm that, despite being in its infancy, has already been widely adopted by major companies such as Google, Microsoft, IBM or Amazon with considerable investment.

Cloud Computing is largely considered the materialization of the Utility Computing concept conceived decades ago, which proposes the leasing of computing resources to remote clients. In the Cloud Computing model this idea is built on three figures [8]: the Infrastructure Provider, which is responsible for leasing resources on demand using virtualization technologies and security and balance policies; the Service Provider, which rents resources to one or more Infrastructure Providers to provide as services new functionalities or even the same resources; and the Services

Clients, who are the final consumers. In particular, the Cloud Computing model is characterized by resources being allocated and released dynamically according to customer needs, which pay for the use made of them, and not necessarily through flat fees or long-term contracts.

The advantages of consuming Software as a Service (SaaS) from Services Providers are that: it eliminates installation and maintenance processes, applications are accessible via standard Internet protocols, configuration and application data can be stored in the cloud, the user pays for specific features, and it adds functionalities to applications to work with other users. All these are properties that would turn intelligent agents into a technology easier to adopt for users.

4 Software Agents as Cloud Computing Services

This section aims to take advantage of the Cloud Computing concept, and particularly regarding SaaS, to facilitate the adoption of intelligent agents in virtual societies where users require representation. When analyzing both fields, we find that some of the problems faced by intelligent agents in its quest to become a more accessible technology are answered by the Cloud Computing model. Specifically, when agents (not just specific functions of them) are offered as software services the complexity that entails developing and updating software agents capable of participating in virtual societies is delegated to Services Providers, users pay for specific agent properties and interact with them using any device with Internet connection, thus participating without hindering mobility. However, agents and settings based on them require features not provided by the technologies traditionally used in service-oriented solutions, such as the ability to control local resources and to engage in complex, bi-directional and asynchronous dialogues. The Agency Services model handles these issues.

In line with Cloud Computing solutions, the new model consists of Agency Services Providers (ASPs), which are companies with the knowledge and the necessary technological infrastructure to provide agents as services that users contract in order to participate in the virtual societies. These virtual environments are initiated by Business Sites in which ASPs participate through a well defined interaction mechanism. Common examples of Business Sites are auction sites, computational grids and intelligent buildings. Clients are users or companies which contract ASP services in order to participate in Business Sites from which they expect to make profit. Clients do not have the necessary knowledge or resources to develop their own solutions, or to do so profitably.

The agent deployed in the infrastructure of the ASP is called Broker Agent and participates in virtual societies conducting planning, negotiation and coordination tasks on behalf of the client which contracts the service. This transfers complex tasks to the cloud. Moreover, a lightweight agent is instantiated on the client side. This is communicated with the Broker Agent and is responsible for simple tasks such as communication with users, monitoring local resources, applying instructions ordered by the broker, reacting to events or sending local information to the broker.

One of the most important properties of the local agent is that, in order not to lose the simplicity that characterizes Cloud Computing, it is remotely and automatically instantiated in the client device by the ASP. An ASP can provide Broker Agents for various virtual societies of various Business Sites; also, a Business Site can accept the participation of brokers from more than one ASP.

Once the client has contracted the services of an ASP, its participation in a virtual society is as follows:

Initiation: The Business Site informs competent ASPs that it is initiating a new business process. In its infrastructure, each ASP deploys a Broker Agent for each client. If necessary, the ASP automatically updates the local agent. The broker tells the local agent a new negotiation process has begun and, if any, the local agent reports on the status of local resources.

Execution: Broker Agent is registered in the virtual society and, according to the state of resources and user preferences, it interacts with the rest of brokers by means of negotiating, planning and coordinating tasks. During the process, the Broker Agent informs the local agent concerning its participation in the virtual society and, if there are local resources, transfers the actions to be applied. Also, the local agent informs the broker of any unforeseen event.

Close: Broker Agent records in the ASP the details of its participation in the virtual society, and tells the local agent that the process has been completed.

The presence of the local agent ensures that the process can monitor local resources, thus eliminating the problem associated with the instantiation of agents on remote servers. The user sets its preferences through a web interface provided by the ASP. This information is forwarded to the Broker Agent when it is instantiated. Thus, in settings with local resources, there is no need for embedded interfaces to communicate with local agents and resources.

The technological challenges posed by intelligent agents presence in the Cloud Computing model involve the ability to conduct bidirectional dialogues initiated by any of the parties, and to deploy agents in local devices which are capable of both monitoring local resources and interacting with the user. To meet these challenges, we propose the use of XMPP [4] for the communication between the broker and local agents, and the Java Network Launching Protocol (JNLP) [6] for remotely transferring and launching the local agent. The mechanism for interaction between Broker Agents and virtual societies is defined by the Business Site and is transparent to the customers, as it is an issue taken on by the ASP.

4.1   Proof of Concept

In this section we test the technological feasibility of our proposal. We seek to confirm that it brings simplicity to the user. With the aim of running the simulation in a cloud environment, we use the Google Application Engine (GAE) [1]. Our example simulates the bargaining process that emerges from the planning for allocating the resources of a computational grid [2]. Users can either offer or demand processing resources. They are represented by Broker Agents that are implemented as GAE applications. The Business Site is also implemented as a GAE application that uses a blackboard to record the requests of the clients and finally clear the allocation. The local agents are Java applications stored in the GAE infrastructure, which users can launch by using a JNLP link. The local agents know which resources are available in the client node by reading randomly properties files compiled with the application. In a real environment, the local agents would use a local interface for monitoring and controlling the resources.

The functionalities of the Business Site are accessible by means of HTTP requests. Broker Agents can invoke the following functions: registering in the business process, sending (demanding) offers (requests), and getting information from the blackboard. Broker Agents receive from clients XMPP messages that contain information about which resources are offered (demanded), what is the maximum (minimum) price for buying (selling) and the desired one. Moreover, Broker Agents provide a HTTP interface that is used by the Business Site to notify them when the best offer changes or the process ends. The Business Site starts sessions periodically, informing competent ASPs. These instantiate a Broker Agent for each customer and try to improve the better offer recorded on the blackboard. Figure 1 illustrates the sequence diagram of the process.

The simulation showed that it is possible to offer software agents as Cloud Computing services by using existing technologies. Moreover, it showed that a JNLP link allows reducing user interaction to downloading a local agent. However, we argue that this is even unnecessary if the local agent is initially installed so that it is run whenever the system loads. Later it can be automatically updated from the ASP.

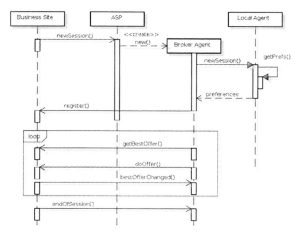

Fig. 1 Sequence diagram illustrating the participation of a client in the test

5   Conclusions and Future Research

The absence of intelligent agents in new virtual societies warns that the community should choose new models that will help implement the advantages tradionally associated with agent-based solutions. The success of the Cloud Computing model is characterized precisely by facilitating access to software, saving the user and developers from all those details that are not related to the application's usefulness. We argue that software agents may be offered as a Cloud Computing service and thus inherit all the virtues of this new paradigm. As discussed, the Agency Services model can be deployed using widely known and proven technologies. In addition, among its properties outstands that, despite adopting service-orientation, it ensures both user autonomy and interaction with local resources.

In future research, we propose the implementation of brokering agents to manage intelligent energy networks that are simulated using a framework [3] developed by our research group.

Acknowledgements. This work has been supported with funding from Agencia Canaria de Investigación, Innovación y Sociedad de la Información of Canary Islands Autonomic Government under project smartgrid.gc, with reference 200801000137.

References

1. Google Application Engine, http://code.google.com/appengine
2. Buyya, R., Abramson, D., Venugopal, S.: The Grid Economy. Proceedings of the IEEE 93, 698–714 (2005)
3. Hernandez, J., Hernandez, M.: TAFAT: An Object Oriented Metamodel in Simulation of Electrical Systems. In: First EIFER Workshop on Simulation of Complex Systems Applied to Energy (2010)
4. Saint-Andre, P.: Streaming XML with Jabber/XMPP. IEEE Internet Computing 9, 82–89 (2005)
5. Sandholm, T.: eMediator: A Next Generation Electronic Commerce Server. Computational Intelligence 18, 656–676 (2002)
6. Schmidt, R.: Java Network Launching Protocol & API Specification (2000)
7. Wurman, P.R., Wellman, M.P., Walsh, W.E.: The Michigan Internet AuctionBot: A Configurable Auction Server for Human and Software Agents. In: Proceedings of the Second International Conference on Autonomous Agents, pp. 301–308 (1998)
8. Zhang, Q., Cheng, L., Boutaba, R.: Cloud computing: state-of-the-art and research challenges. Journal of Internet Services and Applications 1, 7–18 (2010)

Efficient Monitoring of Financial Orders with Agent-Based Technologies

Philippe Mathieu and Olivier Brandouy

Abstract. The execution of orders on stock exchanges is managed by a set of formalized rules based on price and time priority. Nevertheless, orders issued by investors do not show-up directly in the market system : they transit through the brokerage intermediation where they can be arranged in different sequences. We show that the latter operation has a critical impact on investors. In this paper, we propose a decision support system that solves the underlying optimization problem for a given social welfare. We show that the solution cannot be obtained without an agent-based simulation platform that individualizes the consequences of the broker decision in terms of order sequencing at the agent (client) level. In this framework, we study the impact of several social welfares functions and show how the broker can grant his clients with *"just and equitable principles of trade"*.

Introduction

In modern stock exchanges, investors almost never have the ability to route their orders directly to the market; they must use the services of a financial intermediary whose task is to trade securities on behalf of his customers: a broker. Thus, the role of the broker consists in introducing their orders in one of the various trading platforms that are available for the relevant financial commodity. These platforms usually run with electronic order books. Basically, an order book captures the on-going continuous double auction process

Philippe Mathieu
Université Lille 1, Computer Science Dept. & LIFL (UMR CNRS-USTL 8022)
e-mail: philippe.mathieu@lifl.fr

Olivier Brandouy
Sorbonne Graduate School of Business,
Dept. of Finance & GREGOR (EA MESR-U.Paris1 2474)
e-mail: olivier.brandouy@univ-paris1.fr

Y. Demazeau et al. (Eds.): Adv. on Prac. Appl. of Agents and Mult. Sys., AISC 88, pp. 277–286.
springerlink.com

allowing negotiation between buyers and sellers. Electronic order books implement a "price" then "time" priority system. Sell (buy) orders are organized by increasing (decreasing) prices and, in case of equal prices, placed in the queue with respect to the time-stamp indicating when they were introduced in the book.

Consider the situation of a broker having a set of orders $O : \{o_1, o_2, ..., o_n\}$, awaiting to be inserted in a given order book. Whatever the time at which these orders were issued or whoever the customer are, these orders are pending at the broker's desk, and just about to be introduced in the market and processed by the matching algorithm.Nevertheless, nothing is clearly established concerning the "how-to" introducing orders in the book, so to behave smartly and fairly with respect to clients own interests. Nevertheless, a body of literature, both from Finance and Computer Science has tackled related questions such as orders cost of execution (see for example, [8], [2])[1].

This paper investigates this question, shows why it is of main interest both for Computer Scientists and Financial professionals and how it can be solved in an agent-based Decision Support System. We demonstrate that a Social Welfare Function must be defined by the broker and guaranteed throughout the trading process to achieve just and equitable principles of trade he owes to his customers. We also show that a simple "first-in first-out" principle to rule the order flow does not necessarily lead to an optimal situation for broker's clients. To illustrate the research question tackled in this article, let's consider a simplified situation where an order book has two limit orders listed in the Bid and the Ask queues (see Table 1).

Table 1 Initial Order Book

	RAP*	Price	Qty
Ask	2	120	5
Ask	1	110	10
Bid	1	90	10
Bid	2	85	5

* RAP : *rank in the auction process*

We now show that when a set of orders must be introduced in this order book, the sequence along which they are submitted affects the resulting price sequence. For that purpose, we consider the following limit orders:[2]

- Agent "A" wants to sell 25 stocks at a minimum price of 85.
- Agent "B" wants to buy 10 stocks at a maximum price of 95.

---

[1] Most of the time, this question is *officially* solved by a "first-in first-out" principle: orders are treated on the fly. Nevertheless, orders can be re-arranged, and sometimes mixed with the brokers own orders (proprietary trading) in "front running" illegal or parasitic operations [4].

[2] A "limit order" fixes the max (min) acceptable price at which an investor agrees to buy (sell) a financial commodity.

The initial situation of these agents is summarized in Table 2.[3]

If the orders are introduced by the broker in opposite sequence (A *then* B) or (B *then* A), the impact on both agents wealth is clearly different due to priority rules and to the fact that the final portfolio is evaluated *mark-to-market* as well (*i.e* "with the last settled price on the market").

Table 2 Situations for Traders A and B

	Agent	Initial				Final				Δ
		Cash	Stocks	Price	Wealth	Cash	Stocks	Price	Wealth	
(A *then* B)	Ask	375	25	105	3000	2550	0	85	2550	-450
	Bid	3000	0	105	3000	2150	10	85	3000	0
(B *then* A)	Ask	375	25	105	3000	2650	0	85	2650	-350
	Bid	3000	0	105	3000	2050	10	85	2900	-100

Therefore, it is obvious that the sequence chosen by a broker when introducing a series of orders is critical for his clients' wealth. Note again that if a time-stamp constrains the Broker (for example order A arriving first to the desk), this does not really solve the problem since it automatically fixes the order execution (first-in, first-out) without considering any aggregate utility criterion. This could be even more discussed when the time-stamps are close in time. If one desires to respect just and equitable principles of trade this implies to avoid decisions that favor one customer (*in casu* A or B) against the others. In this example, the broker faces a *set of combinations* $\aleph = \{0_1, 0_2\}$ with two *vectors of orders* $0_1 = \{A \text{ then } B\}$ and $0_2 = \{B \text{ then } A\}$ Note that the number of combinations of orders (each one generating a specific sequences 0_j) is fairly large, as soon as n becomes important: $n!$. Thus, the broker has to choose in $\aleph$ a Pareto-optimal vector $O*_j$ for his customers. This issue is clearly related to a social choice problem [1] and the solution necessitates to define a particular Social Welfare Function (here-after SWF).

This paper is organized as follows. In section 1, we introduce the social welfare functions used by the agent-based decision support system. In section 2, we briefly present the decision support system itself. In a last section (3) we propose a limited set of experiment run over artificial data to illustrate the effectiveness of the architecture.

1   Individual and Social Welfare

Individual actions: One should distinguish two different individual actions with respect to the category of agents: those made by investors and those

---

[3] We have fixed an initial arbitrary price for the stocks of 105.

made by the broker. Agent's individual rationality (whatever their role is) is supposed to be standard, *e.g.*, they always prefer having more than less.

1. Investors actions:

   - Investors are represented by a population of n autonomous agents. Each of these agents send one *limit orders* o_i to the broker.
   - A limit order is a 4-uplet o_i : {id, direction, price, quantity} determining the name of the agent, if he wants to sell or buy, at which limit price, and for which amount.

2. Broker actions:

   - The broker must route the resulting set of orders O : $\{o_1, o_2, ..., o_n\}$ to the order book. This necessitates deciding in which sequence these orders will be arranged prior to be introduced in the market system. Any combination of all the elements in O is a vector denoted O_j. As mentioned previously, there are $n! = m$ such combinations.
   - We denote $\aleph$: $\{O_{j,j \in [1,m]} : O_1, O_2, ..., O_m\}$, the set of all possible combinations of orders out of O.

Ideally, investors should obtain as much as they can from their trading activity. Nevertheless, sellers and buyers have clearly opposite interests. Concerning the broker, one should posit that his own aim consists in maximizing the fees he charges his customers with. These fees are based on the market capitalization exchanged by these latter through his services (*i.e.* price × quantity).

1. Investors welfare (w_i):

   - Let $w_{i,t}$ be the wealth of agent i at date t. For the sake of simplicity, we posit a uniform transformation of agent's wealth in terms of cardinal utility (any kind of Von Neumann utility function [7] can be chosen here like $u(w) = log(w)$ or $u(w) = w^\alpha$).
   - $w_{i,t} = p_t.A_i + c_{i,t}$. In this equation, p_t is the price vector for the assets A_i held by agent i (these assets being represented by their quantities); $c_{i,t}$ is agent i cash at time t. Among two states of the world Ω_1 and Ω_2, if $w_{i,t}|\Omega_1 > w_{i,t}|\Omega_2$ then $u_{i,t}|\Omega_1 \succ u_{i,t}|\Omega_2$.
   - $\Delta w_{i,t} = w_{i,t} - w_{i,t-1}$ is the variation of agents individual welfares.

2. Broker welfare (w_B):

   - Let r be the transaction cost applied by the broker and charged to his clients.
   - Any transaction leads to a transfer of cash ($p \times q$) from the buyer to the seller. Therefore, the transaction costs apply on these values cumulated over time. We posit that r is the same for any of the brokers' customers and that fees are charged once for the seller and once for the buyer (see algorithm 1).

```
for (i in 1:n) do
    process o_i ∈ O_i;
    w_B ← w_B + 2 × r × (p × q)
end
```

Algorithm 1. Calculation of Broker's welfare

Social Welfare: We have shown that depending upon the sequence O_i for processing the orders, agents' utilities will be affected accordingly. Based on these individual utilities, a social welfare function (here after SWF) can be defined and used to compute a social welfare measure for each O_i. Let: $U = f(u_{i,i\in[1,n]})$, the collective utility function based on agents individual utilities. This function should ideally respect a series of properties: Extensive developments around these concepts can be found in [6].

Thus, a crucial question here is to determine a computable function U to decide which sequence $0_j^* \in \aleph$ should be chosen in order to ensure a fair treatment of orders among customers.

```
agentsUtilities← NA ;
BestSW← NA ;
BestSeq← NA ;
for (j in 1:m) do
    process O_j ;
    agentsUtilities ← u^j_{i,i∈[1,n]} ;
    sw ← U(agentsUtilities) ;
    if (sw ¿ BestSW) then
        bestSW=sw; BestSeq=O_j
    end
end
O_j^* ← BestSeq ;
return O_j^*
```

Algorithm 2. Algorithm for the choice of $O_j^* \in \aleph$

One can remark from algorithm 2 that U is run over a matrix of m vectors of n values. There is obviously no unique possible function or process U that can solve the question of the fair treatment of orders and each alternative should be considered carefully. For example, the broker can choose several traditional options. These options are based on real-valued functions delivering a "welfare score".

1. Utilitarian SWF.

   In this case, the broker would choose the vector of orders delivering the highest sum of individual utilities : $0_j^* \equiv max_j(\sum_{i:1}^n u^j_{i,i\in[1,n]})$ In the initial example presented in section , this criterion delivers a score of 5550 whatever the sequence. Hence, both are equal and this criterion does not help to choose among these results.

2. Egalitarian SWF (Rawl's "difference principle", see [10]).
   The choice of this rule implies to opt for the sequence of orders delivering
   the maximum of the minimum individual utility amongst the agents $0_j^* \equiv max_j(min_i(u_{i,i\in[1,n]}^j))$. In the basic example, if this SWF is chosen, one
   should decide to process B *then* A (max of min values = 2650).
3. Elitist SWF.
   The sequence of orders delivering the maximum individual utility amongst
   the agents will be chosen : $0_j^* \equiv max_j(max_i(u_{i,i\in[1,n]}^j))$. This approach
   delivers an opposite outcome (A *then* B, max of max values = 3000) with
   respect to the egalitarian approach.
4. Nash SWF.
   This approach proposes to consider the sequence of orders from which the
   product of individual utilities is maximum : $0_j^* \equiv max_j(\prod_{i:1}^n u_{i,i\in[1,n]}^j)$ In
   this last case, B *then* A must be chosen (product of values = 7.685E6).

In our context, these alternative measures can provide a justification for any
of the possible combinations in the order sequence. Nevertheless, one can
imagine that client A or B, will not appreciate these justifications on the
same footage, some appearing fair and others unfair with regard to each
individual point of view. Furthermore, notice that as soon as one uses the
Egalitarian or the Elitist criterion, one client is favored to the detriment of
the other. Nevertheless, this choice could be made by a broker for marketing
reasons or to maximize his own utility w_B. The Utilitarian criterion does
not help to choose in this particular case, even if it is intuitively appealing
in terms of fairness. In summary, in our example, whatever the final decision, A or B will be favored. Note that in the preliminary example, for the
sake of simplicity, we have arbitrarily fixed initial identical endowments for
the agents. In most of the cases, this will not be true. Thus, the result of
the sequence in which orders are entered in the system must be appreciated
"as if" agents were wealth-less (without any stock or cash *prior* to trading).
This "*wealth-less instantiation*" is an initial, necessary step in the process.
It means that an agent willing to sell will have, by construction, negative
stock holdings and positive cash after trading, while an agent willing to buy
will have positive stock holdings and negative cash. Thus, the Utilitarian
and Nash Social Welfare Functions are not adapted because one can imagine
situations in which the post-trading wealth of an agent is negative. [4] This
calculation, based on a "*wealth-less instantiation*", has two straightforward
consequences: i) the Utilitarian criterion is constant whatever the chosen permutation of orders, ii) the Nash criterion cannot be used due to the possibility
of negative individual welfares in some cases. One should therefore consider
alternative SWF.

---

[4] For an application of these criterion in a MAS, see [9].

2   Decision Support System

In this research, we provide a system that allows a broker to simulate within a given order book the execution of his customers' orders in different sequences. Consequently, this decision support system must be able to capture the current state of a real-world market at time t, and to test in a virtual environment where the real-world rules are cloned, the impact of any sequence $O_i \in \aleph$ with regard to the social welfare of the customers population. In doing so, the decision support system extrapolates the impact of the broker possible decision and allow him to select the best possible outcome. Notice again that a social-welfare rule has to be defined prior to any simulation. Thus an agent-based artificial stock market, able to simulate these outcomes, appears vital in this design. Ideally, it not only should allow defining behaviors at the individual level, but also should permit to track the consequences of agents actions at the micro-level, with some realism. To build a powerful decision support system, a full agent-based ASM is definitely necessary since the calculation of social welfare is done from agents' individual welfares. Therefore one should use a system that individualizes the consequences of each agent's action, implements an asynchronous order book and reifies the agents.

The details of the agent-based artificial stock market (here after ABASM) used in this research, named ATOM, can be found in [5]. This platform is validated by both the industrial and scientific worlds: for example, the platform can simulate, using artificial behaviors, the main stylized facts usually considered as necessary for realistic simulations of financial motions (see for example [3]); the "replay-engine" of this platform can use real-world orders and will deliver, in this configuration, the same results as the NYSE-Euronext system . Using this ABASM, a broker will be able to decide, with regard to a social welfare optimization rule, which sequence should be run by the broker in the real-world book. Note again that the choice of a given sequence of orders among all possible sequences cannot be done without tackling its impact at the individual and the collective levels. This imposes to use, in addition to the ABASM itself, a ranking algorithm for each possible sequence of orders.

Ranking Algorithm: The only solution we have found so far is to test all possible combinations of orders. Nevertheless, an initial simplification can be done.

Let b_i and a_i be the sub-vectors respectively gathering the "buy" and "sell" orders $(b_i \bigcup a_i = O)$. Let b'_i and a'_i be the ("buy" and "sell") orders yet present the order book. The subset of orders that are critical to determine the optimal social welfare is :

$$O' = \{\{b_i \geq (min(a_i) \vee min(a'_i))\} \bigcup \{a_i \geq (max(b_i) \vee max(b'_i))\}\} \quad (1)$$

In equation 1, the operator $\geq$ applies on prices appearing in orders. For the next developments, O is supposed to be reduced to O'. In the ranking algorithm 3 presented below, "ATOM" refers to our agent-based artificial stock market presented previously.

Input: Order Book (Real Orders), Waiting Orders, R (rule to optimize)

U(optimal Social Welfare) $\leftarrow -\infty$;
O_i^* (Optimal Sequence of Orders) $\leftarrow \emptyset$;

forall the *possible Permut SEQ of Waiting Orders* **do**
    init ATOM with Order Book;
    execute *SEQ* in ATOM;
    compute SW according to R;
    if *SW > OSW* **then**
        $OSQ \leftarrow SEQ$;

Display OSQ

Algorithm 3. Optimal Sequence of Orders

3   Experiments and Discussions

In this section, we illustrate the effectiveness of the agent-based decision support system using artificial sets of orders that will be used in a context where the broker has to run the sequence of orders in an empty order-book. This means that each of the broker's orders is matched internally (within the set of broker's clients orders). This is a limit case known has "*systematic internalization*". We have chosen to generate artificial data for the sake of simplicity and tractability of the results. The following algorithm is used in both series of simulations. It proposes a process delivering $2 \times k$ orders with price limits and quantities.

```
size=3 ;
for (int k = 1; k ¡= size; k++) do
   | new LimitOrder(ASK, 2^(k−1), 10 − (k − 1))
end
for (int k = 1; k ¡= size; k++) do
   | new LimitOrder(BID, 2^(size−k), 12 − (k − 1))
end
```

Algorithm 4. Automatic Generation of Orders

The complexity of the task is exponential and depends upon the number of orders the broker must deal with. When $k = 3$, one gets 720 potential permutations using the 6 orders (6!). This number of permutation is raised to 40320 when $k = 4$. On a 4-cores computer, this calculation takes around 1.5 seconds, which implies that a heuristic should be found to approximate

Table 3 Broker order set O; orders interpretation : {id, direction, price, quantity}

$k = 3$	{A, Ask, 10, 1},{B, Ask, 9, 2}, {C, Ask, 8, 4}, {D, Bid, 12, 4}, {E, Bid, 11, 2}, {F, Bid, 10, 1}
$k = 4$	{A, Ask, 11, 1}, {B, Ask, 10, 2}, {C, Ask, 9, 4}, {D, Ask, 8, 8}, {E, Bid, 11, 8}, {F, Bid, 12, 4}, {G, Bid, 13, 2}, {H, Bid, 14, 1}

the optimal solution in an acceptable time. Notice for example that with the 4-cores computer, 12 pending orders necessitate nearly 5 hours of computing time; grid computing appears therefore necessary to get a solution at short notice, even if it cannot solve the problem. A realist number of pending orders at the broker's desk – for example 20 – should be arranged optimally in a few seconds, which for the moment, cannot be done. This necessitates to develop an appropriate heuristic. We report the results of this set of experiments in Table 4.

Table 4 Results, *systematic internalization*

	Num of Sol.	Card. of $\aleph$	U	Example of $O*$	w_B	Max w_B	[1]
Utilitarian	720	720	0	$\{A, B, C, D, E, F\}$	160	160*	
Elitist	40	720	12	$\{A, B, E, D, C, F\}$	160	160*	
Egalitarian	56	720	-2	$\{A, B, D, E, F, C\}$	160	160*	

n=3, 6 orders

$* \rightarrow$ sequence in $\aleph$ to obtain this result : $\{A, F, D, B, E, C\}$

n=4, 8 orders

	Num of Sol.	Card. of $\aleph$	U	Example of $O*$	w_B	Max w_B	[1]
Utilitarian	40320	40320	0	$\{A, B, C, D, E, F, G, H\}$	320	338**	
Elitist	288	40320	35	$\{A, B, C, F, G, E, D, H\}$	320	338**	
Egalitarian	152	40320	-4	$\{A, B, C, F, G, D, E, H\}$	320	338**	

$** \rightarrow$ sequence in $\aleph$ to obtain this result :$\{A, G, B, H, E, C, F, D\}$

[1] : Maximum possible welfare for the broker

The left-hand side of the table (column 1 to 4) confirms that different social welfare measures deliver different optimal sequencing O^* for the order set generated by our algorithm (see Algorithm 4). These results clearly illustrate another important issue raised in this research : the broker own interests do not match his clients ones. For example, in the case where 8 orders must be sorted, $\{A, G, B, H, E, C, F, D\}$ is the more interesting sequence for the broker($w_B = 338$). With regard to the individual welfares of his clients, it is never an optimal solution (Utilitarian = 0, Elitist = 4, Egalitarian = -8). In other terms, whatever the social welfare measure, none is compatible with the broker own interests. This result points out a potential conflict of interests between these categories of economic agents (investors and intermediaries), which requires at least some supervision, and probably some regulation.

Conclusion

In stock markets, brokers hold a central position where they have the possibility to influence price dynamics in ordering the flow of orders they receive from their customers. Even if their activity is strictly monitored by professional association, they ultimately could arrange the pending orders from their clients so to capture maximum benefits for themselves (a behavior called "front running" in finance) or, alternatively, arrange these orders in sequences granting fairness amongst their clients. We therefore show in this paper, that since the broker's own interests do not match his clients' ones, order sequencing should be carefully set *before* posting the orders to the market so to match a predefined "fairness among clients" rule. We also defend that only an agent-based decision support system can solve this highly complex task. For the moment, we only propose an algorithm that never scales to find the best orders sequence execution. Nevertheless, the finding of heuristics with appropriate features and lower complexity is mandatory for future extensions.

References

1. Arrow, K.J.: Social Choice and Individual Values, 2nd edn. Yale University Press, New Haven and London (1951)
2. Comerton-Forde, C., Frino, A., Fernandez, C., Oetomo, T.: How broker ability affects institutional trading costs. Accounting and Finance 45(3), 351–374 (2005)
3. Cont, R.: Empirical properties of asset returns: stylized facts and statistical issues. Quantitative Finance 1, 223–236 (2001)
4. Harris, L.: Trading & Exchanges, ch. 11. Oxford University Press, Oxford (2003)
5. Mathieu, P., Brandouy, O.: A generic architecture for realistic simulations of complex financial dynamics. In: Demazeau, Y., Dignum, F., Corchado, J., Perez, J. (eds.) Advances in Practical Applications of Agents and Multi-Agents Systems. AISC, vol. 70, pp. 185–198. Springer, Heidelberg (2010)
6. Moulin, H.: Fair Division and Collective Welfare. MIT Press, Cambridge (2004)
7. von Neumann, J., Morgenstern, O.: Theory of Games and Economic Behavior, 3rd edn. Princeton University Press, Princeton (1953)
8. Nevmyvaka, Y., Kearns, M., Papandreou, A., Sycara, K.: Electronic trading in order-driven markets: Efficient execution, pp. 190–197 (2005)
9. Nongaillard, A., Mathieu, P., Everaere, P.: Nash welfare allocation problems: Concrete issues. In: Intelligent Agent Technology (IAT 2010), pp. 32–39 (2010)
10. Rawls, J.: A Theory of Justice (1971)

An Agent Task Force for Stock Trading

Rui Pedro Barbosa and Orlando Belo

Abstract. In this article the authors present the simulated trading results of a system consisting of 60 intelligent agents, each being responsible for day trading a stock listed on the NYSE or the NASDAQ stock exchange. These agents were implemented according to an architecture that was previously applied to currency trading with interesting results. The performance of the stock trading agents, once integrated in a diversified investment system, showed similar promise. The trading simulation was done using out-of-sample price data for the period between February of 2006 and October of 2010. Throughout this period, the system's performance compared favorably with that of the buy-and-hold strategy, both in terms of return and maximum drawdown. These results indicate that agent technology might be of use for this particular practical application, a conclusion that should interest the investment industry.

Keywords: Intelligent Agent, Stock Trading, Financial Data Mining.

1 Introduction

Financial trading seems like the perfect setting for applying agent technology [1][2]. Most investment analysis, be it technical or fundamental, is nothing more than number crushing, data mining and pattern matching, and these are all tasks at which software agents should be able to excel. This assumption was partially vindicated by the research deriving from the Penn-Lehman Automated Trading Project [3]. This project showed interesting results in what regards to automated stock trading, but for the most part the agents created for this project's trading competitions were more like simple trading bots rather than actual intelligent agents. We believe that, in order to be truly "intelligent", a trading agent must meet at least the following requirements: it should be able to decide when to buy, short sell or close open trades of a given financial instrument; it should be able to perform money and risk management; it should be able to keep learning over time, even as it trades; it should be capable of adapting to changes in market conditions; and finally, it should be smart enough to stop trading when the market becomes less predictable, and to resume trading once the conditions improve. An intelligent agent [4] with these capabilities should be well prepared to trade autonomously and unassisted for an indefinite period of time, hence completely replacing a human trader. In a previous article [5], we proposed an agent architecture that might allow for the creation of trading agents that should more or less meet the

Rui Pedro Barbosa · Orlando Belo
Department of Informatics, University of Minho, Portugal
e-mail: {rui.barbosa,obelo}@di.uminho.pt

Y. Demazeau et al. (Eds.): Adv. on Prac. Appl. of Agents and Mult. Sys., AISC 88, pp. 287–297.

aforementioned requirements. This architecture was applied in the development of a multi-agent currency trading system [6] that showed promising results in a trading simulation encompassing a period of around 2 years. In this article we will describe the use of this same architecture (with a few improvements) in the development of a diversified trading system consisting of 60 stock trading agents, each of which will be trained and configured to autonomously day trade a stock listed on the NYSE or the NASDAQ. Before presenting the actual details of the agent implementation, we will start by describing the proposed agent architecture. The article will end with the reporting of the results obtained by the agent-based trading system in an out-of-sample period comprising around 5 years' worth of data. The system's performance will be compared with that of a simple buy-and-hold strategy, both in terms of return and maximum drawdown. This strategy will be used as the benchmark for our results.

2   The Trading Agent Architecture

As previously mentioned, there are several requirements that a software trading agent needs to meet before it can actually be considered "intelligent". If these requirements are fulfilled, the agent might be able to trade successfully in the long run. The architecture we proposed in the past (figure 1) [5], should allow for the development of this type of agents. It consists of three modules: 1) the prediction module, responsible for forecasting the direction of a stock's price throughout the next trading day (this module is implemented using an ensemble of data mining models); 2) the empirical knowledge module, responsible for deciding how much to invest in each trade (this module is implemented using a case-based reasoning system); and 3) the domain knowledge module, responsible for making the final trading decisions (this module is implemented using an expert system).

Fig. 1 The trading agent architecture.

The prediction module is responsible for deciding if a stock should be bought or short sold, based on the predictions of the data mining models in the ensemble. Each model tries to predict if the stock price will increase or decrease throughout the next trading day (from open to close). Once this is done, the models' predictions are aggregated into a single forecast, which the agent utilizes to pick the direction of the trade: if this forecast dictates that the stock price will increase, then the module's decision is to buy the stock when the market opens, and sell it when it closes; if, on the other hand, it dictates that the price will decrease, the module's decision is to short sell the stock at the open and cover at the close. The algorithm behind the decisions of the prediction module is shown in figure 2. Before each forecast, this component splits the available data into two datasets: the test set, with the most recent n instances, and the training set, with all the rest. The training set is used to retrain each data mining model, after which the retrained models make direction forecasts for the instances in the test set, and trades are simulated accordingly. The results of this trading simulation are subsequently utilized to calculate the profit factors of each model:

$$Overall\ PF = \frac{\sum returns\ of\ profitable\ trades}{\sum |returns\ of\ unprofitable\ trades|} - 1 \tag{1}$$

$$Long\ PF = \frac{\sum returns\ of\ profitable\ long\ trades}{\sum |returns\ of\ unprofitable\ long\ trades|} - 1 \tag{2}$$

$$Short\ PF = \frac{\sum returns\ of\ profitable\ short\ trades}{\sum |returns\ of\ unprofitable\ short\ trades|} - 1 \tag{3}$$

The overall profit factors are utilized to decide if a retrained model should become a part of the ensemble, or if the version prior to retraining should be kept, by choosing the one with the highest profitability. The long and short profit factors are used as the model's vote weights: if the model predicts a price increase, the weight of its vote is its long profit factor, and if it predicts a price decrease, the weight of its vote is its short profit factor. A negative weight is replaced with zero, meaning that the model's prediction is ignored when the models' votes are aggregated to calculate the forecast of the ensemble. The votes' aggregation is accomplished by adding the weights of the votes of the models that predict a price increase, and subtracting the weights of the votes of the models that predict a price decrease. If the resulting value is greater than zero, the final ensemble prediction is that the price will increase, and therefore the stock should be bought; if it is lower than zero, the ensemble prediction is that the price will decrease, hence the stock should be short sold; finally, if it is exactly zero, the ensemble does not make a prediction, and the agent does not trade.

While the prediction module allows the agent to automatically decide when to buy or short sell a financial instrument, that is not sufficient to make it completely autonomous. Besides being able to decide on the direction of each trade (i.e., if it should go long or short), it also needs to be capable of deciding how much to invest (i.e., the size of each trade). More specifically, it should be able to decrease the investment amount or even stop trading when the perceived risk is bigger. That is the purpose of the empirical knowledge module. For each potential trade, it can set the investment amount to three different sizes: if the trade is expected to be profitable, a standard, user-defined trade size is used; if there are doubts regarding the profitability of the trade, half the standard trade size is used; finally, if the trade is expected to be unprofitable, the size is set to zero, which means that the agent will not make the trade. The empirical knowledge module gets its name from the fact that it uses information from previous trades to decide the amount to invest in new trades. The module's main component is a case-based reasoning system. In this system, each case corresponds to a trade that was previously executed by the agent, and contains the following information: the direction predicted by the prediction module, the direction predicted by each model in the prediction module's ensemble, and the return of the trade. The empirical knowledge module tries to capitalize on the bigger profitability associated with certain combinations of models' predictions. For example, our empirical studies show that trades carried out when all the data mining models make the same prediction, i.e., all predict a price increase or all predict a price decrease, are consistently more profitable than those performed when the predictions are mixed. The algorithm governing the empirical knowledge module's decisions is shown in figure 3.

```
Split the available instances in two datasets: the test set with the most
recent n instances, and the training set with all the rest.
For each model in the ensemble:
  - For each instance in the test set, make the model predict its class, and
    use the prediction to simulate a trade.
  - Use the simulation results to calculate the model's overall profit factor
    (equation 1), long profit factor (equation 2) and short profit factor
    (equation 3).
  - Retrain the model with the training set.
  - For each instance in the test set, make the retrained model predict its
    class, and use the prediction to simulate a trade.
  - Use the simulation results to calculate the retrained model's overall
    profit factor (equation 1), long profit factor (equation 2) and short
    profit factor (equation 3).
  - If the overall profit factor of the retrained model is greater than or
    equal to the overall profit factor of the model before retraining: replace
    the model in the ensemble with the retrained model.
  - Else: discard the retrained model.
  - Use the model in the ensemble to make a prediction for the target period.
  - If it predicts a price increase: set the weight of its vote to its long
    profit factor.
  - If it predicts a price decrease: set the weight of its vote to its short
    profit factor.
  - If the weight is a negative number: set it to zero.
Add the weights of the votes of all the models that predict a price increase,
and subtract from this value the weights of the votes of all the models that
predict a price decrease.
If the resulting value is greater than zero: predict a price increase in the
target period.
Else if it is lower than zero: predict a price decrease in the target period.
Else: do not make a prediction for the target period.
Once the target period ends, create a new instance and add it to the
available data.
```

Fig. 2 Pseudo-code of the algorithm used by the prediction module to make the price direction forecast for a given target period.

Both the prediction and the empirical knowledge modules were devised in way that allows the agents to keep learning while they trade. However, there will always be some expert knowledge that the agents will not be able to pick up from practice. The domain knowledge module was created to overcome this problem. As its name implies, its main responsibility is to use domain-specific knowledge to make trading decisions. This module consists of a rule-based expert system in which trading experts can insert rules to regulate the trading activity of the agents. These rules can be related to many different aspects of trading; for example, they can define low liquidity periods during which the agents should not trade, or they can be used to make the agents close open trades when a certain profit or loss is reached. In the proposed agent architecture, the domain knowledge module is responsible for making the final trading decisions, by taking into account the prediction module's recommendations to buy or short sell the financial instrument, the empirical knowledge module's trade size suggestions, and its own expert rules.

Considering the modules' algorithms, and looking at the agent architecture as a whole, we believe that a trading agent that is based on this architecture should be able to:

–   Keep learning new trading patterns as time goes by, as a result of the periodic retraining with new data of the models in the ensemble. This process is essential to the agent's autonomy, because it enables it to update its prediction mechanism without external assistance.

–   Adapt to changes in market conditions, as a result of the continuous reweighting of the models' votes according to their simulated profitability: those that have been more profitable in the recent past increase in weight, while those that have been less profitable decrease in weight.

–   Know when to stay out of the market. All the modules in the agent's architecture can help in this decision: the prediction module will prevent it from trading if the recent past profitability of all the models in the ensemble is negative (because all the vote weights will be set to zero); the empirical knowledge module stops a trade by setting its size to zero, whenever the cases in the database show that similar trades in the past were unprofitable; finally, the domain knowledge module can prevent trades from being open according to the expert's rules (which could, for example, stop the agent from trading based on time restrictions, instrument price levels, or market volatility).

```
Get the ensemble's and the models' price direction forecasts for the target
period from the prediction module.
Retrieve from the database all the cases with the same combination of
predictions.
While the number of retrieved cases is lower than a user-defined minimum:
    remove the last model's prediction from the search and retrieve the cases
    again.
Calculate the overall profit factor of the retrieved cases (equation 1).
If the profit factor is greater than or equal to a user-defined threshold:
    make the trade size equal to the standard size.
Else if it is lower than another user-defined threshold:
    make the trade size equal to zero.
Else:
    make the trade size equal to half the standard size.
Once the trade is closed, inserted a new case in the database.
```

Fig. 3 Pseudo-code of the algorithm used by the empirical knowledge module to make the trade size decision for a given target period.

Obviously, there is no guarantee that the described agent architecture will be able to produce a successful agent for every stock and every trading timeframe combination. Nevertheless, if some of the agents created do possess a statistical trading edge, it is possible that their success could make up for the trading losses of the least competent agents in a diversified multi-agent trading system.

3 The Intelligent Stock Trading Agents

Using the referred architecture, we set out to create 60 stock trading agents. Each agent was configured to day trade a single stock (i.e., to open a position when the market opens every day, and close it at the end of the trading session) according to the following methodology:

- 11 data mining models were placed in the agent's ensemble. These models were randomly selected by an automatic process which trained several hundred models with random parameters and attributes, and then picked the 11 best, according to the performance shown with a very small set of test data. Many different types of classification and regression models were considered, among which artificial neural networks, the naïve Bayes classifier, decision trees and rule learners. As for the attributes used, they were all price and time-based, and included technical analysis indicators like the RSI, the Williams %R and moving averages. The models were trained and tested with the Weka API [7].
- profit factors were calculated using the last 50 instances, hence the models' vote weights for each trade were based on the profitability shown in the previous two and a half months of trading;
- a trade's size was set to zero if similar trades retrieved from the database had a profit factor lower than zero; if the profit factor was between 0 and 1, the trade size was set to half the standard size;
- the following rules were inserted in the agent's expert system: do not trade around Christmas Day or Good Friday (to prevent it from trading in low liquidity days); do not trade if the stock's price is below $10 (because the trading costs can become prohibitively expensive if the stock price is too low, assuming a fixed commission per share,); close a trade if it reaches a profit equal to 2/3 of the average price range in the last 5 days (this is a take-profit rule that instructs the agent to close an open trade once it reaches a reasonable profit). The expert rules were handled by the JBoss Drools engine [8].

The 60 stocks traded by the agents were selected according to two loosely defined requirements: they had to have a high beta, and the market cap of the corresponding companies had to be relatively big. For each of these stocks, we collected as much historical price data as we could find, up to January of 2006. This data was converted into instances, which were utilized to train the agents. Once trained, each agent simulated trades for the out-of-sample period between February of 2006 and October of 2010 (i.e., a total of 1,194 test instances per agent). The cumulative return of the 5 most profitable and the 5 least profitable agents throughout this period is shown in figure 4. As expected, not all the agents achieved an acceptable performance. Overall, the 60 agents made a total of 35,453 trades, 53.8% of which were profitable. These results reflect a commission of $0.01 per stock traded, calculated assuming a standard trade size of $20,000. As a side note, we should point out that the percentage of profitable trades to be expected from a completely random strategy should be, on average, well below 50%, due to these commissions alone.

Since the objective of our work was to create a system with the potential to be profitable in the long run, it was clear we needed to implement some sort of investment diversification. In order to accomplish this, we integrated the 60 intelligent agents into a diversified trading system, and made them share the monetary resources.

The system's main advantage is that it eliminates much of the trading risk associated with the individual agents, because the losses of the worst agents are compensated by the gains of the best ones. The accumulated return achieved with this system in the test period is shown in figure 5 and summarized in table 1. Looking at the difference between the system's gross and net returns, we can see that more than a third of its profit was wasted with trading commissions, which goes to show just how important these expenses are in this type of simulation.

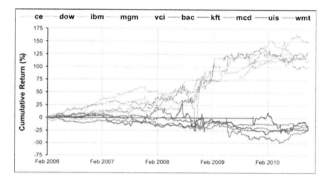

Fig. 4 Individual net cumulative returns of the 5 most profitable and the 5 least profitable stock trading agents in the test period.

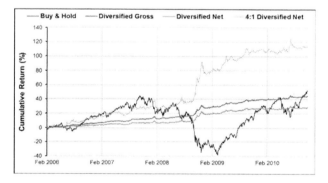

Fig. 5 Gross and net cumulative returns of the diversified trading system.

When compared with the buy-and-hold investment strategy (which implies simply buying equal amounts of the 60 stocks at the beginning of the simulation period, and holding them till the end), the system was not as profitable: it obtained a net return of 28.7%, versus a 51.8% return for the buy-and-hold strategy (calculated using stock prices adjusted for dividends and splits). Nevertheless, if we look at the chart in figure 5, we can see that the system holds a significant advantage over the buy-and-hold strategy, which is the much smaller volatility of its return curve. This advantage is evident in the difference between the two strategies' maximum drawdown (which measures the maximum accumulated loss during the simulation period): the buy-and-hold strategy had an unacceptable 57.5% maximum drawdown, while the system's was just 4.2%. What this means, in practical terms, is that the system's strategy was much safer in the test period. The lower risk implies that it should be better suited for using leverage (i.e., trading on margin with borrowed funds). If, for example, the system was configured to trade with 4:1 leverage (which would simply require quadrupling the agents' standard trade sizes), its net return in the simulation period would increase to 114.6%, with a maximum drawdown of 16.6%. This would be an excellent performance by any standards, and would soundly beat that of the buy-and-hold strategy. Our results hint that the development of a completely autonomous agent-based trading system with the potential to achieve an acceptable performance might be an actual possibility. This corroborates our belief that agent technology can be of use for this type of practical application.

Table 1 Comparison between the cumulative return of the diversified trading system and the buy-and-hold investment strategy.

Strategy		Return (%)	Max Drawdown (%)	Ratio
Buy & Hold		51.8	57.5	0.9
Diversified	Gross	44.5	3.9	11.5
	Net	**28.7**	**4.2**	**6.9**
4:1 Diversified	Gross	178.0	15.5	11.5
	Net	**114.6**	**16.6**	**6.9**

The performance of the described investment system attests the usefulness of the agent architecture we proposed. Nevertheless, there is always the possibility that a different architecture could yield better returns. In particular, it is possible that agents built using a subset of the architecture modules described in the previous section might perform better. In order to test this hypothesis, we implemented the 60 agents utilizing the following subsets: first, the agents were built with just the prediction module; next, we implemented them using the combination between the prediction module and the empirical knowledge module; finally, we combined the prediction module with the domain knowledge module. We then tested the diversified trading system using these three types of agents. Its accumulated return and maximum drawdown, for each simulation run, is summarized in table 2. We can verify that, when the simplest architecture was used, the system was barely profitable, having achieved a return of just 11.0%, with a maximum drawdown of 18.2%. This makes it clear that the agents' forecasting mechanisms can be much improved, be it with better data mining models or better training attributes. Adding the empirical knowledge module to the architecture considerably improved both metrics. An even bigger improvement was obtained when the prediction module was combined with the domain knowledge module. Finally, the use of all three modules resulted in the performance with the best risk-reward ratio, which confirms that all the modules in the proposed architecture were making an important contribution to the system's overall performance.

We should point out that, even though the system that was presented consists of a decentralized group of intelligent autonomous agents, it might not be considered a true multi-agent system due to the lack of inter-agent communication. As is, the agents' only interaction is with other market participants (software and human) through the stock broker. Unlike the previously mentioned multi-agent Forex trading system [6], there is no obvious reason for having the stock trading agents communicate with each other while trading. Nevertheless, there is a particular situation in which the agents might need to report their decisions to another agent. If the base currency of their trading account is different from that in which their stocks are denominated, then it would make sense to hedge the currency exposure whenever a stock trade is opened (because buying/short selling a stock will generate a long/short exposure to the corresponding currency). For this reason, we introduced a new agent in the system, named hedging agent, which is responsible for receiving the investment decisions from the stock trading agents, and making the appropriate currency trades in the Forex market to hedge the currency exposures. This interaction is depicted in the sequence diagram in figure 6 (for simplicity's sake, only one trading agent is shown).

Table 2 Net cumulative return of the diversified investment strategy when the agents are built using different architectures.

Architecture	Return (%)	Max Drawdown (%)	Ratio
Prediction Module	11.0	18.2	0.6
Prediction & Empirical Modules	17.9	8.5	2.1
Prediction & Domain Modules	29.6	11.0	2.7
All Modules	28.7	4.2	6.9

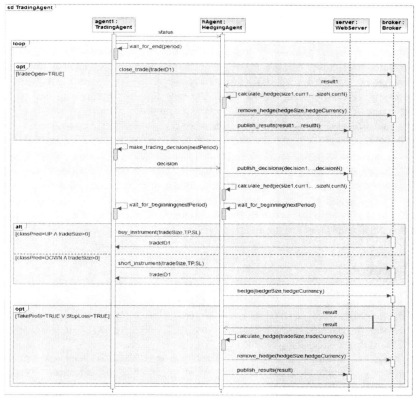

Fig. 6 Sequence diagram describing the interaction between the stock trading agents, the hedging agent and the stock broker.

A typical disclaimer when presenting trading results is that "past performance does not guarantee future returns". Put another way, we cannot be sure that the described system will be able to trade successfully in the future (by successfully we mean profitably with low risk) just because it did so in the past. In order to keep testing it going forward, we made the hedging agent responsible for posting the trading decisions of the 60 agents in a public website, every day before market open. This website has been available at http://ruibarbosa.eu/iquant/iquant_all.html since 2009, allowing its visitors to follow the agents' activity live, and to witness the forward-testing of the trading system. This autonomous system will be kept running continuously over time, so that its performance can be analyzed over the long run.

4 Conclusions and Future Work

The research described in this article leads us to conclude that it may be possible to create an autonomous agent-based trading system that can achieve acceptable results, and that the proposed agent architecture might be a viable option to build this type of systems. Both conclusions should be of interest to the investment industry.

Despite the compelling trading simulation results we obtained, there is still a lot to improve in our stock trading system. First of all, it is likely that using better attributes to train the agents will increase their accuracy, which in turn might make them more profitable. Adding more agents to the system should also improve its performance (in terms of risk-adjusted return), as this would increase the investment diversification, especially if the new agents were configured to trade other types of financial instruments, such as commodity futures or currency pairs. For a real-life trading system, we might also consider a strategy to get rid of the worst agents, possibly replacing them with better ones. We did not do this in our testing, in order to avoid survivorship bias, but it is clear that this measure should vastly improve the system over time. While the system's simulation results do not guarantee that it will be successful in the future, we do believe that they were at least good enough to justify further research on this topic. As is, the system's performance is empirical proof of the usefulness and potential of agent technology for this particular practical application.

References

[1] Vytelingum, P., Dash, R.K., He, M., Sykulski, A., Jennings, N.R.: Trading strategies for markets: A design framework and its application. In: La Poutré, H., Sadeh, N.M., Janson, S. (eds.) AMEC 2005 and TADA 2005. LNCS (LNAI), vol. 3937, pp. 171–186. Springer, Heidelberg (2006)

[2] Wang, H., Mylopoulos, J., Liao, S.: Intelligent agents and financial risk monitoring systems. Communications of the ACM - Robots: Intelligence, Versatility, Adaptivity 45(3), 83–88 (2002)

[3] Kearns, M., Ortiz, L.: The Penn-Lehman Automated Trading Project. IEEE Intelligent Systems 18(6), 22–31 (2003)

[4] Jennings, N., Wooldridge, M.: Applications of Intelligent Agents. In: Agent Technology: Foundations, Applications and Markets, pp. 3–28 (1998)

[5] Barbosa, R., Belo, O.: A Step-By-Step Implementation of a Hybrid USD/JPY Trading Agent. International Journal of Agent Technologies and Systems 1(2), 19–35 (2009)

[6] Barbosa, R.P., Belo, O.: Multi-Agent Forex Trading System. In: Håkansson, A., Hartung, R., Nguyen, N.T. (eds.) Agent and Multi-agent Technology for Internet and Enterprise Systems. Studies in Computational Intelligence, vol. 289, pp. 91–118. Springer, Heidelberg (2010)

[7] Weka Data Mining API, http://www.cs.waikato.ac.nz/~ml/weka/

[8] JBoss Drools Rule Engine, http://www.jboss.org/drools

Contracts Negotiation in Market Environments Taking into Consideration the Strength of Each Economic Player

Łukasz Jankowski, Jarosław Koźlak, and Małgorzata Żabińska

Abstract. The aim of the work is to present the model of the mediator, which takes into consideration the strengths of the different entities in a market. The theoretical background of the mediator is an economic model called Porter's five forces. The mediator is a part of the developed environment for the analysis of strategies of the enterprises functioning on the market and constructing the supply chains. In the paper we present, in a short form, this environment and its functionality and describe the negotiation model and results obtained thanks to it.

1 Introduction

The aim of the work is to develop a system for modelling and analysing the different decision strategies of the environment functioning on the market. To achieve this goal, an environment was developed consisting of companies which negotiate contracts concerning the buying of components with one another, clients ordering final products and additional agents responsible for helping to establish contact.

In this paper, we focus on the problem of negotiation and establishing conditions of the contract. The condition of the contract may consist of parameters, not only especially price but also delivery time, quality of goods, possible condition of breaking contracts or fines if a contract is not fulfilled. In the existing multi-agent systems for market modelling, the market equilibrium model or auction techniques are often used, but they sometimes require long bidding processes to achieve this solution. It is also possible to use bargain models such as Nash's model or Rubinstein's model. In all these cases, it is important to correctly describe both acceptable conditions for every contracting part as well as the negotiation strength of each of them. Because of the complexity of the problem, high level simplified solutions are often used.

Łukasz Jankowski · Jarosław Koźlak · Małgorzata Żabińska
Department of Computer Science, AGH University of Science and Technology,
Al. Mickiewicza 30, 30-059 Kraków
e-mail: l.s.jankowski@gmail.com, kozlak@agh.edu.pl,
        zabinska@agh.edu.pl

Y. Demazeau et al. (Eds.): Adv. on Prac. Appl. of Agents and Mult. Sys., AISC 88, pp. 299–308.
springerlink.com   © Springer-Verlag Berlin Heidelberg 2011

We are going to propose a solution inspired not from game theory but from the concepts originating from economic science. One of the market competition models takes into consideration factors which have influence on the behaviour of the negotiating sides and negotiation results. It is Porter's five forces model [6], which assumes that the competitive strength of companies in a sector depends on 5 factors (4 external and 1 internal) such as: (1) competitive/bargaining strength of suppliers, (2) competitive/bargaining strength of the clients, (3) threats of the arrival of new competitors, (4) threats of appearance of new substitute products, (5) competition inside the sector.

In the presented model we take into consideration the factors 1,2 and 5. We are considering separate negotiation of every product on the separate markets of production components and final products. Particularly, in the components market, the model considers factors 1 and 5 and in the market of final products - factors 2 and 5.

The organisation of the paper is as follows. In section 2, a short state of the art of the problem domain is given. Section 3 and 4 contain a short description of the realised market environment and more detailed description of the negotiation model. Some implementation remarks are in section 5. Section 6 contains the description of the results and section 7 concludes.

2   State of the Art

The features of the problem of the modelling of companies in the market and supply chain management predestines the use of the multi-agent approach. It is associated with several reasons such as the autonomy of the decision making by the given individuals, motivated by the realisation of their goals and also the use of the different techniques of the distributed artificial intelligence such as negotiation techniques, planning and machine learning.

Different multi-agent systems for this problem domain were developed. One of the first such systems, which consisted of different kinds of sellers and buyers, was the one described in [7]. The overview of the multi-agent systems for design and management of supply chains on the market may be found for example in [4].

The popularity and importance of the problem is confirmed by the organised competitions between the agent-producers managed by different algorithms in the market environment defined in [8]. Examples of such agents are Botticelli [1] and Cmieux [2].

3   System Model

The goal of the developed system is to model and analyse different decision strategies taking place on the market environment. It consists of the following main elements: companies, customers of the final products and mediators responsible for the establishment of the contracts between the companies.

Their tasks are as follows:

- Agent Company – represents enterprises, which tries to execute the strategy that is the most advantageous for production, selling and stock management strategies.
- Agent Customer of final product – generates the demand for the final products with given characteristics.
- Agent Mediator – is responsible for finding the buyer–seller pairs which perform the transactions concerning given goods. As fair mediators, they also establish the conditions of the contracts on the basis of information provided by the contracting parties.

The simplified model of the company $Comp_i$ is presented below. The more detailed description of the used market model with more information about components which belong to the company and obtained preliminary results is in [5].

$$Comp_i = (M, R, A, P, S, D, P_D, P_P) \tag{1}$$

where: M – capital with the scheduled incomes and expenses, R – set of resources (components or final goods) with consideration of their states (busy, ordered (with the information about planned arrival time), available), A – asset of production actions, P – a plan, consisting of production plan PP and plan of the component orders PK, S – a set of strategies – rules describing actions, which should be performed in the given states, D – a set of values of decision variables, P_D – demand prediction, P_P – price prediction.

In the next section the negotiation protocol for establishing a contract will be described. The company sends to the mediator the following n-tuple which represent its offer and is perceived as an economic subject by the mediator:

$$s_i = (\phi_i, c_{Ii}, c_{Li}, d_i, R_i, Role_i) \tag{2}$$

where ϕ_i – is an evaluation function which describes how advantageous a conclusion of a contract with given conditions is, c_{Ii} – initial conditions, c_{Li} – limit conditions, d_i – quantity of goods being bought/sold, R_i – kind of the goods, $Role_i$ – role during the negotiations (seller or buyer).

4 Negotiation Model

4.1 Negotiation Protocol

The interactions of Negotiator with agents representing market subjects is shown in fig. 1. Negotiations proceed as follows:

1. Every Client interested in buying a product sends his input to Negotiator. This includes his initial and limit conditions, quantity of product he wants to buy and parameters of his evaluation function.

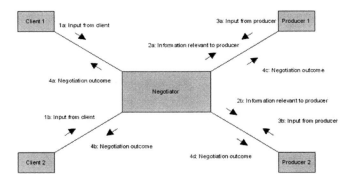

Fig. 1 Interactions between the Negotiator and agents representing market subjects

2. Negotiator passes to producers information they need to form their input to Negotiator. These are: initial offers and quantities of all clients interested in buying the product. Producers do not have any knowledge of clients' limit conditions or the way they assess conditions.
3. Producers send their input to Negotiator. This includes their initial and limit conditions, quantity of product they want to sell and parameters of their evaluation functions.
4. After calculating the outcome of negotiations, Negotiator sends to each agent the results it achieved.

4.2 Negotiation Decisions

Assumptions. The results adequate for every kind of market, showing the balance of strengths of the parts of the market and predictability of the situation, should appear as results of negotiations. To achieve this goal, it is necessary to conduct negotiations parallely with the various partners with a non-binding character of these negotiations.

The aggressiveness of negotiations with given partner (supplier or buyer) should be larger if:

- Other party's strength (factor 1 or 2 of Porter's model depending on the subject being the seller/buyer) is weaker. The number of other partners among which one can choose is larger. The offer of the given partner is less advantageous in comparison with the offers of other partners. The quantity of goods the given partner wants to buy/sell is lower.
- Own strength (factor 5 of Porter's model) is larger. The number of other partners among which the given partner can choose is larger One's offer is less advantageous in comparison with the offers of other partners among which the given partner can choose. The quantity of goods one wants to buy/sell is higher.

The higher number of alternative suppliers/buyers increases the aggressiveness of the subject negotiation, because the subject has a choice among many alternative partners and does not need to make substantial concessions in negotiations with any of them. Disadvantageousness of the partner offer increases the aggressiveness of the subject negotiations, because having a possibility to choose more advantageous solutions, the subject does not care too much to sign a contract with a given partner. On the other hand, negotiating with a partner with a very good offer in comparison to others, the subject would not want to alienate itself and will be ready for making more significant concessions.

Detailed description. The following calculations are all performed by the Negotiator based on information received from buyers and sellers.

We are considering a strictly competitive setting. Subjects perform one of two roles and they have needs specific for their role. The needs of subjects performing different roles are both convergent and opposing. They are convergent because they can only be satisfied by reaching an agreement with one or more subjects that perform a different role. They are opposing because the more the conditions of the agreement are favourable for one party, the more they are unfavourable for the other. The goal is to find the agreements which are likely to be reached considering relative strengths of both groups of subjects, and individual strengths of subjects. In our case the two roles are producers and consumers who meet on the market to sell and buy goods but other applications of this model are conceivable. Each subject comes to negotiations with initial conditions (c_I) and limit conditions (c_L). Initial conditions are the conditions from which the subject would like to begin negotiations. They are more favourable than the limit conditions which are the least favourable conditions under which the subject would accept agreement. Conditions are vectors of variables representing different aspects of the contract. In our case these are price and delivery time. For every aspect of the contract, one group of subjects favours greater values and the other favours lower values. In our case, buyers want both prices and times to be small.

Each subject has an individual evaluation function $\phi : C \to [0; \infty)$ which assesses how favourable conditions are (C is the set of conditions). The evaluation equal 0 means that the agreement under these conditions would be as good as no agreement at all. The evaluation function must have such a quality that $\phi(c_L) = 0$.

Solution is reached as a result of not binding, bilateral negotiations of every subject performing one role with every subject performing the other role. Fig. 2 1 shows the state diagram of each such negotiation. The negotiation becomes "In progress" after determining negotiation space, initial negotiated quantity and the number of steps of negotiation.

Each bilateral negotiation begins with defining the negotiation space. It is done by each subject stating its initial offer. Initial offers are usually equal to initial conditions. It is however possible that the initial conditions of one subject are unacceptable for another subject (they are worse than its limit conditions). In such a case, initial offer is equal to the limit conditions of the other subject.

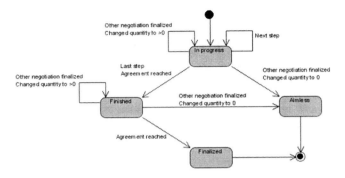

Fig. 2 States of the negotiation

The quantity of goods that are the object of each negotiation can be different. It is always the minimum of the quantity not yet sold by the producer and quantity not yet bought by the consumer. When any party satisfies part of its needs the quantities of all its negotiations are updated. During negotiations subjects make concessions until they reach an agreement. The agreement is reached in a number of steps. The number of these steps is a random variable having geometric distribution. It is realised at the beginning of negotiation. During each step both parties make some concessions and after the last step the agreement is reached.

Negotiations are not binding (fig. 2). This means that reaching an agreement (state "Finished") does not necessary imply making a transaction (state "Finalized"). For the negotiations to be finalised and the transaction to be made, an agreement must be reached and both parties must at the same time declare that they want to trade under the agreed conditions. Subject declares it wants to finalise negotiation if it decides that in order to satisfy all its needs and to do it under best conditions possible it would have to finalise this negotiation.

At some point at least one subject satisfies all its needs by finalising negotiations. Then the quantities of all other of its negotiations are updated to 0 and negotiations became "Aimless" regardless of the state they were in. Algorithm ends when all negotiations became either finalised or aimless.

An additional mechanism is used in order to guarantee that the results are Pareto efficient. It is conceivable that all negotiations are finished and there are some negotiations that are neither finalised nor aimless and none of them can be finalised. Such a deadlock would occur if there were no negotiations in which both parties would declare they want to finalise. Deadlocks are resolved by randomly forcing declarations to finalise.

Subjects base their behavior on one negotiation on the state of this and all other negotiations. We introduce a measure of how aggressively a subject negotiates which we shall call determination. The more determined a subject is in bilateral negotiations, the more it favours reaching any agreement with the other party than forcing other party's concessions possibly reaching a better agreement but risking the other party's withdrawal. Determination can increase or decrease during negotiation and it can be different in different bilateral negotiations at the same time.

Determination is the speed of making concessions. If the conditions previously offered by i to j are c_{ij} and the conditions previously offered by j to i are c_{ji}, and there are k steps left until the end of negotiation then the next offered conditions are $c'_{ij} = c_{ij} + \frac{X_{ij}}{k}(c_{ji} - c_{ij})$ $c'_{ji} = c_{ji} + \frac{1-X_{ij}}{k}(c_{ij} - c_{ji})$ where redX_{ij} is a random variable having $Beta(\frac{1}{D_{ji}}, \frac{1}{D_{ij}})$ probability distribution. If k=1 then $c'_{ij} = c'_{ji}$. Determination is the relative decline in subject's prospect of satisfying its needs after the other party would take its offer "of the table". Hence determination equals $D_{ij} = \frac{(P_i - P'_{ij})}{P_i}$ where P_i is the current prospect of satisfying i's needs considering the state of all its negotiations and P'_{ij} is the prospect of satisfying i's needs considering the state of all its negotiations if negotiation with j would fail.

The measure of prospect is chosen based on its desired properties. A withdrawal of the other party in one of negotiations leads to a not better prospect than before. In case of the withdrawal of the other party in one of the negotiations the decrease of prospect is the greater, the better the conditions were proposed by it. In case of the withdrawal of the other party in any negotiations, the decrease of prospect is the greater, the bigger part of needs this negotiation could have satisfied.

In case of the withdrawal of the other party in one of negotiations the decrease of prospect is the greater, the lesser part of needs can be satisfied by those of the other negotiations that have better current conditions. Suppose subject i is negotiating with N partners. Let q_{ij} be the quantity of goods that are the object of negotiation between i and j. Let c_{ji} be the conditions that j currently proposes to i. Let d_i (demand) be the amount of goods subject i needs to buy or sell. Let $s_i = \Sigma_{(j=1)}^N q_{ij}$ be the total quantity of goods that are objects of all negotiations (supply). Let $c_i^{(n)}$ be the conditions that are n-th the most favourable in terms of ϕ_i and let $q_i^{(n)}$ be the quantity that is the subject of the same negotiation.

Then the prospect equals

$$P_i = \int_0^{s_i} w_i(x)o_i(x)dx \tag{3}$$

where $o_i(x)$ is the offers function of subject i and $w_i(x)$ is the weights function for subject i. Offers function is defined as

$$o_i(x) = \phi_i(c_i^{(n)}) \text{ for } 0 \leq x \leq s_i \text{ where n is such that } \Sigma_{k=0}^{n-1} q_i^{(k)} < x \leq \Sigma_{k=0}^n q_i^{(k)} \tag{4}$$

Weights function is defined as

$$w_i(x) = \frac{\exp\left(-\frac{x}{d_i}\right)}{d_i} \tag{5}$$

The weights function is used to properly model the fact that the influence of other party's withdrawal on prospect is the weaker the bigger part of d_i can be satisfied by finalising other contracts on terms better than those offered by this party.

5   Realisation

The simulation environment was implemented in Java language and using JADE platform [3]. On the realisation level the following kinds of agents are also used:

- Base Agent - offers basic functions used by each agent,
- Data Agent- responsible for picking configuration data from database and storing statistics about the running of the system,
- Coordinator Agent- registers created agents and provides information about offers of companies for given kinds of goods,
- Simulator Agent - responsible for the synchronisation of activities, verification of whether the system is ready to pass to the next step and move into simulation steps (steps representing consecutive days).

6   Results

A number of features of this negotiation model have been confirmed. Three of them are worth mentioning in particular. These are: the results given for different market structures, the results given depending on supply/demand relation on the market and the results for individual subjects depending on their size.

To assess the influence of market structure on results, a number of simulations were ran. In each of them there were 10 buyers who wanted to purchase 10 units each and were ready to pay $c_L=20$ at most. Their initial offer was $c_I=0$. Sellers were willing to sell at $c_L=10$ at least and their initial offer was $c_I=20$. In each simulation the total supply was equal to 60. The number of sellers was different in every simulation ranging from a monopoly to 60 buyers producing 1 item each. The results are shown in fig. 3a.

It was found that the closer the market structure is to monopoly, the better results the few producers receive and the closer it resembles perfect competition, the more the power shifts to the buyers side of market.

To assess the influence of supply/demand relation on results in a number of simulations the amounts produced were changed from 1 to 9 per producer. In each simulation there were 10 producers who wanted to buy 10 units each ($c_I=0,c_L=20$) and 10 producers ($c_I=20,c=10$). It was found that the bigger the total supply is, the smaller is the market price. It is a result of two factors. Firstly, the bigger a single producer is, the bigger its own determination in negotiations is, which leads to worse results for it. Secondly, the bigger the amount offered, the lower the determination of buyers is, which also leads to worse results for the producers. The results are shown in fig. 3b.

To assess the influence of individual size on negotiation outcomes one simulation was repeated a number of times. There was one buyer who wanted to buy 55 units ($c_I=0,c_L=20$). There were 10 sellers who wanted to sell amounts from 1 to 10 units ($c_I=20,c_L=10$). Each producer had initial capital of 5000. It costed 10 to produce one unit of product. Producers suffered storage costs of 1 per turn for each unit in storehouse. Their constant costs were equal 5. The results are shown in fig. 3c.

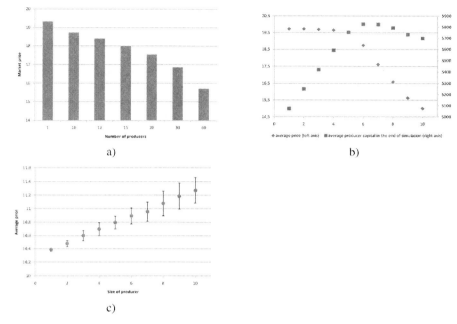

Fig. 3 Influence of market structure (a) and supply (b) on average negotiation outcome. Influence of the size of a producer on his average negotiation outcome (c).

It was found that the bigger the producer is, the better negotiation outcomes he can achieve. This is a result of two opposing factors. Firstly, the bigger a subject is, the greater the determination of others to negotiate with it is, which leads to better results. Secondly, the bigger a subject is, the greater its own determination in all negotiations is, which leads to worse results. As tests have shown, of these two factors the first is stronger.

It is worth noting that although a bigger size of single producer allows him to negotiate better conditions, bigger sizes of all producers lead to worse conditions for all of them. As the model is symmetrical, all of above properties apply to the buyers as well.

7 Conclusion

In the paper the negotiation protocol which takes into consideration the strength of the economic subjects in the market and which uses a mediator agent for establishing a contract was presented. The presented results obtained during the tests of the modules confirmed their expected features.

An important feature of the model is that the results adequate for different kinds of the markets are achieved thanks to micro-level decisions and without the knowledge of the participants regarding the kind of market. The mechanisms of the appropriate adjustments of the aggression level in the negotiations according to the participant advantages is sufficient to achieve it. Individual decisions of the self-interest subjects lead to the results expected for the given kinds of markets, which is the situation present in the real markets.

The subsequent work will consist of the use of the mediator module in the developed environment for the modelling and optimisation of the companies strategies in the market. We are especially focusing on the analysis of the stock management strategies and level of risks allowed by the enterprises during the negotiations as well as the different techniques of future state predictions.

Acknowledgements. We would like to thanks J. Marszałek and P. Kolarz, the former students of Department of Computer Science AGH-UST for the development of the initial version of the system. The presented work has been financed by a grant from the Polish Ministry of Science and Higher Education number N N516 366236.

References

1. Benisch, M., Greenwald, A., Grypari, I., Lederman, R., Naroditskiy, V., Tschantz, M.: Botticelli: A supply chain management agent. In: Proceedings of AAMAS 2004, New York, USA, pp. 1174–1181 (2004)
2. Benisch, M., Sardinha, A., Andrews, J., Sadeh, N.: Cmieux: adaptive strategies for competitive supply chain trading. SIGecom Exch. 6(1), 1–10 (2006)
3. Java Agent DEvelopment Platform (2008), http://jade.tilab.com
4. Moyaux, T., Chaib-draa, B., D'Amours, S.: Supply chain management and multiagent systems: An overview. In: Chaib-draa, B., Muller, J. (eds.) Multiagent Based Supply Chain Management. SCI, vol. 28, Springer, Heidelberg (2006)
5. Nawarecki, E., Koźlak, J.: Building multi-agent models applied to supply chain management. Control and Cybernetics 39(1 spec. iss.), 149–176 (2010)
6. Porter, M.: How competitive forces shape strategy. Harvard Business Review (1979)
7. Swaminathan, J., Smith, S., Sadeh, N.: Modeling supply chain dynamics: A multiagent approach. Decision Sciences 29(3) (1998)
8. TAC SCM Game Description (2008), http://www.sics.se/tac/

Bid Definition Method for Electricity Markets Based on an Adaptive Multiagent System

Tiago Pinto, Zita Vale, Fátima Rodrigues, Hugo Morais, and Isabel Praça

Abstract. Electricity markets are complex environments with very particular characteristics. MASCEM is a market simulator developed to allow deep studies of the interactions between the players that take part in the electricity market negotiations. This paper presents a new proposal for the definition of MASCEM players' strategies to negotiate in the market. The proposed methodology is multi-agent based, using reinforcement learning algorithms to provide players with the capabilities to perceive the changes in the environment, while adapting their bids formulation according to their needs, using a set of different techniques that are at their disposal. Each agent has the knowledge about a different method for defining a strategy for playing in the market, the main agent chooses the best among all those, and provides it to the market player that requests, to be used in the market.

This paper also presents a methodology to manage the efficiency/effectiveness balance of this method, to guarantee that the degradation of the simulator processing times takes the correct measure.

1 Introduction

The recent restructuring of the energy markets, characterized by an enormous increase of the competition in this sector, led to relevant changes, and consequently new challenges in the participating entities operation [1]. In order to overcome these challenges, it is essential for the professionals to understand fully the principles of the markets, and how to evaluate their investments under such a competitive environment [1].

This necessity for understanding those mechanisms and how the involved players' interaction affects the outcomes of the markets, and thus, the revenues of the investments, contributed to the need of using simulation tools, with the purpose of taking the best possible results out of each market context for each participating entity. Multi-agent based software is particularly well fitted to analyze dynamic and adaptive systems with complex interactions among its constituents [2, 3, 4].

Tiago Pinto · Zita Vale · Fátima Rodrigues · Hugo Morais · Isabel Praça
GECAD – Knowledge Engineering and Decision-Support Research Center
Institute of Engineering – Polytechnic of Porto (ISEP/IPP)
Rua Dr. António Bernardino de Almeida, 431 4200-072 Porto
e-mail: {tmp,zav,mfc,hgvm,icp}@isep.ipp.pt

Y. Demazeau et al. (Eds.): Adv. on Prac. Appl. of Agents and Mult. Sys., AISC 88, pp. 309–316.
springerlink.com © Springer-Verlag Berlin Heidelberg 2011

To explore and study such approaches is the main goal of our research, and for that we use the multi-agent system MASCEM (Multi-Agent System for Competitive Energy Markets) [4, 5] that provides us with a simulation of the electricity market, considering all the most important entities that take part in such operations.

Players in MASCEM are implemented as independent agents, with their own capability to perceive the states and changes of the world, and acting accordingly. These agents are provided with biding strategies, to try taking the best possible advantage from each market context.

This paper presents a new methodology for the negotiating agents' bidding definition. The proposed methodology is developed as a multiagent system, and uses reinforcement learning algorithms, which allow players to quickly perceive changes in their environment, and so granting them the means to adapt their bids according to their needs. This adaptation is performed using a set of different agents, each containing the knowledge to perform a distinct methodology based on different approaches and techniques. The variety in approaches and large range of considered techniques asset the player with solutions to overcome different obstacles, and environment changes that occur during the course of the market.

2 Multiagent System for Strategy Formulation

In order to provide MASCEM's [4, 5] negotiating players with competitive advantage in the market it is essential to endow them with strategies capable of dealing with the constant market changes, allowing adaptation to the competitors' actions and reactions. For that it is necessary to have adequate forecast techniques to analyse the market data properly. The way prices are predicted can be approached in several ways, namely through the use of statistical methods, data mining techniques [6], neural networks (NN) [6], support vector machines (SVM), or several other methods [7, 8]. There is no method that can be said to be the best for every situation, only the best for particular cases.

To take advantage of the best characteristics of each technique, we decided to create a method that integrates several distinct technologies and approaches. This method is implemented as a multiagent system. There is one agent performing each distinct algorithm, detaining the exclusive knowledge of its execution. This way the system can be executing all the algorithms in parallel, preventing as possible the degradation of the method's performance. As each agent gets its answer, sends it to the main agent, which chooses the most appropriate answer among all.

The communications between the agents of this system are managed by an adaptation of a Prolog [9] Facilitator [5], this way making it possible for this system to be integrated in MASCEM, with the communications treated separately from the rest of the simulation, guaranteeing the independence and parallelism between the distinct groups of agents. The multiagent system for bidding strategies' integration with MASCEM is presented in figure 1.

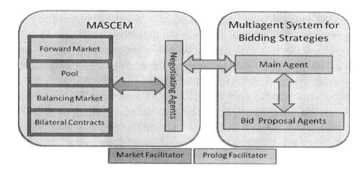

Fig. 1 Multiagent system for bidding strategies integration with MASCEM

As presented in Fig.1, the main entity of this system is the Main Agent,, which is the agent responsible for executing the reinforcement learning algorithms, which allow that in each moment and in each circumstance the technique that presents the best results for every actual scenario is chosen as the simulator's response. So, given as many answers to each problem as there are algorithms, the reinforcement learning algorithm will choose the one that is most likely to present the best answer according to the past experience of their responses and to the present characteristics of each situation, such as the week day, the period, and the particular market context that the algorithms are forecasting.

The other entities present in this system, the *Bid Proposal Agents*, are: the *NN Agent,* which uses a feed-forward neural network; the *AMES Agent,* performing an adaptation of the AMES bidding strategy [2]; the *Composed Goal Directed Agent;* the *Adapted Derivative Following Agent,* which uses an adaptation of the strategy proposed by Greenwald [8]; the *Market Price Following Agent;* the *Game Theory Agent,* [7]; and finally the *Statistical approaches based Agents.*

This system's and its entities detailed description can be found in [10].

2.1 Mechanism for Dynamic Efficiency Adaptation

Since there are many algorithms running simultaneously, some requiring higher processing times than others and some achieving a higher level of effectiveness in their forecasts, it becomes necessary to build a suitable mechanism to manage the algorithms efficiency in order to guarantee the minimum degradation of the previous implementation performance, i.e. the MASCEM simulator's processing time without considering this system's integration.

To adjust this mechanism to each simulation purpose, the user is able to define the importance that efficiency and effectiveness will have. If the user chooses 100% importance for effectiveness, in case of the simulation being used for decision support for the real market, and so the most important issue being the quality of the forecasts, and not how long they take to be achieved, since market bids are asked only once every 24 hours, then all the algorithms and agents will be turned on, and contribute to the simulation. In the other extreme, if the user chooses 100% importance for efficiency, in case of the simulation being with the purpose of studying other issue of the market that MASCEM offers, that not the forecasts, and so, preferring a much faster simulation, with no degradation of the simulator times, then all the strategy agents that require more time to process than the MASCEM simulation without them, are excluded, remaining only the faster ones.

Choosing anywhere in between demands a careful and smart analysis of the efficiency/effectiveness balance. This means that increasing the importance of the efficiency doesn't traduce into a linear exclusion of the agents that are taking longer to provide an answer. Rather a conjugation of this aspect with the effectiveness of each method, so an agent requiring a little more time than other but giving best responses, can be excluded later than the other. This is done using two fuzzy sets [11], one for the effectiveness, characterizing the difference between each forecast and the real verified value, and one variable for the efficiency, doing the same to represent the difference between the processing times of each method and MASCEM's without considering this system integration. The fuzzy sets are defined as in Fig. 2.

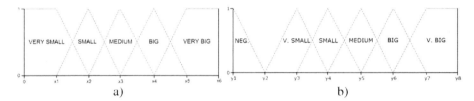

Fig. 2 Fuzzy Sets and respective membership functions definition, for the characterizatrion of: a) the effectiveness, b) the efficiency

Both fuzzy sets are dynamic, being actualized in every run, to guarantee the suitability of the membership functions in each situation. Defining the intervals as static values could lead to situations where all, or a big part, of the methods are being classified in the same interval of difference, and so, this methodology losing its logic. This way, the values are always adapting to the characteristics that are being observed, making sure that a Very Big value is actually a very big value considering the values that were registered so far, and a Medium value is actually a value in the average of the registered values.

To consistently adapt these intervals, in case of the effectiveness, as presented in figure 2 a), the variable $x6$ is always the maximum value of difference verified so far, $x3$ the average value between all the forecast methods, and $x1$, $x2$, $x4$ and $x5$are defined accordingly to the other two values.

In case of the efficiency, figure 2 b), $y1$ represents the minimum value of processing time verified among all the methods, $y2$ the processing time of the

MASCEM simulator without considering the integration of this multiagent system, y4 the average of the processing times of all the methods, and y6 the maximum time that a method took so far to achieve an answer. This way all the values between y1 and y2 will be determined as Negative, and the corresponding methods will never need to be excluded by this method. The methods taking all the values above will depend on the percentage of importance for effectiveness/efficiency. The fuzzy confusion matrix that joins these two fuzzy sets will determine if each method will be excluded or not, combining the two fuzzy values, and depending on that value of importance.

3 Case Study

In this section we present three simulations undertaken using MASCEM, referring to the same 14 consecutive days, starting from Wednesday, 29th October. The data used in this case study has been based on real data extracted from OMEL [12].

These simulations involve 7 buyers and 5 sellers (3 regular sellers and 2 VPPs). This group of agents was created with the intention of representing the Spanish reality, reduced to a smaller summarized group, containing the essential aspects of different parts of the market, in order to allow a better individual analysis and study of the interactions and potentiality of each of those actors.

For the first simulation we will consider a different bidding definition for each agent. In the second and third simulations, all strategies remain the same except for Seller 2, which will be our test subject. In the first simulation Seller 2 will use the Neural Network as strategy for the bid definition (*NN Agent* method). In the second Seller 2 will use a statistical approach, a regression on the data of the last 5 business days. Finally, in the third simulation, Seller 2 will use the proposed multiagent strategy for bid definition, presented in Section 3. The selected reinforcement learning algorithm for the *Main Agent* for this simulation has been the revised Roth-Erev, with equal value of the algorithms weight, and a past experience weight value of 0.4, a small value to grant higher influence to the most recent results, so that it can quickly learn and catch new tendencies, since the market is always changing fast.

After the simulations, we can compare the profits obtained by Seller 2 using each of the three strategies. Seller 2 power supply will remain constant at 550MW for each period throughout the simulations.

The other players' bids definition and further details of this case-study can be consulted in www.mascem.com.

Since the reinforcement learning algorithm treats each period of the day as a distinct case, we have to analyse the development of the performance in each period individually. Figure 3 presents the evolution of Seller 2 profits in the first period of the day, along the 14 days, using each of the three considered methods.

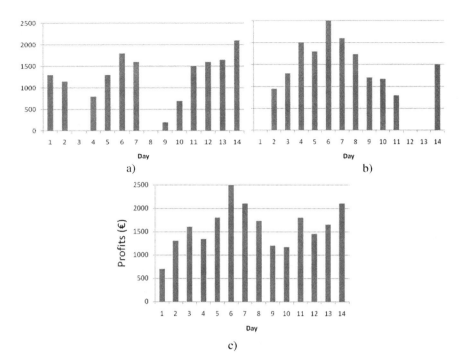

Fig. 3 Profits obtained by Seller 2 in the first period of the considered 14 days, using: a) the regression on the data of the last 5 business days, b) the Neural Network, c) the proposed strategy for bid definition

Comparing the charts presented in figure 3, we can see that the third simulation was clearly the most profitable for Seller 2 for this period.

In the first day the profit using the proposed strategy is located below the value of the profit using the regression. That is because the confidence values for all algorithms have been initialized equally, and so the selection of the answer is made randomly. The selected agent's answer originated a low profit.

In the second day, after the reinforcement learning algorithm is updated, it chooses the agent's answer that had the greater success in the first day, obviously the one that got the best reward, and we can see that the profit is located above both the other comparison strategies, as well as it happens on the third day.

In the fourth day its value is below the results of the NN; the algorithm is still selecting the other agent that got the best results in the first 3 days. In the fifth day the value is equal to the NN, and it follows that trend until day 10, in spite of its lower values in days 9 and 10. In day 11 the chosen agent's answer is no longer the *NN Agent*, as its value is higher than the ones of the other two strategies. Other algorithm passed the NN Agent as the best result of confidence as a result of the NN worst results in days 9 and 10.

In the last two days the selected response is the regression, catching its high value tendency still on time to get two good final results. This is due to the low weight value of past experience, otherwise this algorithm would not be chosen, for its weaker performance in the first days.

4 Conclusions and Future Work

This paper proposed a new methodology for bids definition in electricity markets. This methodology uses a multiagent system based on reinforcement learning algorithms to provide players with the capabilities to perceive the changes in the environment, adapting their bids according to their needs and to the dynamic environment in which they are acting. The proposed method is integrated in MASCEM, an electricity market simulator developed by the authors' research centre.

The results obtained with the proposed methodology are clearly encouraging, since we can observe that the proposed method achieves better results than the individual strategies that it has been compared to. The proposed method proved being able to catch the good result trends and reacting quickly to the decreasing tendencies. This is a very important result, for it demonstrates that several agents with distinct characteristics can be combined, and all give their contribution in the right time, making the object player in fact more intelligent, being able to learn, adapt and take the most advantage out of the environment.

The inclusion of further agents using other algorithms with new approaches can surely bring even more success to this strategy, by allowing a greater coverage of unexpected situations. This is our next step, along with the refinement of the reinforcement learning techniques, with the inclusion of clustering to group periods that present similar tendencies.

References

1. Rothwell, G., Goméz, T.: Electricity Economics: Regulation and Deregulation. In: IEEE Series on Power Engineering, pp. 15–42 (2003)
2. Sun, J., Tesfatsion, L.: Dynamic Testing of Wholesale Power Market Designs: An Open-Source Agent-Based Framework. Computational Economics 30(3), 291–327 (2007)
3. Koritarov, V.: Real-World Market Representation with Agents: Modeling the Electricity Market as a Complex Adaptive System with an Agent-Based Approach. IEEE Power & Energy Magazine, 39–46 (2004)
4. Praça, I., et al.: MASCEM: A Multi-Agent System that Simulates Competitive Electricity Markets. IEEE Intelligent Systems 18(6), 54–60 (2003); (Special Issue on Agents and Markets)
5. Pinto, T., et al.: Multi-Agent Based Electricity Market Simulator With VPP: Conceptual and Implementation Issues. In: 2009 IEEE PES General Meeting (2009)
6. Azevedo, F., Vale, Z., Oliveira, P.: A Decision-Support System Based on Particle Swarm Optimization for Multi-Period Hedging in Electricity Markets. IEEE Transactions on Power Systems 22(3), 995–1003 (2007)

7. Bompard, E., et al.: A game theory simulator for assessing the performances of competitive electricity markets. IEEE St. Petersburg Power Tech (2005)
8. Greenwald, A., Kephart, J.: Shopbots and Pricebots. In: Proceedings of the Sixteenth International Joint Conference on Artificial Intelligence, IJCAI, Stockholm (1999)
9. http://www.lpa.co.uk (accessed on January 2010)
10. Vale, Z., et al.: MASCEM - Electricity markets simulation with strategically acting players. IEEE Intelligent Systems 26(2) (2011); (Special Issue on AI in Power Systems and Energy Markets)
11. Gomide, F., Pedrycz, W.: Notions and Concepts of Fuzzy Sets. Fuzzy Systems Engineering: Toward Human-Centric Computing (2007)
12. Operador del Mercado Ibérico de Energia, http://www.omel.es (accessed on September 2010)

Author Index

CPSIA information can be obtained at www.ICGtesting.com
233870LV00002B/40/P